Verilog Digital System Design

RT Level Synthesis, Testbench and Verification

Zainalabedin Navabi, Ph.D.

Professor of Electrical and Computer Engineering
Northeastern University
Boston, Massachusetts

W0193367

Second Edition

McGraw Hill Education (India) Private Limited

NEW DELHI

McGraw Hill Education Offices
New Delhi New York St Louis San Francisco Auckland Bogotá Caracas
Kuala Lumpur Lisbon London Madrid Mexico City Milan Montreal
San Juan Santiago Singapore Sydney Tokyo Toronto

McGraw Hill Education (India) Private Limited

Special Indian Edition 2008

Adapted in India by arrangement with The McGraw-Hill Companies, Inc., New York

Sales territories: India, Pakistan, Nepal, Bangladesh, Sri Lanka and Bhutan

Verilog Digital System Design 2/e

ISBN (13): 978-0-07-025221-9
ISBN (10): 0-07-025221-1

Managing Director: *Kaushik Bellani*

Manager—Production: *Sohan Gaur*
Manager—Sales & Marketing: *S Girish*
Senior Product Manager—Science, Technology and Computing: *Rekha Dhyani*
General Manager—Production: *Rajender P Ghansela*

Published by McGraw Hill Education (India) Private Limited,
P-24, Green Park Extension, New Delhi 110 016 and printed at
India Binding House, Noida, (U.P.)

Cover Printer: India Binding House

Cover Designer: Kapil Gupta

To my mother, Sadri Kheradmand (Navabi),
who inspired me to pursue a life of science
and engineering.

Contents

Preface   xiii

Chapter 1. Digital System Design Automation with Verilog   1

   1.1   Digital Design Flow   2
         1.1.1   Design entry   3
         1.1.2   Testbench in Verilog   4
         1.1.3   Design validation   4
         1.1.4   Compilation and synthesis   7
         1.1.5   Postsynthesis simulation   10
         1.1.6   Timing analysis   10
         1.1.7   Hardware generation   10
   1.2   Verilog HDL   10
         1.2.1   Verilog evolution   11
         1.2.2   Verilog attributes   11
         1.2.3   The Verilog language   13
   1.3   Summary   13
       Problems   13
       Suggested Reading   14

Chapter 2. Register Transfer Level Design with Verilog   15

   2.1   RT Level Design   15
         2.1.1   Control/data partitioning   16
         2.1.2   Data part   16
         2.1.3   Control part   17
   2.2   Elements of Verilog   18
         2.2.1   Hardware modules   18
         2.2.2   Primitive instantiations   19
         2.2.3   Assign statements   20
         2.2.4   Condition expression   20
         2.2.5   Procedural blocks   20
         2.2.6   Module instantiations   21
   2.3   Component Description in Verilog   22

v

		2.3.1 Data components	22
		2.3.2 Controllers	29
	2.4	Testbenches	33
		2.4.1 A simple tester	33
		2.4.2 Tasks and functions	34
	2.5	Summary	34
		Problems	35
		Suggested Reading	35

Chapter 3. Verilog Language Concepts   37

	3.1	Characterizing Hardware Languages	37
		3.1.1 Timing	37
		3.1.2 Concurrency	39
		3.1.3 Timing and concurrency example	40
	3.2	Module Basics	41
		3.2.1 Code format	41
		3.2.2 Logic value system	41
		3.2.3 Wires and variables	42
		3.2.4 Modules	42
		3.2.5 Module ports	43
		3.2.6 Names	43
		3.2.7 Numbers	44
		3.2.8 Arrays	46
		3.2.9 Verilog operators	48
		3.2.10 Verilog data types	54
		3.2.11 Array indexing	58
	3.3	Verilog Simulation Model	59
		3.3.1 Continuous assignments	61
		3.3.2 Procedural assignments	65
	3.4	Compiler Directives	71
		3.4.1 `timescale	71
		3.4.2 `default-nettype	71
		3.4.3 `include	71
		3.4.4 `define	71
		3.4.5 `ifdef, `else, `endif	72
		3.4.6 `unconnected-drive	72
		3.4.7 `celldefine, `endcelldefine	72
		3.4.8 `resetall	72
	3.5	System Tasks and Functions	72
		3.5.1 Display tasks	73
		3.5.2 File I/O tasks	73
		3.5.3 Timescale tasks	74
		3.5.4 Simulation control tasks	74
		3.5.5 Timing check tasks	74
		3.5.6 PLA modeling tasks	74
		3.5.7 Conversion functions for reals	75
		3.5.8 Other tasks and functions	75
	3.6	Summary	76
		Problems	76
		Suggested Reading	80

Chapter 4. Combinational Circuit Description   81

4.1   Module Wires   81
    4.1.1   Ports   81
    4.1.2   Interconnections   82
    4.1.3   Wire values and timing   82
    4.1.4   A simple testbench   84
4.2   Gate Level Logic   85
    4.2.1   Gate primitives   85
    4.2.2   User defined primitives   87
    4.2.3   Delay formats   88
    4.2.4   Module parameters   90
4.3   Hierarchical Structures   93
    4.3.1   Simple hierarchies   93
    4.3.2   Vector declarations   95
    4.3.3   Iterative structures   96
    4.3.4   Module path delay   99
4.4   Describing Expressions with Assign Statements   102
    4.4.1   Bitwise operators   102
    4.4.2   Concatenation operators   104
    4.4.3   Vector operations   104
    4.4.4   Conditional operation   105
    4.4.5   Arithmetic expressions in assignments   108
    4.4.6   Functions in expressions   109
    4.4.7   Bus structures   110
    4.4.8   Net declaration assignment   111
4.5   Behavioral Combinational Descriptions   112
    4.5.1   Simple procedural blocks   113
    4.5.2   Timing control   113
    4.5.3   Intra-assignment delay   116
    4.5.4   Blocking and nonblocking assignments   116
    4.5.5   Procedural if-else   118
    4.5.6   Procedural case statement   120
    4.5.7   Procedural for statement   122
    4.5.8   Procedural while loop   123
    4.5.9   A multilevel description   124
4.6   Combinational Synthesis   125
    4.6.1   Gate level synthesis   127
    4.6.2   Synthesizing continuous assignments   128
    4.6.3   Behavioral synthesis   129
    4.6.4   Mixed synthesis   132
4.7   Summary   132
    Problems   132
    Suggested Reading   134

Chapter 5. Sequential Circuit Description   135

5.1   Sequential Models   135
    5.1.1   Feedback model   136
    5.1.2   Capacitive model   136
    5.1.3   Implicit model   136

5.2   Basic Memory Components   137
      5.2.1   Gate level primitives   137
      5.2.2   User defined sequential primitives   139
      5.2.3   Memory elements using assignments   140
      5.2.4   Behavioral memory elements   142
      5.2.5   Flip-Flop timing   149
      5.2.6   Memory vectors and arrays   151
5.3   Functional Registers   157
      5.3.1   Shift registers   157
      5.3.2   Counters   161
      5.3.3   LFSR and MISR   163
      5.3.4   Stacks and queues   167
5.4   State Machine Coding   171
      5.4.1   Moore machines   171
      5.4.2   Mealy machines   174
      5.4.3   Huffman coding style   176
      5.4.4   A more modular style   180
      5.4.5   A ROM based controller   181
5.5   Sequential Synthesis   181
      5.5.1   Latch models   183
      5.5.2   Flip-flop models   184
      5.5.3   Memory initialization   185
      5.5.4   General sequential circuit synthesis   186
5.6   Summary   186
      Problems   187
      Suggested Reading   189

Chapter 6. Component Test and Verification   191

6.1   Testbench   191
      6.1.1   Combinational circuit testing   192
      6.1.2   Sequential circuit testing   194
6.2   Testbench Techniques   195
      6.2.1   Test data   196
      6.2.2   Simulation control   197
      6.2.3   Limiting data sets   198
      6.2.4   Applying synchronized data   199
      6.2.5   Synchronized display of results   200
      6.2.6   An interactive testbench   201
      6.2.7   Random time intervals   204
      6.2.8   Buffered data application   205
6.3   Design Verification   206
6.4   Assertion Verification   207
      6.4.1   Assertion verification benefits   208
      6.4.2   Open verification library   208
      6.4.3   Using assertion monitors   209
      6.4.4   Assertion templates   216
6.5   Text Based Testbenches   219
6.6   Summary   220
      Problems   220
      Suggested Reading   221

Chapter 7. Detailed Modeling   223

7.1   Switch Level Modeling   223
    7.1.1   Switch level primitives   224
    7.1.2   The basic switch   225
    7.1.3   CMOS gates   226
    7.1.4   Pass gate logic   230
    7.1.5   Switch level memory elements   234
7.2   Strength Modeling   241
    7.2.1   Strength values   242
    7.2.2   Strength used in resolution   244
    7.2.3   Strength reduction   247
7.3   Summary   250
    Problems   250
    Suggested Reading   251

Chapter 8. RT Level Design and Test   253

8.1   Sequential Multiplier   253
    8.1.1   Shift-and-add multiplication process   254
    8.1.2   Sequential multiplier design   256
    8.1.3   Multiplier testing   261
8.2   von Neumann Computer Model   265
    8.2.1   Processor and memory model   265
    8.2.2   Processor model specification   266
    8.2.3   Designing the adding CPU   267
    8.2.4   Design of datapath   268
    8.2.5   Control part design   269
    8.2.6   Adding CPU Verilog description   270
    8.2.7   Testing adding CPU   275
8.3   CPU Design and Test   281
    8.3.1   Details of processor functionality   281
    8.3.2   SAYEH datapath   283
    8.3.3   SAYEH Verilog description   287
    8.3.4   SAYEH top-level testbench   298
    8.3.5   Sorting test program   304
    8.3.6   SAYEH hardware realization   304
8.4   Summary   306
    Problems   306
    Suggested Reading   307

Appendix A. List of Keywords   309

Appendix B. Frequently Used System Tasks and Functions   311

Appendix C. Compiler Directives   319

Appendix D. Verilog Formal Syntax Definition   321

Appendix E. Verilog Assertion Monitors   345

Index   375

Preface

This book is on the IEEE Standard Hardware Description Language based on the Verilog® Hardware Description Language (Verilog HDL), IEEE Std 1364–2001. The intended audiences are engineers involved in various aspects of digital systems design and manufacturing and students with the basic knowledge of digital system design. The emphasis of the book is on using Verilog HDL for the design, verification, and synthesis of digital systems. We will discuss Register Transfer (RT) level digital system design, and discuss how Verilog can be used in this design flow.

In the last few years RT level design of digital systems has gone through significant changes. Beyond simulation and synthesis that are now part of any RTL design process, we are looking at testbench generation and automatic verification tools. As with any book on Verilog, this book covers digital design and Verilog for simulation and synthesis. However, to ready design engineers for designing, testing, and verifying large digital system designs, the book contains material for testbench development and verification. The subjects of testbench and verification are introduced in Chapter 1. Chapter 2 onwards we concentrate on Verilog for design and synthesis. This will teach the readers efficient Verilog coding techniques for describing actual hardware components. When all of Verilog from a design point of view is presented, we turn our attention to test and verification. Chapter 6 covers testbench development techniques and use of assertion verification monitors for better analysis of a design. Toward the end of the book we put together our coding techniques for synthesis and testbench development, and present several RT level designs from design specification to verification.

Embedded in the presentation of the language, the book provides a review of digital system design and computer architecture concepts. This review is useful for relearning these concepts as demanded by new design methodologies and hardware description language based design tools. For practicing engineers the flow of the book, which starts from introductory material and advances into complex digital design concepts, provides a self-sufficient learning tool. The material is suitable for an upper division undergraduate or a first year graduate course. For a one-semester course on the Verilog HDL language and its use in a digital system design environment, the book can be used in its entirety. The book can also be used as a supplement for graduate and undergraduate digital system design and computer organization courses.

Overview of the Chapters

Chapter overviews are presented below. This material is intended to help a reader concentrate on parts of the book that he or she finds suit able to his or her needs best. Chapters 1 and 2 are introductory, and contain material with which many readers may already be familiar. It is, however, recommended that these chapters not be completely omitted, even by experienced readers. The Verilog language is presented in Chapter 3 and includes the details of language syntax and semantics. The next two chapters (4 and 5) concentrate on Verilog for describing hardware from a design point of view. This is followed by a chapter on testing. Together, Chapters 4, 5, and 6 cover use of Verilog for design and test of digital systems. Chapter 7, which is on detailed modeling, is useful for VLSI designers. The last example in Chapter 8 is a complete processor that is modeled for synthesis and a complete testbench is developed for it.

Chapter 1 gives an overview of digital design process and the use of hardware description languages in this process. Simulation, synthesis, formal verification, and assertion verification are discussed in this chapter.

Chapter 2 shows various ways hardware components can be described in Verilog. The purpose of this chapter is to give the reader a general overview of the Verilog language.

Chapter 3 discusses the complete Verilog language structure. The focus of the chapter is more on the linguistic issues and not on modeling hardware components. A general understanding of the language is necessary before it can be used for hardware modeling. Writing Verilog for describing hardware is discussed in the chapters that follow this chapter.

Chapter 4 starts with gates and ends with high-level Verilog constructs for description of combinational circuits. Concurrency and timing will be discussed in the examples of this chapter. Except for specification of timing parameters, codes discussed in this chapter are synthesizable. A section in this chapter presents rules for writing synthesizable combinational circuits.

Chapter 5 discusses modeling and description of sequential circuits in Verilog. The chapter begins with models of memory and shows how they can be specified in Verilog. Registers, counters, and state machines are discussed in this chapter. A section in this chapter presents rules for writing synthesizable sequential circuits.

Chapter 6 is on writing testbenches in Verilog. The previous two chapters discussed Verilog from a hardware design point of view, and this chapter shows how components described as such can be tested. We talk about data generation, response analysis, and assertion verification.

Chapter 7 covers switch level modeling and detailed representation of signals in Verilog. This material is geared more for those using Verilog as a modeling language and less for designers. VLSI structures can be described by Verilog constructs discussed here.

Chapter 8 shows complete RTL design flow, from problem specification to test. We show several complete examples that take advantage of material of Chapters 4, 5, and 6 for description, simulation, verification, and synthesis of digital systems. Examples in this chapter take advantage of text IO facilities of Verilog for storing test data and circuit responses.

Appendix A contains Verilog keywords. Appendix B lists commonly used system tasks and briefly describes each task. Appendix C lists Verilog compiler directives and explains their use. Appendix D presents the standard IEEE Verilog HDL syntax. Language constructs terminals and nonterminals are presented here in a formal grammar representation. Appendix E presents the OVL assertion monitors. After a brief description of each assertion monitor its parameters and arguments are explained.

Suggested Reading Flow

The book teaches the Verilog language for RT level design, simulation, verification, and synthesis of digital systems. For a complete comprehension of these issues, or for a complete one-semester graduate course, the book is recommended in its entirety. However, for specific needs and requirements or for an undergraduate course on automated design methodologies, parts of the book can also be used. The following paragraphs present several such uses.

For a hardware designer interested in learning about synthesis, Chapters 4 and 5 are the most important ones. For such users, Chapter 3 can be used as a reference, and Chapter 6, which is on testbench development, can be studied as needed. When the designer is ready to consider complete systems, Chapter 8 is recommended.

Chapter 2 is introductory and provides an overview of the language. For a student using Verilog in a lower-level undergraduate course, this chapter is a good starting point for learning the language. More complex parts of the language can then be learned as needed.

Chapter 8 can be used for learning computer organization concepts and the use of Verilog in description of these structures. Readers familiar with Verilog can use their knowledge to learn the inter-workings of CPU structures, instruction execution, and testing large systems.

The flow of the book is such that it provides a complete knowledge of Verilog using the same flow as that used in teaching hardware design in most 4-year Computer Engineering programs. The following outlines indicate various applications of the book for beginners, undergraduate students, graduate students, designer engineers, modelers, and system designers.

1. General introduction for a lower-level undergraduate course or an entry level design engineer:
   - *Chapters 1–2.*   Design flow and Verilog overview
   - *Chapters 4–5.*   Combinational and sequential circuits for synthesis

2. Advanced logic design for a senior-level course or an advanced design engineer with some familiarity with design flow and Verilog syntax:
   - *Chapters 1–2.*   A review of Verilog-based design
   - *Chapter 3.*   Language semantics and constructs
   - *Chapters 4–5.*   Combinational and sequential circuits for synthesis
   - *Chapter 6.*   Test methods

3. Advanced system design for a senior-level course or an advanced system design engineer with some familiarity with design flow and Verilog syntax:
   - *Chapters 1–2.*   A review of Verilog-based design
   - *Chapter 3.*   Use as reference as needed
   - *Chapters 4–5.*   Combinational and sequential circuits for synthesis
   - *Chapter 6.*   Test methods
   - *Chapter 8.*   Top-down design of systems

4. Advanced modeling and system design for a graduate-level course or an advanced VLSI design engineer:
   - *Chapters 1–2.*   A review of Verilog-based design
   - *Chapter 3.*   Use as reference as needed
   - *Chapters 4–5.*   Combinational and sequential circuits for synthesis
   - *Chapter 6.*   Test methods
   - *Chapter 7.*   Switch level and CMOS modeling
   - *Chapter 8.*   Top-down design of systems

5. Parallel with undergraduate Computer Engineering program:
   - Use Chapters 1 and 2 early in a digital logic design course
   - Use Chapters 4 and 5 in a digital logic design course in parallel with discussion of combinational and sequential circuits
   - Use Chapter 6 in a technical elective design course

- Use Chapter 7 in the senior-level VLSI course
- Use Chapter 8 in the Junior or Sophomore computer architecture course

Code Examples

Among many tasks involved in the preparation of the manuscript, for a book describing a language that is as example oriented as this book, selecting appropriate set of examples and presenting them to the reader are of special importance. For every design example presented in this book, a testbench is generated and the design has been tested. With every example, there is a logic design concept and there are several Verilog constructs and features that are covered. The set of examples is chosen to present the complete Verilog language for synthesis. These examples start with using simple Verilog constructs and progressively move into more complex ones. Parallel with the flow of language constructs, the book starts with using simple logic design concepts, such as using basic gates for combinational circuits, and moves into advanced logic design concepts such as queues and processors.

The CD accompanying this book includes simulation, synthesis, and device programming software tools. Verilog description of the examples of this book and their testbenches are also included on this CD. For the instructors using this book in an educational setting, solutions for the end of chapter problems and Power Point lecture slides can be obtained from the author or the publisher.

Acknowledgments

Guidelines, comments, reviews, and support of many people helped the development of this book, and the author wishes to thank them. The style used for presenting the material is based on simple examples that cover a certain topic and discussing the issues that the example covers. As with the other books that I have written, I have used guidelines and writing philosophy of the late Professor Fredrick J. Hill of the University of Arizona, with whom I worked many years as a student and a research associate. My students and colleagues were particularly helpful in the development of this book. In the past 15 years, my students at the University of Tehran, Northeastern University and National Technological University have been very helpful in bringing up ideas for more illustrative examples. Many examples come from exam and homework questions that these students had to struggle with.

At the start of this writing project, my associate, Ms. Fatemeh Asgari assumed responsibility for managing the preparation of the manuscript. Organizing the efforts for manuscript preparation, managing the timing of this task with my many other tasks has been a very challenging task for her. Her crystal ball always told the truth about how bad I would miss my deadlines. Students at the University of Tehran, Armin Alaghi, Najmeh Fakhraie, Amirali Ghofrani, Aida Hasani, and Mahsan Rofouei, were very helpful in completion of this project. They helped reviewing the manuscript, coding, preparing the artwork, and suggesting ways of improving the flow of the book for different levels of audiences.

Most of all, I thank my wife, Irma Navabi, for help encouragement and understanding of my working habits. Such an intensive work could not be done if I did not have support of my wife and my two sons, Aarash and Arvand. I thank them for this and other scientific achievements I have had.

Zainalabedin Navabi, Ph.D.
Boston, Massachusetts
navabi@ece.neu.edu

Digital System Design Automation with Verilog

As the size and complexity of digital systems increase, more computer-aided design (CAD) tools are introduced into the hardware design process. Early simulation and primitive hardware generation tools have given way to sophisticated design entry, verification, high-level synthesis, formal verification, and automatic hardware generation and device programming tools. Growth of design automation tools is largely due to hardware description languages (HDLs) and design methodologies that are based on these languages. Based on HDLs, new digital system CAD tools have been developed and are now widely used by hardware designers. At the same time research for finding better and more abstract hardware languages continues. One of the most widely used HDLs is the Verilog HDL. Because of its wide acceptance in digital design industry, Verilog has become a must-know for design engineers and students in computer-hardware-related fields.

This chapter presents tools and environments that are based on Verilog and are available to a hardware designer for automating his or her design process, and hence improving the final product's time to market. We discuss steps involved in taking a hierarchical, high-level design from a Verilog description of the design to its implementation in hardware. Processes and terminologies are illustrated here. We discuss available electronic design automation (EDA) tools that are based on Verilog, and talk about their role in an automated design environment. The last section of this chapter discusses some of the properties of Verilog that make this language a good choice for designers and modelers of hardware.

1.1   Digital Design Flow

For the design of a digital system using an automated design environment, the design flow begins with specification of the design at various levels of abstraction and ends with generating netlist for an application specific integrated circuits (ASIC), layout for a custom IC, or a program for a programmable logic devices (PLD). Figure 1.1 shows steps involved in this design flow.

In the design entry phase, a design is specified as a mixture of behavioral Verilog code, instantiation of Verilog modules, and bus and wire assignments. A design engineer is also responsible for generating testbenches

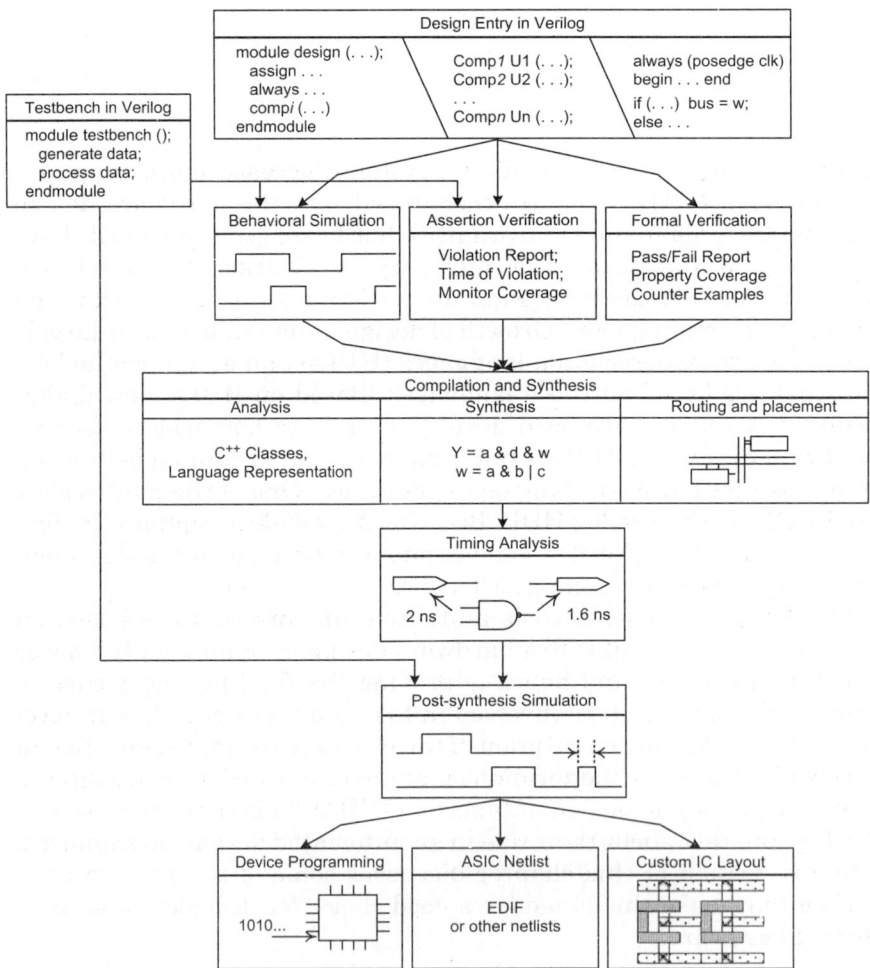

Figure 1.1   FPLD Design Flow

for his or her design for verification of the design and later for verifying the synthesis output. Design verification can be done by simulation, assertion verification, formal verification, or a mix of all three. After performing this design validation phase (this is called the presynthesis verification), this design is taken through the synthesis process to translate it into actual hardware of a target device. Here, target device refers to the specific field programmable logic device (FPLD) that is being programmed, the ASIC that is being manufactured by an outside source, or the custom IC that is being fabricated. After the synthesis process and before the actual hardware is generated, another simulation, which is referred to as postsynthesis simulation, is done. This simulation can take advantage of the same testbench generated for the Verilog model of the system before it is synthesized. This way, the behavioral model of the design and its hardware model are tested with the same data. The difference between pre- and postsynthesis simulations is in the level of details obtained from each simulation.

The sections that follow elaborate on each of the blocks shown in Fig. 1.1. Most Verilog based EDA environments provide blocks shown in this figure.

1.1.1 Design entry

The first step in the design of a digital system is the design entry phase. In this phase, the design is described in Verilog in a top-down hierarchical fashion. A complete design may consist of components at the gate or transistor level, behavioral parts describing high-level functionality of a hardware module, or components described by their bussing structure.

Because high-level Verilog designs are usually described at the level that specifies system registers and transfer of data between registers through busses, this level of system description is referred to as register transfer level (RTL). A complete design described as such has a clear hardware correspondence. Verilog constructs used in an RT level design are procedural statements, continuous assignments, and instantiation statements.

Verilog *procedural statements* are used for high-level behavioral descriptions. A system or a component is described in a procedural fashion similar to the way processes are described in a software language. For example, we can describe a component by checking its input conditions, setting flags, waiting for events to occur, monitoring handshaking signals, and issuing outputs. Describing a system procedurally, Verilog **if-else**, **case** and other software-language-like constructs can be used.

Verilog *continuous assignments* are statements for representing logic blocks, bus assignments, and bus and input/output interconnect specifications. Combined with boolean and conditional operations, these language constructs can be used for describing components and systems in terms of their register and bus assignments.

Verilog *instantiation statements* are for using lower-level components in an upper-level design. Instead of describing behavior, functionality, or bussing of a system, we can describe a system in Verilog in terms of its lower-level components. These subcomponents can be as small as a gate or a transistor, or as large as a complete processor.

1.1.2   Testbench in Verilog

A system designed in Verilog must be simulated and tested for functionality before it is turned into hardware. In this simulation pass, design errors and incompatibility of components used in the design can be detected. Simulating a design requires generation of test data and observation of simulation results. This process can be done by use of a Verilog module that is referred to as a testbench. A Verilog testbench uses high-level constructs of this language for data generation, response monitoring, and even handshaking with the design. Inside the testbench, the design that is being simulated is instantiated. The testbench together with the design forms a simulation model used by a Verilog simulation engine.

1.1.3   Design validation

An important task in any digital design is design validation. Design validation is the process that a designer checks his or her design for any design flaws that may have occurred in the design process. A design flaw can happen due to ambiguous problem specifications, designer errors, or incorrect use of parts in the design. Design validation can be done by simulation, assertion verification, or formal verification.

1.1.3.1   Simulation.   Simulation for design validation is done before a design is synthesized. This simulation pass is also referred to as behavioral, RT level, or presynthesis simulation. At the RT level a design includes clock-level timing but no gate and wire delays are included. Simulation at this level is accurate to the clock level. Timing of RT-level simulation is at the clock level and does not usually consider hazards, glitches, race conditions, setup and hold violations, and other detailed timing issues. The advantage of this simulation is its speed compared with simulations at the gate or transistor levels.

Simulation of a design requires test data, and usually Verilog simulation environments provide various methods for application of these data to the design being tested. Test data can be generated graphically using waveform editors, or through a testbench. Figure 1.2 shows two alternatives for defining test input data for a simulation engine. Outputs of simulators are in the form of waveforms (for visual inspection) and text for large designs for machine processing.

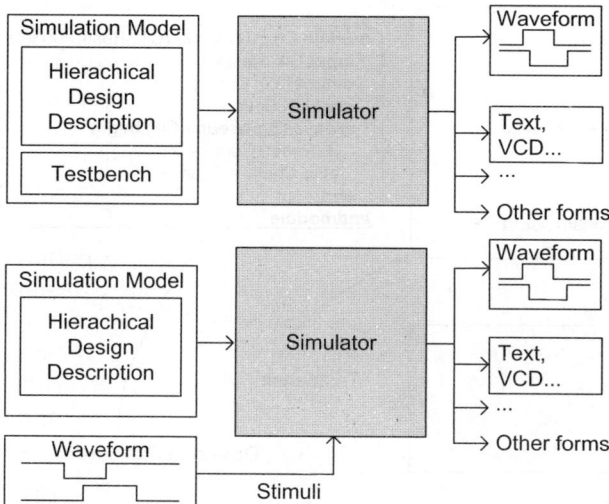

Figure 1.2 Using a Testbench or a Waveform Editor for Simulation

For simulating with a Verilog testbench, the testbench instantiates the design under test, and as part of the code of the testbench it applies test data to the instantiated circuit. Figure 1.3 shows a Verilog code of a counter circuit, its testbench, and its simulation results in form of a waveform. As shown here, simulation validates the functionality of the counter circuit being tested. With every clock pulse the counter is incremented by 1. Note in the timing diagram that the counter output changes with the rising edge of the clock and no gate delays and propagation delays are shown. Simulation results show the correct functionality of the counter regardless of the clock frequency.

Obviously, an actual hardware component behaves differently. Based on the timing and delays of the parts used, there will be a nonzero delay between the active edge of the clock and the counter output. Furthermore, if the clock frequency applied to an actual part is too fast for propagation of values within the gates and transistors of a design, the output of the design becomes unpredictable.

The simulation shown here is not provided with the details of the timing of the hardware being simulated. Therefore, potential timing problems of the hardware that are due to gate delays cannot be detected. This is typical of a presynthesis or high-level behavioral simulation. What is being verified in Fig. 1.3 is that our counter counts binary numbers. How fast the circuit works and what clock frequency it requires can only be verified after the design is synthesized.

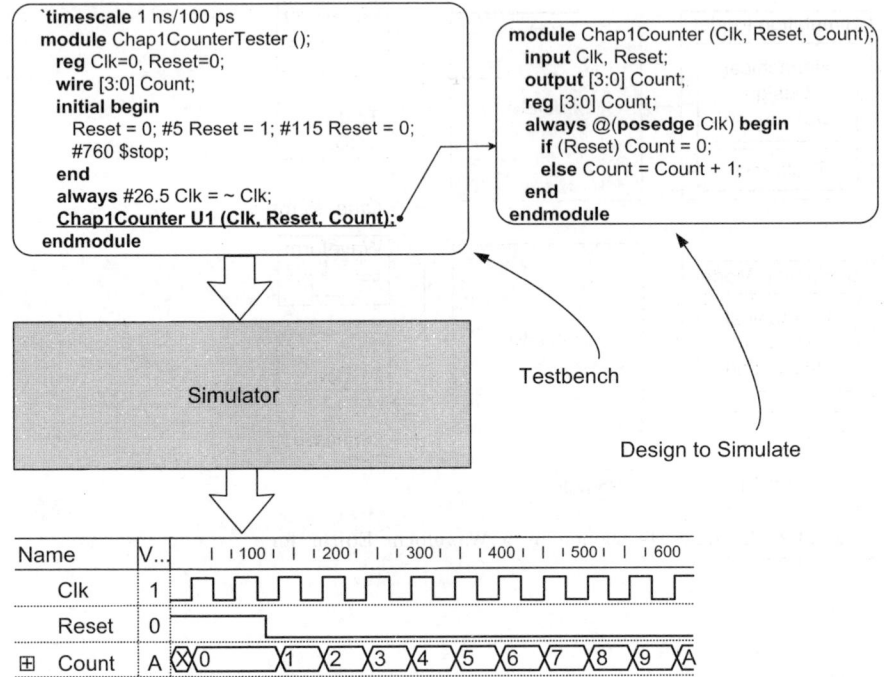

```
`timescale 1 ns/100 ps
module Chap1CounterTester ();
  reg Clk=0, Reset=0;
  wire [3:0] Count;
  initial begin
    Reset = 0; #5 Reset = 1; #115 Reset = 0;
    #760 $stop;
  end
  always #26.5 Clk = ~ Clk;
  Chap1Counter U1 (Clk, Reset, Count);
endmodule
```

```
module Chap1Counter (Clk, Reset, Count);
  input Clk, Reset;
  output [3:0] Count;
  reg [3:0] Count;
  always @(posedge Clk) begin
    if (Reset) Count = 0;
    else Count = Count + 1;
  end
endmodule
```

Figure1.3   Verilog Simulation with a Testbench

1.1.3.2 Assertion verification.

Instead of having to inspect simulation results manually or by developing sophisticated testbenches, assertion monitors can be used to continuously check for design properties while the design is being simulated. Assertion monitors are put in the design being simulated by the designer. The designer decides that if the design functions correctly, certain conditions have to be met. These conditions are regarded as design properties, and assertion monitors are developed by designer to assert that these properties are not violated. An assertion monitor fires if a design property put in by the designer is violated. This alerts the designer that the design is not functioning according to the designer's expectation. Open verification library (OVL) provides a set of assertion monitors for monitoring common design properties. Designers can use their own assertions and use them in conjunction with their testbenches.

1.1.3.3 Formal verification.

Formal verification is the process of checking a design against certain properties. When a design is completed, the designer develops a set of properties reflecting correct behavior of his or her design. A formal verification tool examines the design to make sure that the described properties hold under all conditions. If a situation is found that the property will not hold, the property is said to have been violated. Input conditions that make a property fail are regarded as the property's counter examples. Property coverage indicates how much of the complete design is exercised by the property.

1.1.4   Compilation and synthesis

Synthesis is the process of automatic hardware generation from a design description that has an unambiguous hardware correspondence. A Verilog description for synthesis cannot include signal and gate level timing specifications, file handling, and other language constructs that do not translate to sequential or combinational logic equations. Furthermore, Verilog descriptions for synthesis must follow certain styles of coding for combinational and sequential circuits. These styles and their corresponding Verilog constructs are defined under Verilog for RTL synthesis.

In the design process, after a design is successfully entered and its presynthesis simulation results have been verified by the designer, it must be compiled to make it one step closer to an actual hardware on silicon. This design phase requires specification of the hardware that the design is to be realized in. For example, we have to specify a specific ASIC, or a field programmable gate array (FPGA) part as our "target hardware." When the target hardware is specified, technology files of that hardware (ASIC, FPGA, or custom IC) with detailed timing and functional specification become available to the compilation process. The compilation process, translates various parts of the design to an intermediate format (analysis phase), links all parts together, generates the corresponding logic (synthesis phase), places and routes components of the target hardware, and generates timing details.

Figure 1.4 shows the compilation process and a graphical representation for each of the compilation phase outputs. As shown, the input of this phase is a hardware description that consists of various levels of Verilog, and its output is a detailed hardware for programming an FPLDor manufacturing an ASIC.

1.1.4.1   Analysis.   A complete design that is described in Verilog may consist of behavioral Verilog, bus and interconnection specifications, and wiring of other Verilog components. Before the complete design is turned into hardware, the design must be analyzed and a uniform format must be generated for all parts of the design. This phase also checks the syntax and semantics of the input Verilog code.

1.1.4.2   Generic hardware generation.   After obtaining a uniform presentation for all components of a design, the synthesis pass begins its

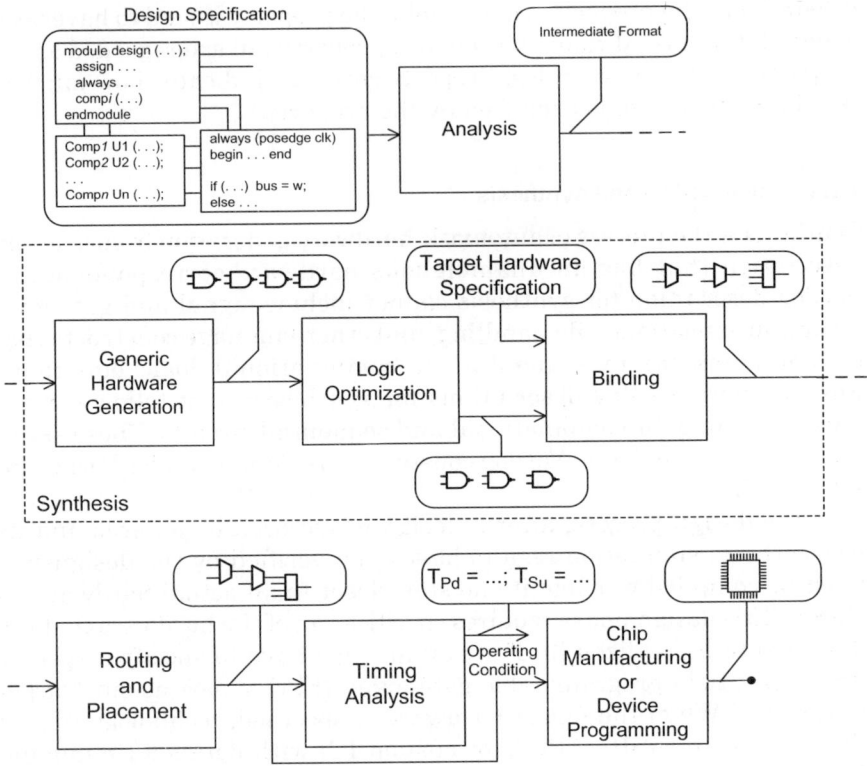

Figure 1.4   Compilation and Synthesis Process

operation by turning the design into a generic hardware format, such as a set of boolean expressions or a netlist of basic gates.

1.1.4.3   Logic optimization.

The next phase of synthesis, after a design has been converted to a set of boolean expressions, is the logic optimization phase. This phase is responsible for reducing expressions with constant input, removing redundant logic expressions, two-level minimization, and multilevel minimization that include logic sharing.

This is a very computationally intensive process, and some tools allow users to decide on the level of optimization. Output of this phase is in the form of boolean expressions, tabular logic representations, or primitive gate netlists.

1.1.4.4   Binding.

After logic optimization, the synthesis process uses information from target hardware to decide exactly what logic elements and cells are needed for the realization of the circuit that is being

designed. This process is called binding and its output is specific to the
FPLD, ASIC, or custom IC being used.

1.1.4.5 Routing and placement. The routing and placement phase
decides on the placement of cells of the target hardware. Wiring inputs
and outputs of these cells through wiring channels and switching areas
of the target hardware are determined by the routing and placement
phase. The output of this phase is specific to the hardware being used
and can be used for programming an FPLD or manufacturing an ASIC.

An example of a synthesis run is shown in Fig. 1.5. In this figure, the
counter circuit used in the simulation run of Fig. 1.3 is being synthesized.
In addition to the Verilog description of the design, the synthesis tool
shown requires specification of the target hardware to synthesize to.
The output of the synthesis tool is a list of gates and flip-flops available

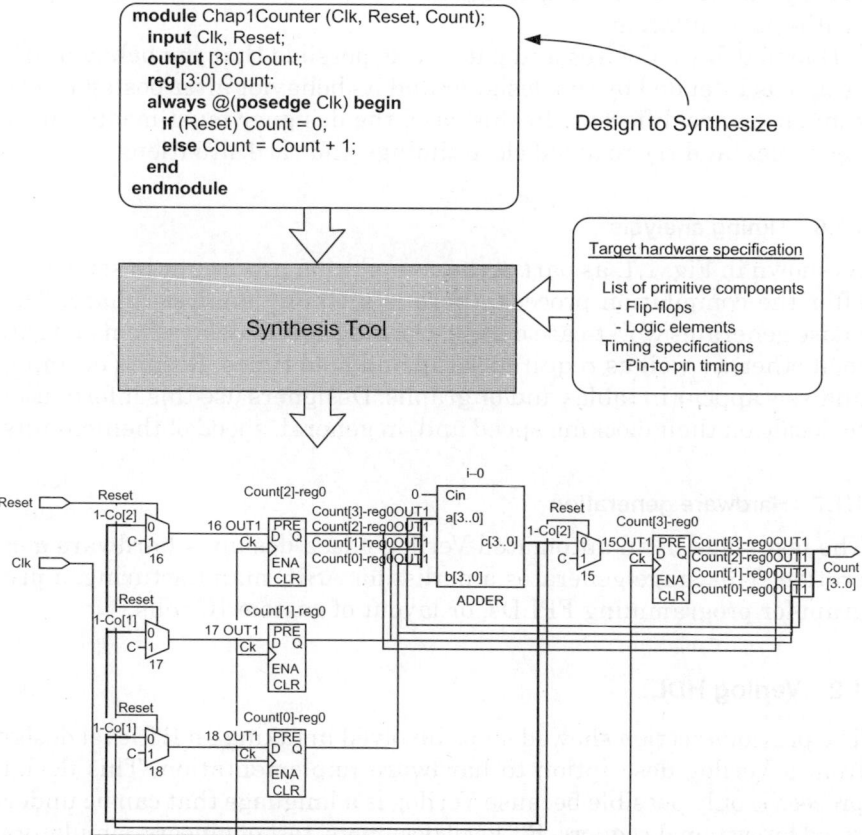

Figure 1.5 An Example Synthesis Run

in the target hardware, and their interconnections. A graphical representation of this output that is automatically generated by the synthesis tool of Altera's Quartus II is shown in Fig. 1.5.

1.1.5   Postsynthesis simulation

After synthesis is done, the synthesis tool generates a complete netlist of target hardware components and their timings. The details of gates used for the implementation of the design are described in this netlist. The netlist also includes wiring delays and load effects on gates used in the postsynthesis design. The netlist output is made available in various netlist formats including Verilog. Such a description can be simulated and its simulation is referred to postsynthesis simulation. Timing issues, determination of a proper clock frequency and race, and hazard considerations can only be checked by a postsynthesis simulation run after a design is synthesized. As shown in Fig. 1.1, the same testbench testing the original Verilog design before synthesis can be used for postsynthesis simulation.

Due to delays of wires and gates, it is possible that the behavior of a design as intended by the designer and its behavior after postsynthesis simulation are different. In this case, the designer must modify his or her design and try to avoid close timings and race situations.

1.1.6   Timing analysis

As shown in Fig. 1.1, as part of the compilation process, or in some tools after the compilation process, there is a timing analysis phase. This phase generates worst-case delays, clocking speed, delays from one gate to another, as well as required setup and hold times. Results of timing analysis appear in tables and/or graphs. Designers use this information to decide on their clocking speed and, in general, speed of their circuits.

1.1.7   Hardware generation

The last stage in an automated Verilog-based design is hardware generation. This stage generates a netlist for ASIC manufacturing, a program for programming FPLDs, or layout of custom IC cells.

1.2   Verilog HDL

The previous section showed steps involved in taking an RT level design from a Verilog description to hardware implementation. This design process is only possible because Verilog is a language that can be understood by system designers, RT level designers, test engineers, simulators, synthesis tools, and machines. Because of this important role in design,

Verilog has become an IEEE standard. The standard is used by users as well as tool developers.

1.2.1 Verilog evolution

Verilog was designed in early 1984 by Gateway Design Automation. Initially the original language was used as a simulation and verification tool. After the initial acceptance of this language by electronic industry, a fault simulator, a timing analyzer, and later in 1987, a synthesis tool was developed based on this language. Gateway Design Automation and its Verilog-based tools were later acquired by Cadence Design System. Since then, Cadence has been a strong force behind popularizing the Verilog hardware description language.

In 1987 VHDL became an IEEE standard hardware description language. Because of its Department of Defense (DoD) support, VHDL was adapted by the U.S. government for related projects and contracts. In an effort for popularizing Verilog, in 1990, OVI (Open Verilog International) was formed and Verilog was placed in public domain. This created a new line of interest in Verilog for the users and EDA vendors.

In 1993, efforts for standardization of this language started. Verilog became the IEEE standard, IEEE Std. 1364-1995, in 1995. Already having simulation tools, synthesizers, fault simulation programs, timing analyzers, and many of their design tools developed for Verilog, this standardization helped further acceptance of Verilog in electronic design communities.

A new version of Verilog was approved by IEEE in 2001. This version that is referred to as Verilog-2001 is the present standard used by most users and tool developers. New features for external file access for read and write, library management, constructs for design configuration, higher abstraction level constructs, and constructs for specification of iterative structures, are some of the features added to this version of Verilog. Work on improving this standard continues in various IEEE sponsored study groups.

1.2.2 Verilog attributes

Verilog is a hardware description language for describing hardware from transistor level to behavioral. The language supports timing constructs for switch level timing simulation and at the same time, it has features for describing hardware at the abstract algorithmic level. A Verilog description may consist of a mix of modules at various abstraction levels with different degrees of detail.

1.2.2.1 Switch level. Features of the language that make it ideal for switch level modeling and simulation includes primitive unidirectional and bidirectional switches with parameters for delay and charge storage. Circuit delays may be modeled as propagation delay, rise and fall delay, and line delays. The charge storage feature at this level of abstraction in Verilog makes this language capable of describing dynamic complimentary metal oxide semicondutor (CMOS) and metal oxide semiconductor (MOS) circuits.

1.2.2.2 Gate level. Gate level primitives with predefined parameters provide a convenient platform for netlist representation and gate level simulation. For more detailed and special purpose gate simulations, gate components may be defined at the behavioral level. Verilog also provides utilities for defining primitives with special functionalities. A simple 4-value logic system is used in Verilog for signal values. However, for more accurate logic modeling, Verilog signals also include 16 levels of strength in addition to the four values.

1.2.2.3 Pin-to-pin delay. A utility for timing specification of components at the input/output level is provided in Verilog. This utility can be used for back annotation of timing information in original predesigned descriptions. Moreover, the pin-to-pin language facility enables modelers to fine-tune timing behavior of their models based on physical implementations.

1.2.2.4 Bussing specifications. Bus and register modeling utilities are provided in Verilog. For various bus structures, Verilog supports predefined wire and bus resolution functions using its 4-value logic value system. Combination of bus logic and resolution-functions enable modeling of most physical bus types. For register modeling, high-level clock representation and timing-control constructs can be used for representation of registers with various clocking and resetting schemes.

1.2.2.5 Behavioral level. Procedural blocks of Verilog enable algorithmic representations of hardware structures. Constructs similar to those in software programming languages are provided for describing hardware at this level.

1.2.2.6 System utilities. System tasks in Verilog provide designers with tools for testbench generation, file access for read and write, data handling, data generation, and special hardware modeling. System utilities for reading memory and programmable logic array (PLA) images provide convenient ways of modeling these components. Verilog display and I/O tasks can be used to handle all inputs and outputs for data application and simulation. Verilog allows random access to files for read and write operations.

1.2.2.7 PLI. Programming language interface (PLI) of Verilog provides an environment for accessing Verilog data structures using a library of C-language functions.

1.2.3 The Verilog language

The Verilog HDL satisfies all requirements for design and synthesis of digital systems. The language supports hierarchical description of hardware from system to gate or even switch level. Verilog has strong support at all levels for timing specification and violation detection. Timing and concurrency required for hardware modeling are specially emphasized.

In Verilog a hardware component is described by the *module_declaration* language construct. Description of a module specifies a component's input and output list as well as internal component busses and registers. Within a **module**, concurrent assignments, component instantiations, and procedural blocks can be used to describe a hardware component.

Several modules can hierarchically be instantiated to form other hardware structures. Leaves of a hierarchical design specification may be modules, primitives, or user defined primitives. For simulating a design, it is expected that all leaves of the hierarchy are individually compiled.

Many Verilog tools and environments exist that provide simulation, fault simulation, formal verification, and synthesis. Simulation environments provide graphical front-end programs and waveform editing and display tools. Synthesis tools are based on a subset of Verilog. For synthesizing a design, target hardware, e.g., specific FPGA or ASIC, must be known.

1.3 Summary

This chapter gave an overview of mechanisms, tools, and processes used for taking a design from the design stage to a hardware implementation. This overview contained information that will become clearer in the chapters that follow. This chapter also provided the reader with the history of Verilog evolution. With this standard HDL, the efforts of tool developers, researchers, and software vendors have become more focused, resulting in better tools and more uniform environments. The next chapter presents an overview of Verilog.

Problems

1.1 Study Altera's FPGA design environment and see their simulation and synthesis environments. How do you compare Altera's environment with the simulation and synthesis environments discussed in this chapter?

1.2   Search for several commercial formal verification tools and generate a report of their input formats, capabilities, and their verification utilities.

1.3   Study Accellera's OVL library and discuss how this library helps the design automation process.

1.4   Study SystemC and discuss tools available for this language.

1.5   Study the VHDL hardware description language and discuss tools available for this language.

Suggested Reading

Accellera, *Open Verification Library: Assertion Monitor Reference Manual*, www.accellera.org, v1.0, 2005.

Bening, L., and H. D. Foster, *Principles of Verifiable RTL Design Second Edition–A Functional Coding Style Supporting Verification Processes in Verilog*, 2d ed. Springer, Boston, MA, 2001, ISBN: 0792373685.

Brown, S., and Z. Vranesic, *Fundamentals of Digital Logic with Verilog Design*, McGraw-Hill, New York, 2002, ISBN: 0-07-283878-7.

IEEE Std 1364-2001, *IEEE Standard Verilog Language Reference Manual*, SH94921-TBR (print) SS94921-TBR (electronic), ISBN 0-7381-2827-9 (print and electronic), 2001.

IEEE Std 1076-2002, *IEEE Standard VHDL Language Reference Manual*, SH94983-TBR (print) SS94983-TBR (electronic), ISBN 0-7381-3247-0 (print) 0-7381-3248-9 (electronic), 2002.

Lam, W. K., *Hardware Design Verification: Simulation and Formal Method-Based Approaches*, Prentice Hall PTR, New Jersey, 2005, ISBN: 0131433474.

Navabi, Z., *Digital Design and Implementation with Field Programmable Devices*, Kluwer Academic Publishers, Boston, MA, 2005, ISBN: 1-4020-8011-5.

Navabi, Z., *Verilog Computer-Based Training Course*, CBT CD with hardcopy User's manual, McGraw-Hill, New York, 2002, ISBN 0-07-137473-6.

2

Register Transfer Level Design with Verilog

The intent of this chapter is to present an overview of Verilog and the design styles in which this language is used. Various concepts of a language, be it a software or a hardware language, are interdependent. A general knowledge of the language is therefore needed before more detailed features of the language can be discussed. This chapter discusses register transfer level (RTL) design of digital systems and shows how Verilog is used for description, testing, simulation, and synthesis of various RT level components of a design. With this presentation, we will also give an overview of Verilog and set the stage ready for more elaborate discussion of Verilog constructs in the chapters that follow.

In the sections that follow we will first discuss RT level design and how a complete system is put together at this abstraction level. The section that follows this introductory material presents basic structures of Verilog such as modules, ports, and utilities for test and verification of design components. The rest of this chapter discusses coding of a complete RT level design in Verilog. This part serves as an overview of the complete Verilog HDL language.

2.1 RT Level Design

Design of small hardware components can usually be done by describing the hardware for synthesis and synthesizing and implementing the design by appropriate computer aided design tools. On the other hand, a large design requires proper planning, architectural design, and partitioning before its various parts can be written in Verilog for synthesis. Taking a high-level description of a design, partitioning it, coming up with an architecture for it (i.e., designing its bussing structure), and then describing and implementing various components of this architecture is referred to as RT level design.

2.1.1 Control/data partitioning

The first step in an RT level design is the partitioning of the design into a data part and a control part. The data part consists of data components and the bussing structure of the design and the control part is usually a state machine generating control signals that control the flow of data in the data part.

Figure 2.1 shows a general sketch of an RT level design that is partitioned into its data and control parts. We will use this diagram to discuss the two partitions and at the same time show how Verilog may be used for describing an RTL circuit.

2.1.2 Data part

The data part of an RTL design consists of the interconnection of data components that are, registers, combinational logic units, register files, and busses that interconnect them. The data part, which we also refer to as the data path, has external data inputs and outputs, as well as control inputs and outputs from and to the control part. Figure 2.2 shows partial code of the data part of Fig. 2.1 described in Verilog. This partial code shows ports of the *DataPath* **module** and indicates that within this module various data components are specified. Control signals are inputs to the data part and are sent to the data components and busses. This code shows the module header including its name and its ports. Following the header, inputs and outputs, and their dimensions are declared. Texts that are followed by // are comments.

A data component has certain control signals that control its clocking and/or its functionalities.

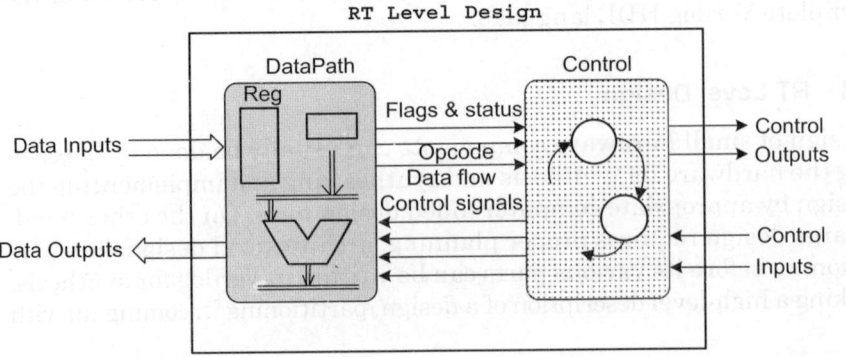

Figure 2.1   Control/Data Partitioning

```
module DataPath
    (DataInput, DataOutput, Flags, Opcodes, ControlSignals);

    input   [15:0] DataInputs;
    output  [15:0] DataOutputs;
    output  Flags, ...;
    output  Opcodes, ...;
    input   ControlSignals, ...;
    // instantiation of data components
    // ...
    // interconnection of data components
    // bussing specification
endmodule
```

Figure 2.2   *DataPath* Module

A module describing a typical data component shows how the component uses its input control signals to perform various operations on its data inputs. Figure 2.3 shows a partial code of a data component.

Busses in the data part of an RTL design have control signals that select their sources and routing of data from one data component to another. The data part has output signals going to the control part that provide flags and status of the data.

2.1.3   Control part

The control part of an RTL design takes control inputs from the data part and external control inputs and depending on its state makes decisions as to when and what control signals to issue.

The control part, which we also refer to as the control unit, consists of one or more state machines that keep the state of the circuit, make decisions based on the current data and data status, and control how data is routed and what operations are performed on the data in the data part.

```
module DataComponent (DataIn, DataOut, ControlSignals);
    input   [7:0] DataIn;
    output  [7:0] DataOut;
    input   ControlSignals;
    // Depending on ControlSignals
    // Operate on DataIn and
    // Produce DataOut
endmodule
```

Figure 2.3   Partial Verilog Code of a Data Component

```
module ControlUnit
        (Flags, Opcodes, ExternalControls, ControlSignals);
    input   Flags, ...;
    input   Opcodes, ...;
    input   ExternalControls, ...;
    output  ControlSignals;
    // Based on inputs decide :
    // What control signals to issue,
    // and what next state to take
endmodule
```

Figure 2.4   Outline of a Controller

Partial Verilog module of Fig. 2.4 shows an outline of tasks handled by the control unit of an RTL design.

2.2   Elements of Verilog

Constructs of the Verilog language are designed for describing hardware modules and primitives. This section presents basic constructs of the language for describing a hardware module.

2.2.1   Hardware modules

The Verilog hardware description language (HDL) is used to describe hardware modules of a system and complete systems. Therefore, the main component of the language, which is a *module*, is dedicated for this purpose. As shown in Fig. 2.5, a module description consists of the keyword **module**, the name of the module, a list of ports of the hardware module, the module functionality specification, and the keyword **endmodule**. Following a module name and its list of ports, usually variables, wires, and module parameters are declared. After the declarations, statements in a module specify its functionality. This part defines how output ports react to changes on the input ports.

```
module module-name
    List of ports;
    Declarations
    ...
    Functional specification of module
    ...
endmodule
```

Figure 2.5   **Module** Specifications

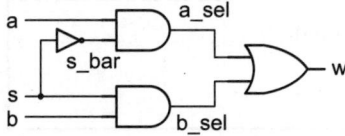

Figure 2.6 A Multiplexer Using Basic Gates

As in software languages, there is usually more than one way a module can be described in Verilog. Various descriptions of a component may correspond to descriptions at various levels of abstraction or to various levels of detail of the functionality of a module. One module description may be at the behavioral level of abstraction with no timing details, while another description for the same component may include transistor-level timing details. A module may be part of a library of predesigned library components and include detailed timing and loading information, while a different description of the same module may be at the behavioral level for input to a synthesis tool. It must be noted that descriptions of the same module need not behave in exactly the same way nor is it required that all descriptions describe a behavior correctly. In a fault simulation environment, faulty modules may be developed to study various failure forms of a component.

In the sections that follow we show a small example and several alternative ways it can be described in Verilog. This presentation is to serve as an introduction to various forms of Verilog constructs for the description of hardware.

2.2.2 Primitive instantiations

Verilog uses different constructs for describing a module with different levels of detail. Verilog basic logic gates are called primitives and for describing a component using these primitives, a construct called primitive instantiation is used. See for example the multiplexer of Fig. 2.6 that is made of AND and OR gates. This structure can be described in Verilog as shown in Fig. 2.7.

```
module MultiplexerA (input a, b, s, output w);
    wire a_sel, b_sel, s_bar;
    not U1 (s_bar, s);
    and U2 (a_sel, a, s_bar);
    and U3 (b_sel, b, s);
    or  U4 (w, a_sel, b_sel);
endmodule
```

Figure 2.7 Primitive Instantiations

```
module MultiplexerB (input a, b, s, output w);
    assign w = (a & ~s) | (b & s);
endmodule
```

Figure 2.8 Assign Statement and Boolean

The first line of this code contains the name of the module, *MultiplexerA*, and its input and output ports. Following this line, intermediate wires are declared. The rest of this code consists of instantiation of **not**, **and**, and **or** gates. These instantiations are done according to the diagram of Fig. 2.6, and their wirings are as indicated in this diagram.

2.2.3 Assign statements

Instead of describing a component using primitive gates, boolean expressions can be used to describe the logic, and Verilog **assign** statements can be used for assigning results of these expressions to various outputs. Our simple multiplexer example can be described as shown in Fig. 2.8.

The statement shown in the body of the *MultiplexerB* module continuously drives *w* with its right-hand side expression.

2.2.4 Conditional expression

In cases where the operation of a unit is too complex to be described by boolean expressions, conditional expressions can be used. Our multiplexer example is described in Fig. 2.9 using an **assign** statement and a conditional operation.

Because conditional expressions mimic if-then-else behavior of software languages, they are very effective in describing complex functionalities. Furthermore, the nesting capability of the conditional operator makes it useful in describing a behavior in a very compact way.

2.2.5 Procedural blocks

In cases where the operation of a unit is too complex to be described by assignment of boolean or conditional expressions, higher-level procedural

```
module MultiplexerC (input a, b, s, output w);
    assign w = s ? b : a;
endmodule
```

Figure 2.9 Assign Statement and Condition Operator

```
module MultiplexerD (input a, b, s, output w);
    reg w;
    always @ (a, b, s) begin
        if (s) w = b;
        else w = a;
    end
endmodule
```

Figure 2.10   Procedural Statement

constructs should be used. Verilog's main construct for procedural specification of hardware is the **always** statement used in the example of
Fig. 2.10.

The example shown in Fig. 2.10 is still another Verilog code for our
multiplexer example discussed previously. In this code, an **always** statement, which is the main procedural body of Verilog, encloses an **if-else**
statement that assigns a or b to w depending on the value of s.

2.2.6 Module instantiations

Still another way of describing a component is by describing its subcomponents and instantiating and wiring these lower-level components
to form the intended upper-level design. Verilog's construct for this
application is called *module_instantiation*, an example of which is shown
in Fig. 2.11.

In this Figure, module *ANDOR* is first defined. Then in *MultiplexerE*,
the four-input *ANDOR* module and an inverter are instantiated to form
the intended 2-to-1 multiplexer. The diagram of Fig. 2.12 corresponds
to the Verilog code of Fig. 2.11.

```
module ANDOR (input i1, i2, i3, i4, output y);
    assign y = (i1 & i2) | (i3 & i4);
endmodule
//
module MultiplexerE (input a, b, s, output w);
    wire s_bar;
    not U1 (s_bar, s);
    ANDOR U2 (a, s_bar, s, b, w);
endmodule
```

Figure 2.11   Module Instantiation

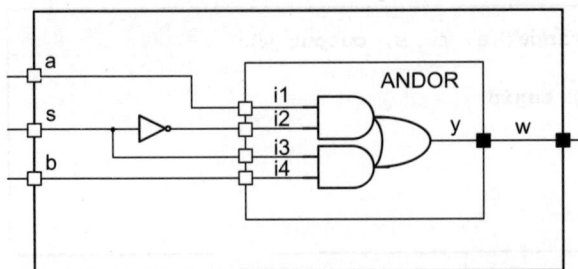

Figure 2.12 Multiplexer Using *ANDOR*

2.3 Component Description in Verilog

As discussed in Sec. 2.1, an RT level design consists of data and control parts. The data part consists of instantiation and wiring of various data components. With the brief introduction to Verilog in the previous section, we are now ready to take a longer step towards giving an overview of this language, by describing simple RT level components.

2.3.1 Data components

Data components generally consist of multiplexers for bus specifications, registers for data storage, flip-flops for flags, and combinational logic units for arithmetic and/or logical operations on data. In what follows, we will show small examples illustrating how such components are coded in Verilog.

2.3.1.1 Multiplexer.   As discussed in the previous section, there are many ways a multiplexer can be described in Verilog. We use an **assign** statement for describing an octal 2-to-1 multiplexer. The multiplexer selects its 8-bit *data0* or *data1* inputs depending on its *sel* input. Figure 2.13 shows the Verilog code for this multiplexer.

As shown in Fig. 2.13, the name of the multiplexer module is *Mux8*. The description begins with the `timescale directive. This directive defines the module's time unit. The *1ns/100ps* used in this example indicates that

```
`timescale 1ns/100ps
module Mux8 (input sel, input [7:0] data1, data0,
             output [7:0] bus1);
   assign #6 bus1 = sel ? data1 : data0;
endmodule
```

Figure 2.13   Octal 2-to-1 Mux

all timing values are in *ns*, and the time precision is *100ps*, or *0.1ns*. This means that we can specify time values with one fractional digit (0.1ns). Following the `timescale directive, the first line of the code specifies the name of the module and its ports. Four input and output ports of *Mux8* are named as *sel*, *data1*, *data0*, and *bus1*. The header also specifies the size of module ports and their modes (input or output). For size specification, Verilog uses square brackets enclosing a vector's dimensions. Since *sel* is a scalar, no size is specified for it, and one is assumed for its number of bits. Following the header, other declarations such as intermediate wires or timing parameters used in the module description must be declared. Our *Mux8* example does not require such declarations.

Following the declarations, the main body of a Verilog module describes the operation of the module. In this part, a module may be described in terms of its subcomponents, its register and bus structure, or its behavior. In the *Mux8* example, an **assign** statement is used to specify output values for various input combinations. This statement specifies a 6-ns delay for all values assigned to *bus1*. The right-hand side of this statement selects *data1* or *data0* depending on whether the *sel* value is binary **1** or not. Signals, such as *bus1*, to which assigning is done are presumed to be driven by their right-hand side at all times. Such signals are considered wire and do not need to hold any value.

2.3.1.2 Flip-flop. Flip-flops are used in the data part of a design for flags and data storage. A multi-bit flip-flop is a register, of which the Verilog style of coding is very similar to that of a flip-flop. Figure 2.14 show the Verilog code of a 1-bit flip-flop with a synchronous *reset* input and a *din* data input. The flip-flop triggers on the falling edge of its *clk* input.

As in Fig. 2.13, the first line in Fig. 2.14 specifies the time unit and its precision. Also as in the description of *Mux8*, the first line after the module

```
`timescale 1ns/100ps

module Flop (reset, din, clk, qout);
    input reset, din, clk;
    output qout;
    reg qout;
    always @ (negedge clk) begin
        if (reset) qout <= #8 1'b0;
        else qout <= #8 din;
    end
endmodule
```

Figure 2.14 Flip-Flop Description

definition in the *Flop* code specifies the input and output ports of the flip-flop. In this example we are using a different format for declaration of inputs and outputs of the *Flop* module. Declarations shown here specify which ports are inputs and which are considered as outputs of the module. An additional declaration specifies that *qout* is a signal that has the capability of holding its values. This becomes clearer in the following paragraph.

The part of the code in Fig. 2.14 that begins with the **always** keyword specifies the values assigned to *qout* in response to changes of *clk* and other flip-flop inputs. As specified by the statement following the @ sign, the body of this **always** statement is executed at the negative edge of the *clk* signal. At such times, if *reset* is true, *qout* receives *1'b0* (1-bit binary 0); otherwise, *qout* receives *din*. Value assignments to *qout* take place only on the negative edge of the clock. Therefore, in order for this output to hold its value between clock edges, it has been declared as a **reg**.

Notice that the assignment to *qout* uses an arrow, while in the previous examples an = sign was used. This assignment is called a nonblocking assignment and assignments using an equal sign are called blocking. The use of nonblocking assignments in descriptions of sequential circuits is a usual practice in Verilog.

In all Verilog descriptions, a delay value is specified by an integer following a # sign. In Fig. 2.14, the 8 ns delay value specified on the right-hand side of assignments to *qout* specifies the time delay between evaluation of the right-hand side expression and its assignment to *qout*.

A software-like procedural coding style is used for describing the *Flop* model. In this description we are only concerned with assigning appropriate values to circuit outputs. Neither the structure of the circuit nor the details of the hardware in which data flows are of any concern.

2.3.1.3 Counter.

Counters are used in the data part of an RTL design for registering data, accessing memory, or queues and register stacks. Figure 2.15 shows Verilog code for a 4-bit modulo-16 counter. The counter has a synchronous *reset* and a 4-bit *count* output. With every negative edge

```
`timescale 1ns/100ps

module Counter4 (input reset, clk, output [3:0] count);
    reg [3:0] count;
    always @ (negedge clk) begin
        if (reset) count <= #3 4'b00_00;
        else count <= #5 count + 1;
    end
endmodule
```

Figure 2.15   Counter Verilog Code

of the *clk* input, the counter counts up one place. When the counter reaches **1111**, it rolls back and starts counting from **0000** with the next clock edge.

The Verilog code for the counter begins with the module name and port list. Input and output declarations as well as declaration of *count* as a 4-bit **reg** follow the module heading. The signal *count* is to hold values between activations of assignment of values to this signal, and therefore it is declared as **reg**. This variable keeps the count of our up-counter at all times. As in the description of *Flop*, an **always** statement that becomes active on the negative edge of *clk* encloses the statements specifying the behavior of the counter. Following the keyword **begin** that follows this statement, an **if-else** statement increments *count* if *reset* is not active. The **else** part of this statement, that is taken when *reset* is not active, increments *count* by 1. When *count* reaches **1111**, the next count taken, by treating *count* as an unsigned number, is **10000**. However, since the left-hand side of *count+1* is the 4-bit *count* variable, Verilog truncates the next count to **0000**. This is why when count reaches its upper limit, it rolls back to 0.

The description of Fig. 2.15 shows a delay of 3 ns when *count* is reset and a delay of 5 ns when this variable is incremented. As discussed, delay values are numbers that follow the sharp-sign (#) symbol. As in the case of the flip-flop example, nonblocking assignments are used for assignment of values to register outputs.

2.3.1.4 Full adder. In the data part of an RT level design, full adders are used for building carry-chain adders. A full adder is a combinational circuit with two data inputs (*a* and *b*) and one carry input (*cin*). The outputs of this circuit are *sum* and carry-out (*cout*). Figure 2.16 shows the Verilog code for this circuit. As shown, the body of *full_adder* module encloses two **assign** statements for *sum* and *cout* outputs of the circuit. Each statement represents a logic block driving the corresponding left-hand side output signal. All changes on *sum* occur after 5 ns from the time that one of its inputs change. Similarly, changes on *cout* occur after 3 ns from the time that *a*, *b*, or *cin* change. Because we are only using one delay value for every output, the specified value is considered as t_{PLH} (low-to-high propagation time), and t_{PHL} (high-to-low propagation time).

```
`timescale 1ns/100ps

module full adder (input a, b, output sum, cout);
    assign #5 sum = a ^ b ^ cin;
    assign #3 cout = (a & b)|(a & cin)|(b & cin);
endmodule
```

Figure 2.16 Full adder Verilog Code

In our example we are using two **assign** statements in the statement part of the *full_adder* module. This part of a module is considered as a concurrent body of Verilog. The order of statements in this section is not important and all statements are sensitive to their sensitivity list, meaning that they execute when an event occurs on any of their right-hand side signals. Sensitivity lists are discussed further on.

2.3.1.5 Shift-register.   Another structure that is used as a data component is a register with or without various shift capabilities. Here we show a shift-register with two mode inputs *m[1:0]* that form a 2-bit number. When *m* is 0, the shifter does nothing (retains its old value), for values of *m* = 1 and *m* = 2 it shifts its contents right and left, respectively, and for *m* = 3 it loads its parallel inputs into the register. This latter mode is its normal register mode.

Figure 2.17 shows the Verilog code for this shift-register. As with other examples we have discussed, inputs and outputs of this circuit are declared in the upper part of the module. The mode input is *m* and is declared as a 2-bit vector. The shift-register parallel data input and output are 8 bits.

The four modes of operation of this circuit are handled inside an **always** block by a **case** statement with four case alternatives. The last case alternative defaults to all values not specified before it.

The actual shifting of the contents of the shift-register is done by the use of the concatenation operator that uses a pair of curly brackets for concatenating all the bits that it is bracketing. For shifting *ParOut* to the right, the serial left input (*sl*) is concatenated to the left of bits 7 to 1 of *ParOut*. This way, *sl* moves into position 7 of *ParOut* and bits 7 down

```
`timescale 1ns/100ps

module ShiftRegister8 (input sl, sr, clk, input [7:0] ParIn,
                       input [1:0] m, output reg [7:0] ParOut);
    always @ (negedge clk) begin
        case (m)
            0: ParOut <= ParOut;
            1: ParOut <= {sl, ParOut [7:1]};
            2: ParOut <= {ParOut [6:0], sr};
            3: ParOut <= ParIn;
            default: ParOut <= 8'bX;
        endcase
    end
endmodule
```

Figure 2.17   An 8-bit Universal Shift Register

to 1 move into positions 6 down to 0 of this register. Similarly, for shifting *ParOut* to the left, the serial right (*sr*) input is concatenated to the right of bits 6 down to 0 of this register. This causes *sr* to be clocked in bit 0 of *ParOut*, and *ParOut[6:0]* to be clocked into *ParOut[7:1]*, causing a left shift of this register.

2.3.1.6 ALU.   In our next example of an RT level component, we discuss the Verilog coding of an 8-bit 4-function arithmetic and logic unit (ALU). ALUs with various functionalities are used in the data parts of many RTL designs for performing arithmetic and/or logical operations on their vector inputs.

Our ALU example here has a 2-bit *mode* input that selects one of its four (add, subtract, AND, and OR) functions. The mode input takes values 0, 1, 2, or 3 to specify the function performed by the ALU.

The Verilog code of this example is shown in Fig. 2.18. The inputs and outputs of the ALU are declared in the module header. The *input [7:0]* declaration applies to all signals that follow it, up to the next declaration. Therefore, this declaration applies to *left* and *right* inputs of ALU. Also shown in the header of the module, is the declaration of *ALUout* both as **output** and as **reg**. Since we will be assigning *ALUout* within a procedural block the **reg** declaration is required. Following the module header, an **always** statement that encloses a **case** statement describes the ALU. The signals enclosed in parenthesis following the @ sign, which follows the **always** keyword, are referred to as the sensitivity list of the **always** block. This means that an event on any of these signals causes the **always** statement to execute once. Because *ALU8* is a combinational circuit, all its inputs must appear in the sensitivity list of the **always**

```
`timescale 1ns/100ps

module ALU8 (input [7:0] left, right, input [1:0] mode,
             output reg [7:0] ALUout);
    always @(left, right, mode) begin
        case (mode)
            0: ALUout = left + right;
            1: ALUout = left - right;
            2: ALUout = left & right;
            3: ALUout = left | right;
            default: ALUout = 8'bX;
        endcase
    end
endmodule
```

Figure 2.18   An 8-bit ALU

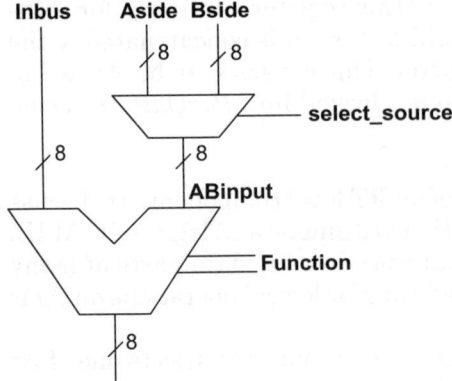

Inbus   Aside Bside

select_source

ABinput

Function

Figure 2.19 Partial Hardware
Using *Mux8* and *ALU*

Outbus

block that describes it. The **case** statement in the body of the **always** statement is similar to that used in the shift-register example. Assignments made to *ALUout* are of the blocking type (using an equal sign) that is a common practice in Verilog for assignments made to non-register outputs. Finally the **default** alternative puts all **X**s on *ALUout* if *mode* contains anything but **1**s and **0**s. The *8'bX* format translates to binary *X* expanded to 8 bits.

2.3.1.7 Interconnections.   Many data path components and the wiring of the complete data part itself require lower-level component interconnections. To illustrate Verilog mechanism for this, we use our *Mux8* and *ALU8* examples to form the partial hardware shown in Fig. 2.19.

The partial hardware shown selects *Aside* or *Bside* depending on *select_source* and puts it on the *ABinput* side of the ALU (this is connected to the actual *ALU8* right port). The *Inbus*, which is a local signal in the scope of this partial hardware, is connected to the *left* input of *ALU8*. The ALU output (*ALUout*) connects to *Outbus* in this hardware, and its *mode* input connects to the local 2-bit *function* signal. The partial Verilog code of Fig. 2.20 is the code that corresponds to the diagram of Fig. 2.19.

```
ALU8 U1 ( .left(Inbus), .right(ABinput),
          .mode(function), .ALUout(Outbus) );
Mux8 U2 ( .sel(select_source),
          .data1(Aside), .data0(Bside) );
```

Figure 2.20   Partial Verilog code of Fig. 2.19

```
ALU8 U1 ( Inbus, ABinput, function, Outbus );
Mux8 U2 ( select_source, Aside, Bside );
```

Figure 2.21   Ordered Port Connection

This partial code shows interconnections described above. *ALU8* and *Mux8* modules are instantiated and *U1* and *U2* instance names are used for these modules. Following instance names, sets of parenthesis enclose port connections to the instantiated modules. The port connection format shown here begins with a dot (.), followed by the name of the actual port of the module, and then it is followed by a set of parenthesis that encloses the expression or local signal that connects to the named port. An alternative port connection (shown in Fig. 2.21) format is to exclude the actual ports of the instantiated components and only list the local signals in the same order as their connecting ports.

2.3.2 Controllers

Data components are put together in the data part of an RT level design, and the controller is wired to the data part to control its flow of data. Controllers can be as easy as one flip-flop, handshaking handlers, or as complex as several concurrent state machines. Figure 2.22 shows an outline of a controller circuit. The inputs to the controller determine its next states and its outputs. The controller monitors its inputs and makes decisions as to when and what output signals to assert. Controllers keep past history of circuit data by switching to appropriate states.

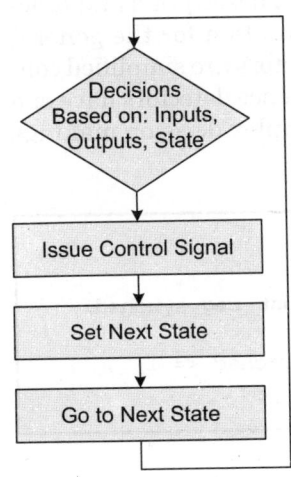

Figure 2.22   Controller Outline

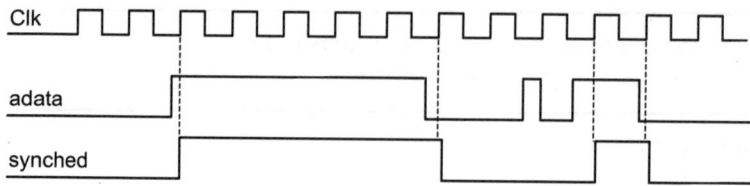

Figure 2.23   Synchronizing *adata*

This section presents two examples to illustrate some of the features of Verilog for describing state machines.

2.3.2.1   Synchronizer.   When inputs to a clocked circuit are generated by a system that is run by a different clock, or by an external asynchronous circuit, a synchronizer is used to synchronize incoming data with a given clock. Figure 2.23 shows an asynchronous data signal (*adata*) and an output that is synchronized with the positive-edge of the *clk* signal.

Figure 2.24 shows the Verilog module for generating the *synched* signal of Fig. 2.23. The module header declares inputs and outputs of this circuit, and within this module an **always** block handles the synchronization. As shown, the flow into the **always** block begins when the positive edge of *clk* is detected. At this time if *adata* is **0**, the output *synched* signal remains **0** for the next clock period. If a **1** is detected on *adata* on the rising edge of the clock, *synched* becomes **1** and remains **1** for at least one clock period, at which time it may be set to **0** if *adata* is **0**. Since the flow into the **always** block only begins on the rising edge of the *clk*, it is guaranteed that changes of *synched* only occur with this clock edge. The description shown assumes that only **0** and **1** values can appear on the *adata* input.

2.3.2.2   Sequence detector.   While being simple in description and functionality, a sequence detector is a good representation for the general class of controllers. In other words, sequence detectors are simplified controllers. Instead of many inputs and outputs, sequence detectors have one or two input and output lines, and instead of complex decision makings

```
`timescale 1ns/100ps

module Synchronizer (input clk, adata, output reg synched);
    always @ (posedge clk)
        if (adata == 0) synched <= 0; else synched <= 1;
endmodule
```

Figure 2.24   A Simple Synchronization Circuit

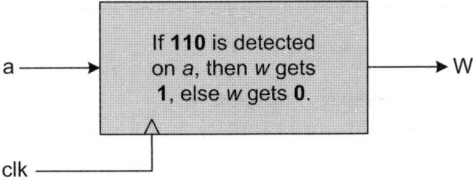

Figure 2.25 State Machine Description

and input conditions, sequence detectors generally search for a sequence of **1**s and **0**s on their input. We will present a Verilog description for the simple sequence detector shown in Fig. 2.25.

A Moore machine sequence detector, the pseudo-code of which is shown in Fig. 2.27, searches on it's a input for the **110** sequence. When this sequence is detected in three consecutive clock pulses, the output (w) becomes **1** and stays **1** for a complete clock cycle. The state machine for this detector is shown in Fig. 2.26. States of the machine are named $s0$, $s1$, $s2$, and $s3$. The $s0$ state is the reset state and $s3$ is the state in which the **110** sequence is detected. The reset input of this machine resets it to its $s0$ state. Starting in this state, it takes at least three clock periods for the machine to get to the $s3$ state, in which output becomes **1**.

The Verilog behavioral description of this machine is shown in Fig. 2.27. The list of ports are $a, clk, reset$, and w. A parameter declaration in this description defines constants $s0, s1, s2$, and $s3$ to be used for the names of the states of this machine. Assigning binary values assigned to these parameters, can be regarded as making state assignments.

The two-bit **reg** declaration declares $current$ as the variable that holds the current state of the machine. This variable corresponds to a 2-bit register, which represents the state variables of this machine.

The main flow of the state machine is implemented by an **always** block that is sensitive to the positive edge of the clock. In this statement an **if-else** statement checks for $reset$, and in case of the absence of a **1** on this inputs, state transitions are taken care of by a **case** statement.

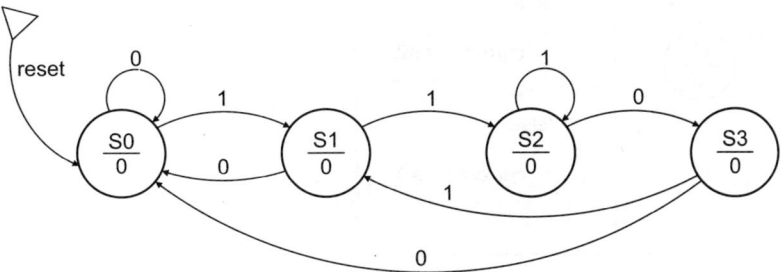

Figure 2.26 Sequence Detector State Machine

```
`timescale 1ns/100ps

module Detector110 (input a, clk, reset, output w);
    parameter [1:0] s0=2'b00, s1=2'b01, s2=2'b10, s3=2'b11;
    reg [1:0] current;

    always @ (posedge clk) begin
        if (reset) current = s0;
        else
            case (current)
                s0: if (a) current <= s1; else current <= s0;
                s1: if (a) current <= s2; else current <= s0;
                s2: if (a) current <= s2; else current <= s3;
                s3: if (a) current <= s1; else current <= s0;
            endcase
    end

    assign w = (current == s3) ? 1: 0;

endmodule
```

Figure 2.27 Verilog Code for **110** Detector

The four case alternatives of the **case** statement each correspond to a state of the state machine. Figure 2.28 shows state *s2*, its next state, and the corresponding Verilog code in the **always** block of Fig. 2.27.

The body of the *Detector110* module has an **assign** statement that assigns a **1** to the *w* output when the machine reaches the *s3* state. This statement is outside of the **always** block and represents a combinational

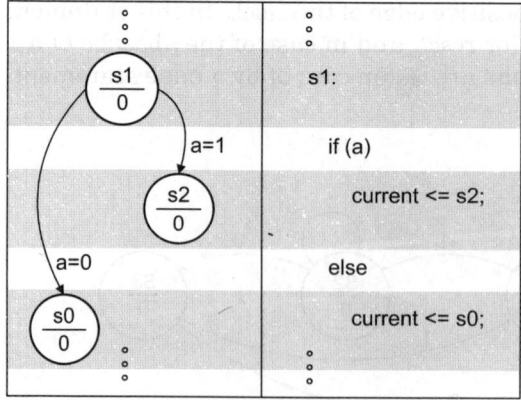

Figure 2.28 State Transitions and Corresponding Verilog Code

circuit that drives the *w* output. This statement runs concurrent with the **always** block and is evaluated every time *current* changes.

2.4 Testbenches

Although the main application of Verilog is accurate representation of hardware for simulation and synthesis, we cannot ignore the role of testing our designs for design and functional verification. Verilog has language constructs for application of test data to a design, as well as constructs for monitoring responses a circuit generates.

2.4.1 A simple tester

We develop a simple test bench for our *Detector110* example of the previous section to illustrate the kinds of Verilog constructs used for this purpose.

As shown in Fig. 2.29, like any other Verilog description a testbench description begins with the **module** keyword. Unlike other descriptions, a testbench does not have input or output ports. In the body of the module shown, variables that are input to our unit under test (UUT) are declared as **reg** and its outputs are declared as **wire**. The testbench shown instantiates the *Detector110* module and uses *UUT* for its instance name. Connections of *aa, clock, rst,* and *ww* local signals to the

```
`timescale 1ns/100ps

module Detector110Tester;
    reg aa, clock, rst;
    wire ww;
    Detector110 UUT (aa, clock, rst, ww);
    initial begin
        aa = 0; clock = 0; rst = 1;
    end
    initial repeat (44) #7 clock = ~clock;
    initial repeat (15) #23 aa = ~aa;
    initial begin
        #31 rst = 1;
        #23 rst = 0;
    end
    always @ (ww) if (ww == 1)
        $display ("A 1 was detected on w at time = %t", $time);
endmodule
```

Figure 2.29 Testbench for *Detector110*

ports of *Detector110* are done according to the order in which these ports appear in the port list of the detector circuit.

We are using **initial** statements to drive test values into the variables that are connected to the inputs of the detector. An **initial** statement is a procedural statement that runs once and stops when it reaches its last statement. All **initial** blocks in a module, start at time 0 and run concurrently.

In the body of our testbench, an **initial** statement is used for initializing the input signals. Following this statement, another **initial** statement repeats 44 times of complementing the *clock* input with 7 ns delay values. This generates a periodic signal on *clock*. Signal *aa* is also assigned a periodic signal, but with a different frequency. The last **initial** statement shown waits 31 ns before assigning a **1** to *rst* and another 23 ns before assigning a **0** to this variable. This is intended to cover several clock edges in order to reset the machine.

The last statement in the body of the *Detector110tester* module is an **always** statement that reports the times at which the *ww* variable becomes **1**. Recall that this variable is connected to the output of our sequence detector. The **always** block wakes up when *ww* changes. If the change on *ww* has caused it to become **1**, a statement displaying a note and the time of occurrence of *ww* will be issued. This note will appear in the simulation environment's window that is often referred to as the "console" or "transcript". The **$display** task is a Verilog system task.

2.4.2   Tasks and functions

The testbench of Fig. 2.29 shows utilization of the **$display** system task. In addition to system tasks for input, output, display, and timing checks, Verilog allows definition and utilization of user defined tasks and functions.

A task can represent a sub module within a Verilog module. A task begins with the **task** keyword and for its header uses a format that is very similar to that of a module. The body of a task can only consist of procedural statements like **if-else** and **case**.

Functions can also be used for corresponding to hardware entities, or they may be used for writing structured codes, in much the same way as they are used in software languages. Typical applications of functions include representation of boolean functions, data and code conversion, and input and output data formatting. Generally, any time the final value of a process is used on the right-hand side of an expression, a function can be used to simplify the expression.

2.5   Summary

This chapter gave an overview of Verilog and its use for design and test of RT level description. After discussing components of an RT level design, we used small examples to illustrate such components and at the same time showed Verilog coding of hardware modules. The descriptions presented in this part were all synthesizable and had a one-to-one hardware correspondence. To make the discussion of the language complete, and show alternatives of the use of this language in a design, we showed how testbenches could be developed in Verilog. In this part several new constructs of Verilog were presented.

Problems

2.1 Write a behavioral description for a 4-to-1 multiplexer. The multiplexer has $s1$ and $s0$ select inputs and four data inputs.

2.2 Using an **assign** statement describe a majority circuit with three inputs and one output. When the majority of the inputs are **1**, the output of the majority circuit becomes **1**.

2.3 Write a Verilog description for a **101** sequence detector with an a input and a w output.

2.4 Write a testbench for the **101** detector of Prob. 2.3. Make sure you apply an input sequence to make the output become **1**.

Suggested Reading

IEEE Std 1364-2001, *IEEE Standard Verilog Language Reference Manual*, SH94921-TBR (print) SS94921-TBR (electronic), ISBN 0-7381-2827-9 (print and electronic), 2001.

Navabi, Z., *"Digital Design and Implementation with Field Programmable Devices"*, Kluwer Academic Publishers, Boston, MA, 2005, ISBN: 1-4020-8011-5.

Navabi, Z., *"Verilog Computer-Based Training Course"*; CBT CD with hardcopy User's manual, McGraw-Hill, New York, 2002 ISBN 0-07-137473-6.

3

Verilog Language Concepts

Because Verilog is a language for description of hardware, it has features that are conceptually different from those of software languages. Two main features that characterize hardware languages are timing and concurrency. Timing is associated with values that are assigned to hardware carriers, while concurrency refers to simultaneous operation of various hardware components. Because of these features, data types and operators take new definitions in hardware languages and must be looked at differently from those of software languages. This chapter defines timing and concurrency and then it discusses Verilog language constructs that make it an efficient language for design and test of hardware modules.

3.1 Characterizing Hardware Languages

Timing and concurrency are the main characteristics of hardware description languages. These features are instrumental in the correct description of hardware components at various levels of abstraction.

3.1.1 Timing

Transfer of values between hardware components or within a component is done through wires or busses. Variables in Verilog may be used for representation of actual wires, and because of delays associated with the transfer of values through wires, variable assignments in Verilog can include timing specification. Consider, for example, the AND-OR circuit of Fig. 3.1. This circuit can be simply described by a boolean equation as shown below. (Note that, **&** is AND, **|** is OR, and ~ is NOT).

```
assign w1 = a & b | a & ~b;
```

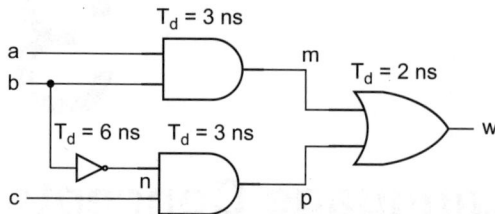

Figure 3.1 An AND-OR circuit

The problem with this representation is that it does not consider gate delays, and glitches that may appear on w due to different delay paths from b to w. These delays and glitches will not be seen on the output of this expression ($w1$).

For a more accurate representation of this circuit, assignments that incorporate timing as well as value assignments must be used. Verilog allows the use of gate and wire delays for this purpose. The partial code shown below is a more accurate representation of Fig. 3.1.

```
assign #6 n = ~b;
assign #3 m = a & b;
assign #3 p = n & c;
assign #2 w2 = m | p;
```

An **assign** statement drives the signal on the left-hand side of the equal sign with the boolean expression on its right. A sharp-sign (#) followed by a number specifies the delay of the left-hand side signal.

Although this description is more complex than the previous representation of this circuit, it represents details of the timing behavior that the single **assign** statement does not. For example, consider the waveform shown in Fig. 3.2. When b changes from **1** to **0** the final value of the circuit output remains at **1**. However, because of the inverter delay in forwarding a **1**, a 6 ns glitch appears on this output.

Name	V...	. 85 .	. 90 .	. 95 .	. 100	. 105	. 110	. 115 .	. 120	. 125 .
a	1									
c	1									
b	1					5 ns				
w1	1									
w2	1					6 ns				

Figure 3.2 Output Glitch

As shown in this figure, *w1*, which is the result of the single **assign** statement, does not show the delay, while *w2* shows both propagation delays and glitches that may occur on the *w* output of circuit of Fig. 3.1.

From the above discussion, we conclude that accurate representation of hardware requires handling of timing, and a language for modeling hardware must have constructs for doing so. The Verilog HDL allows many schemes for incorporating timing into description of hardware. Constructs of this language are tailored to have such timing specifications.

3.1.2 Concurrency

Like timing, concurrency is an essential feature of any language for description of hardware. When a software programmer develops code for performing a certain task, he or she thinks of this task in a sequential manner. The software developed this way will have a top down sequential flow. On the other hand, when a hardware designer or modeler is to describe a hardware system, he or she thinks of this hardware as interconnections of components. The functionality of the overall system is achieved by concurrently active components communicating through their input and output ports. The functionality of each component may be described by concurrent subcomponents or described by a program in a sequential manner.

We refer to concurrency as the way the simulation of components or constructs appears to the user. Obviously, Verilog is a language for which simulators have been developed on single-processor platforms, and true concurrency in the execution of thousands of components cannot exist. Through the use of concurrent constructs, timing of interconnecting signals, and order of simulation of constructs or components, a Verilog simulator makes us (the users) *think* that such execution is being done concurrently.

Going back to our example of the AND-OR circuit, the gates of the circuit of Fig. 3.1 are concurrently active. This means that we cannot decide on a pre-determined order in which these gates perform their operations. Instead, while hand simulating this circuit, we evaluate a gate only when its input changes. Similarly, the four **assign** statements that we discussed for representing this circuit are regarded as concurrent. The order in which these statements appear in a concurrent body of Verilog is not important. As in its corresponding hardware, each statement only evaluates its boolean expression when an event occurs on one of its right-hand side signals.

For example, if the **assign** statement driving *w2* appears first in the list of statements shown, the result will be no different than if it appears last, as it is now. In Verilog, the body of a module is referred to as a concurrent body, and encloses statements that are concurrently active.

Although individual hardware components of a system are concurrent, for describing behavior of a component, it is sometimes easier for designers to describe them behaviorally in a procedural fashion. For this reason, Verilog allows the use of procedural blocks that can enclose procedural statements like **if-else** and **case** statements.

3.1.3 Timing and concurrency example

As an example of a design that uses concurrent statements with timing, consider the full-adder description of Fig. 3.3. The module header declares inputs and outputs of this circuit and in the concurrent body of the module two **assign** statements drive s (sum) and co (carryout) outputs of this circuit. The s and co outputs have 3 ns and 4 ns propagation delays respectively. When an input changes, e.g., a, the right-hand sides of both expressions are evaluated. The new value of s is scheduled into this output for 3 ns later, and that of co for 4 ns. Because of the delay values, if an input change causes both outputs to change, s changes before co does, even though the **assign** statement of co appears before that of s.

As an example of a procedural body of Verilog consider the testbench of the *Full_adder* **module** shown in Fig. 3.4. After declaration and initialization of variables, wires, and constants, the body of this module uses an instantiation statement to instantiate the full adder that is the module under test. Following this statement, and concurrent with it, an **always** block applies test data to the ports of the circuit being tested.

An **always** block is a concurrent structure on the outside, but has a procedural body. In the body of this block an **if**-statement checks the simulation time and if it exceeds *tlimit* it stops the simulation. If the time has not reached this limit, in a sequential flow, after a wait of 17 ns input a is complemented. This is followed by a wait of 13 ns and a wait of 19 ns before complementing ci and b inputs. After new values have been given to the inputs of *Full_adder*, the **always** block returns to its beginning and repeats its process. For this example, this process stops at 539 ns, allowing 11 iterations.

This example has shown how timing is used in a procedural body of Verilog. Procedural bodies are used for description of testbenches or for describing a hardware component whose behavior is too complex to be described with simple boolean equations.

```
`timescale 1ns/100ps

module Full_adder (input a, b, ci, output co, s);
    assign #4 co = a & b | a & ci | b & ci;
    assign #3 s = a ^ b ^ ci;
endmodule
```

Figure 3.3 Full adder Concurrent Description

```
`timescale 1ns/100ps

module FulladderTester;
    reg a = 0, b = 0, ci = 0;
    wire co, s;
    parameter tlimit = 500;
    Fulladder MUT (a, b, ci, co, s);
    always begin
        if ($time >= tlimit) $stop;
        else begin
            #17;
            a = ~a;
            #13;
            ci = ~ci;
            #19;
            b = ~b;
        end
    end
endmodule
```

Figure 3.4   Full adder Tester Procedural Description

3.2   Module Basics

The previous section discussed some of the main concepts of the Verilog language. To prepare for description of hardware, this section shows how modules are developed, and how names, numbers, and operators are used. We discuss conventions, lexical issues, and code formal in Verilog. The standard IEEE std 1364-2001 has a complete presentation of these topics.

3.2.1   Code format

Verilog code is free format, with spaces and new lines serving as separators. Source text is case-sensitive, i.e., identifiers using lowercase or uppercase characters are distinguished from each other. The language uses certain keywords, all of which must use lowercase characters.

Comments may appear anywhere in a Verilog source text. A comment designator starting with // makes the rest of the line, up to a new-line character, a comment. The symbols /* and */ bracket a section of code as a comment, and they go across new-line characters.

3.2.2   Logic value system

Bit type, or bits of vectors or arrays, of Verilog wires and variables take the 4-value logic value system. Values in this system are 0, 1, Z, and X.

The **0** value represents forcing **0** like a direct pull to the ground, or a resistive **0**, or a capacitive **0**. A resistive **0** is generated when there is a large resistance between a line and a forcing **0** value. A capacitive **0** is when a line is float; but has a capacitance that has a zero charge.

The **1** value represents forcing **1**, resistive **1**, and a capacitive **1**. These are defined similar to various modes of the **0** value. For example a forcing **1** is defined as the logic value driven by a supply voltage.

The **Z** value represents an undriven, high-impedance value. This is the electrical float which causes no current flow to either supply or ground voltage. Both **Z** and **z** are acceptable forms of this logic value.

The **X** value represents a conflict in multiple driving values, an unknown, an uninitialized value, a short between two opposing values (**0** and **1**), or a bus contention. Driven wires and Verilog variables assume **X** for their initial values. Figure 3.5 shows several examples for the four values of Verilog's logic value system. Both **X** and **x** are acceptable forms of this logic value.

3.2.3 Wires and variables

Verilog has two main data types, **net** and **reg**. A **net** represents a wire driven by a hardware structure or output of a gate. A **reg** represents a variable that can be assigned values in a behavioral description of a component in a Verilog procedural block. A later section on data types elaborates on this topic.

3.2.4 Modules

A module is the main structure for definition of hardware components and testbenches. Modules begin with the **module** keyword and end with **endmodule**. Immediately following the **module** keyword, port list of the module appears enclosed in parenthesis. Declaration of mode, type, and size of ports can either appear in the port list or as separate declarations. The example of Fig. 3.3 demonstrates the former method, whereas in the *FlipFlop* description of Fig. 3.6 names only ports that are

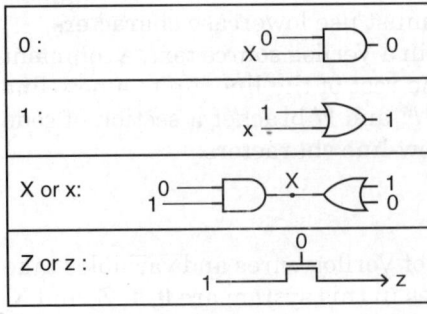

Figure 3.5 Logic Values and Examples

```
`timescale 1ns/100ps

module FlipFlop (preset, reset, din, clk, qout);
    input preset, reset, din, clk;
    output qout;
    reg qout;
    always @ (posedge clk) begin
        if (reset) qout <= #7 0;
            else if (preset) qout <= #7 1;
                else qout <= #8 din;
            end
endmodule
```

Figure 3.6 Separate Port Declarations Statements

listed in the port list, and declared as separate **input** and **output** ports inside the body of the *FlipFlop* module.

The body of a **module** consists of the specification of the operation of the hardware the **module** is representing. The *Full_adder* of Fig. 3.3 is described by two concurrent **assign** statements, while the *FlipFlop* of Fig. 3.6 uses a single **always** procedural block to describe its operation.

A testbench **module** has no ports. It instantiates the module under test (MUT) and through the use of concurrent statements or procedural blocks applies data to the ports of MUT. Multiple modules can be tested with the same testbench.

3.2.5 Module ports

Inputs and outputs of a module must be declared as **input**, **output**, or **inout**. By default, all declared ports are regarded as **net**s, and the default **net** type is used for the ports. For example, if defaults are not changed, an input or an output automatically assumes the **wire** type **net**. Ports declared as **output** may be declared as **reg**. This way they can be assigned values inside procedural blocks. (e.g., in an **always** block). However, an **inout** port can be used only as a **net**. To assign values to an **inout** port in procedural bodies, a **reg** corresponding to the port must be declared and used. Values of this **reg** type variable can then be assigned to the **inout** port using a continuous assignments. For an output, a **reg** specification can follow the **output** keyword in the port list of the module.

3.2.6 Names

A stream of characters starting with a letter or an underscore forms a Verilog identifier. The **$** character and the underscore are allowed in an identifier. A stream of special characters may be used as an identifier if preceded by a backslash character. Verilog uses keywords that are all formed by streams of lowercase characters. In our examples, we use bold type for Verilog codes for keywords.

System tasks and functions are part of the Verilog standard. The names of these utilities begin with a $ character. An example system task is **$display**, which is used for formatted output. System tasks and functions will be discussed later in this chapter.

The Verilog language defines a number of compiler directives that will be discussed later. Compiler directive names are preceded by the ` (back single quote) character. An example is the `**timescale** directive, which defines the time unit for a Verilog code in a source text. The following are valid names for identifiers.

```
aname , name1 , _name , Name ,
 Name$ , name55 , _55name , setup ,
 _$name.
```

And, the following are Verilog keywords or system tasks.

$display default $setup
begin tri1 small

3.2.7   Numbers

Constants in Verilog are integer or real. Specification of integers can include **X** (or **x**) and **Z** (or **z**) in addition to the standard **0** and **1** logic values.

Integer formats provide various ways for representing bit streams. Integers may be sized or unsized. A sized integer begins with the number of equivalent bits, followed by the single quote character ('), a base specifier, and the digits of the number in the specified base. The base specifier is a single lower or uppercase character, **b**, **d**, **o**, or **h** for binary, decimal, octal, and hexadecimal bases. The general format for integers is:

number_of_bits **`base_identifier** *digits*

Digits in the decimal (**d**) system are **0** through **9**. For hexadecimal, octal, and binary systems, in addition to their standard digits, **X** and **Z** (both upper and lowercase) characters are also allowed. Hexadecimal and octal **X** and **Z** digits expand to 4 or 3 bits of **X** and **Z** respectively. A number without the *number_of_bits* specification is regarded as an unsized number.

Optionally, the **base_identifier** can be preceded by the single character S (or s) to indicate a signed quantity. A simple integer without the base specification is treated as a signed number, whereas a number that has its base specified is only treated as a signed number if the optional **s** precedes its base specification. This designator does not change bit pattern of a number, but only its interpretation.

A plus or minus operator can be used on the left of the number specification. A minus sign in this position is treated as a unary operator and changes the sign of the number. The underscore character (_) can be used anywhere in a number for grouping its bits or digits for readability purposes. This and other formats described above are used in the examples listed in Table 3.1.

More examples of constants are shown in Fig. 3.7. As shown, variables a through l are declared and initialized. Following their initializations, **$displayb** tasks display their binary values. Display results are shown in the comments that follow the statements.

Real constants in Verilog use the standard format as described by IEEE std 754-1985. This is the IEEE standard for double precision floating-point numbers. Examples for real number representations are: 1.9, 2.6E9, 0.1e-6, 315.96-12.

TABLE 3.1   Number Representation Examples

Number representation	Binary equivalent	Explanation
4'd5	0101	Decimal 5 is interpreted as a 4-bit number.
8'b101	00000101	Binary 101 is turned into an 8-bit number.
12'h5B_3	010110110011	Binary equivalent of hex; underscore is ignored.
-8'b101	11111011	This is the 2's complement of the number in the above example.
10'o752	0111101010	This is the octal 752 with a 0 padded to its left to make it a 10-bit number.
8'hF	00001111	Hexadecimal F is expanded to 8 bits by padding zeros to its left.
12'hXA	xxxxxxxx1010	Hexadecimal XA is expanded to 12 bits by extending the left X.
12'shA6	Signed 111110100110	This is an 8-bit number treated as a 2's complement signed number.
-4'shA	Signed 0110	The 2's complement (because of the minus sign) 4-bit A is regarded as a signed number.
596	1001010100	This is a positive constant.

```
`timescale 1ns/100ps

module NumberTest;
    reg [11:0] a = 8'shA6;    initial $displayb ("a=", a);
    // a=111110100110
    reg [11:0] b = 8'sh6A;    initial $displayb ("b=", b);
    // b=000001101010
    reg [11:0] c = 'shA6;     initial $displayb ("c=", c);
    // c=000010100110
    reg [11:0] d = 'sh6A;     initial $displayb ("d=", d);
    // d=000001101010
    reg [11:0] e = -8'shA6;   initial $displayb ("e=", e);
    // e=000001011010
    reg [11:0] f = -'shA6;    initial $displayb ("f=", f);
    // f=111101011010
    reg [11:0] g = 9'shA6;    initial $displayb ("g=", g);
    // g=000010100110
    reg [11:0] h = 9'sh6A;    initial $displayb ("h=", h);
    // h=000001101010
    reg [11:0] i = -9'shA6;   initial $displayb ("i=", i);
    // i=111101011010
    reg [11:0] j = -9'sh6A;   initial $displayb ("j=", j);
    // j=111110010110
    reg [11:0] k = 596;       initial $displayb ("k=", k);
    // k=001001010100
    reg [11:0] l = -596;      initial $displayb ("l=", l);
    // l=110110101100
endmodule
```

Figure 3.7 Integer Constants

3.2.8 Arrays

Verilog allows declaration and usage of multidimensional arrays for **net**s or **reg**s. The following declares *a_array* as a two-dimensional array of 8-bit words. This array contains 1024 × 512 8-bit words.

```
reg [7:0] a_array [0:1023][0:511];
```

In an array declaration, the address range (or ranges, for multi-dimensional arrays) of the elements of the array comes (or come, for multi-dimensional arrays) after the name of the array. Range specifications are enclosed in square brackets. The size and range specification of the elements of an array come after the **net** type (e.g., **wire**) or **reg** keyword.

In the absence of a range specification before the name of the array, an element size of one bit is assumed. Figure 3.8 shows several examples of array and vector declarations and their corresponding graphical representations. Array indexing will be discussed in a later section.

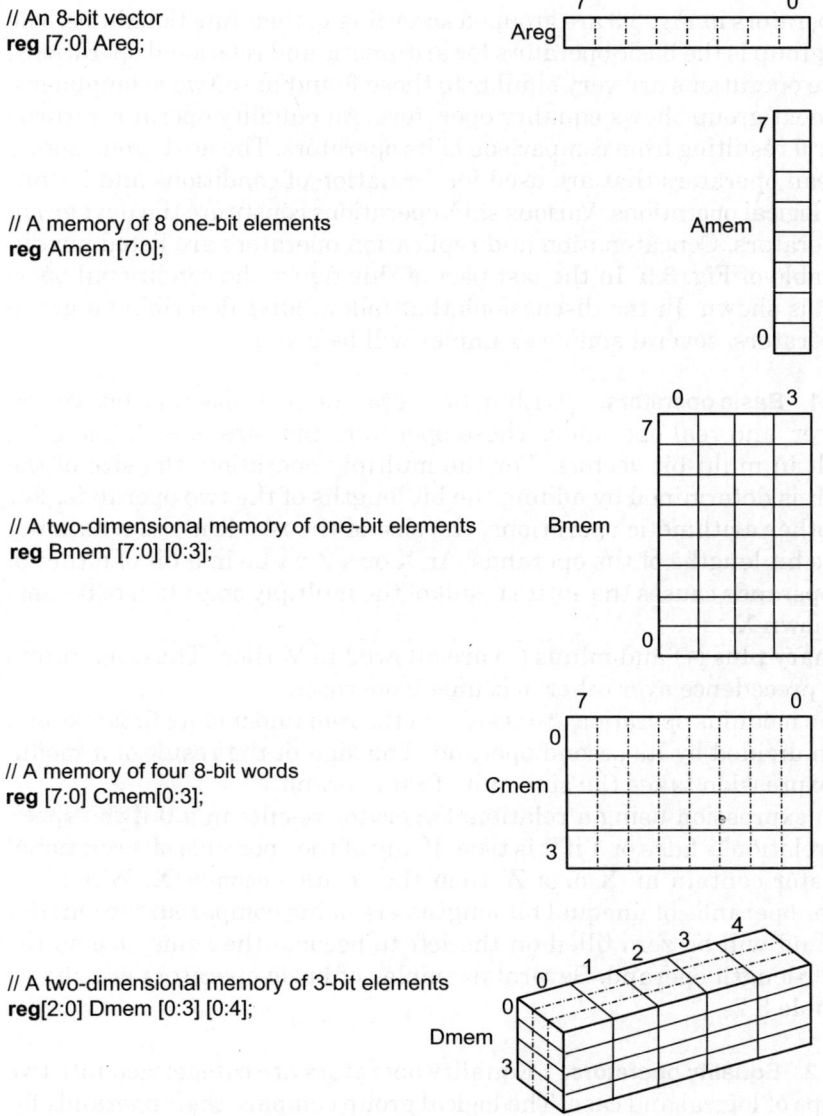

```
// An 8-bit vector
reg [7:0] Areg;
```

```
// A memory of 8 one-bit elements
reg Amem [7:0];
```

```
// A two-dimensional memory of one-bit elements
reg Bmem [7:0] [0:3];
```

```
// A memory of four 8-bit words
reg [7:0] Cmem[0:3];
```

```
// A two-dimensional memory of 3-bit elements
reg[2:0] Dmem [0:3] [0:4];
```

Figure 3.8 Array structures

3.2.9 Verilog operators

Boolean operations are the most common type of operations for describing functions of hardware components at the gate or even RT level. In addition, there are operations for the behavioral or functional description of hardware. Most operations found in software languages, are also supported in Verilog. Figure 3.9 shows Verilog operators and a brief description of each.

Operators in Fig. 3.9 are grouped according to their functionalities. The first group is the basic operators for arithmetic and relational operations. These operations are very similar to those found in software languages. The next group shows equality operators. An equality operator returns a 1 or 0 resulting from comparison of its operators. The next group shows boolean operators that are used for formation of conditions and for bitwise logical operations. Various shift operations constitute the next group of operators. Concatenation and replication operators are listed next in the table of Fig. 3.9. In the last part of this figure the conditional operation is shown. In the discussions that follow, after describing a group of operators, several simple examples will be given.

3.2.9.1 Basic operators.
Arithmetic operators in Verilog take bit, vector, integer, and real operands. These operators that are +, −, *, /, and **, result in multi-bit vectors. For the multiply operation, the size of the result is determined by adding the bit lengths of the two operands. For the other arithmetic operations, the size of the result is the maximum of the bit lengths of the operands. An **X** or a **Z** value in a bit of either of the operands causes the entire result of the multiply operation to become unknown **X**.

Unary plus (+) and minus (−) are allowed in Verilog. These operators take precedence over other arithmetic operators.

The modulus operation (%) results in the remainder of its first operand when divided by its second operand. The sign of the result of a modulus expression takes the sign of its first operand.

An expression using a relational operator results in a **0** if the specified relation is false or **1** if it is true. If any of the operands of a relational operator contain an **X** or a **Z**, then the result becomes **X**. When two vector operands of unequal bit lengths are being compared, the smaller operand will be zero filled on the left to become the same size as the larger-length operand. Several examples of basic operators are shown in Table 3.2.

3.2.9.2 Equality operators.
Equality operators are categorized into two groups of logical and case. The logical group compare their operands for equality (==) or inequality (!=), and return a one-bit result, **0**, **1**, or **X**. An **X** ambiguity arises when an **X** or a **Z** occurs in one of the operands.

Basic	OPERATION	DESCRIPTION	RESULT
Arithmetic	+ − * / **	Basic arithmetic	Multi-bit
Relational	> >= < <=	compare	One-bit

Equality	OPERATION	DESCRIPTION	RESULT
Logical	== ! =	Equality not including Z, X	One-bit
Case	=== ! ==	Equality including Z, X	One-bit

Boolean	OPERATION	DESCRIPTION	RESULT
Logical	&& ‖ !	Simple logic	One-bit
Bit-wise	~ & | ^ ^~ ~^	Vector logic operation	One-bit
Reduction	& ~& |~| ^ ^~ ~^	Perform operation on all bits	One-bit

Shift	OPERATION	DESCRIPTION	RESULT
Logical right	>>n	Zero-fill Shift n places	Multi-bit
Logical left	<<n	Zero-fill Shift n places	Multi-bit
Arithmetic right	>>>n		Multi-bit
Arithmetic left	<<<n		Multi-bit

Concat	OPERATION	DESCRIPTION	RESULT
Concatenation	{ }	Join bits	Multi-bit
Replication	{{ }}	Join & replicate	Multi-bit

Condition	OPERATION	DESCRIPTION	RESULT
Concatenation	? :	If-then-else	Multi-bit

Figure 3.9 Verilog Operators

On the other hand case-equality and case-inequality operators (===
and !==) consider **X** and **Z** values in comparing their operands. The
result of these operands is always **0** or **1**. Examples of the use of these
operations are shown in Table 3.3.

TABLE 3.2   Examples of Basic Operators

Example	Results in
25 * 8'b6	150
25 + 8'b7	32
25 / 8'b6	4
22 % 7	1
81b10110011 > 8'b0011	1
4'b1011 < 10	0
4'b1Z10 < 4'b1100	X
4'b1x10 < 4'b1100	X
4'b1x10 <= 4'b1x10	X

3.2.9.3 Boolean operators.   Operators performing boolean operations are logical, bit-wise, and reduction. The logical boolean operators (**&&**, **||**, and **!**) are connectives used for formation of conditions. These operators return one bit results, regardless of the size of their operands. The result of the **&&** operator is **1** if none of the operands is **0**. The result of the **||** operator is **1** if at least one of the operands is non zero, and the **!** operator complements its operand. If an **X** or a **Z** appears in an operand of a logical operator, an **X** will result.

Bitwise operators consists of **&**, **|**, **^**, **^~**, and **~** for bitwise AND, OR, XOR, XNOR, and NOT, respectively. For the XNOR operation, **~^** is also a valid symbol. All bitwise operators operate on scalars and vectors. In operations on vectors of differing sizes, the result becomes the larger of the two vectors. All four values of the standard Verilog logic value system are valid bit values for bitwise operators. Figure 3.10 shows the results of bitwise operators operating on two 1-bit operands for four bitwise operators. The complement operator **~**, not shown in this figure, results in **1** and **0** for **0** and **1** inputs and **X** for **X** and **Z** inputs.

A reduction operation is referred to as one that performs a certain bitwise operation on bits of a vector and reduces it by a single bit. There are six reduction operations in Verilog. These operations are **&**, **~&**, **|**, **~|**, **^**, and **~^** (or **^~**) for AND, NAND, OR, NOR, XOR, and XNOR. Applying a

TABLE 3.3   Examples of Equality Operators

Example	Results in
8'b10110011 == 8'b10110011	1
8'b1011 == 8'b00001011	1
4'b1100 == 4'b1Z10	0
4'b1100 != 8'b100X	1
8'b1011 !== 8'b00001011	0
8'b101X === 8'b101X	1

& / ^ / ~		0		1		X		Z	
0		0	0	0	1	0	x	0	x
		0	1	1	0	x	x	x	x
1		0	1	1	1	x	1	x	1
		1	0	0	1	x	x	x	x
X		0	x	x	1	x	x	x	x
		x	x	x	x	x	x	x	x
Z		0	x	x	1	x	x	x	x
		1	x	x	x	x	x	x	x

Figure 3.10 Bit-by-bit Bitwise and Reduction Operators

reduction operator to a vector performs bitwise operations on a pair of bits, starting in the least-significant position and repeating until all bits have been covered. For every pair of bits, the table in Fig. 3.10 determines the result of the operation. Complement reduction operations (~&, ~|, and ~^) perform reduction first and then complement the result. Examples of logical, bit-wise, and reduction operators are shown in Table 3.4.

3.2.9.4 Shift operators.   Logical and arithmetic shift operators are provided in Verilog for shifting their operands *n* places to the right or left. Such an operator shifts its first operand by the number specified by its second operand. Logical shift operators (>>, and << for shift right and left) fill the vacated bit positions with zeros.

Fill values for arithmetic shift operators depend on the type of their results being signed or unsigned. For unsigned results, arithmetic shift operators behave like the logical operators and fill their vacated bit positions with zeroes. If an arithmetic shift operator is used for producing a signed result (determined by the left-hand side, or other expressions), vacated bit positions due to shifting will be filled with the most significant bit (sign bit) of the first operand of the shift operator. Several examples are shown in Table 3.5.

TABLE 3.4 Logical, Bit-Wise, and Reduction

Example	Results in
8'b01101110 && 4'b0	0
8'b01101110 || 4'b0	1
8'b01101110 && 8'b10010001	1
! (8'b10010001)	1
8'b01101110 & 8'bxxzz1100	8'b0xx01100
8'b01101110 | 8'bxxzz1100	8'x11x1110
~& (4'b0xz1)	1
~| (4'b0xz1)	0

TABLE 3.5   Shift Operators

Example	Results in
8'b0110_0111<<3	8'b0011-1000
8'b0110_0111<<1'bz	8'bxxxx-xxxx
Signed_LHS = 8'b1100-0000>>>2	8'b1111-0000

3.2.9.5 Concatenation operators.   An important operation in hardware modeling is concatenation. This operation is used for formation of vectored sources or destinations from smaller vectors or scalars. The notation used for this operator is a pair of curly brackets ({...}) enclosing all scalars and vectors that are being concatenated.

The concatenation operator may be used on the left-hand side of an assignment. For example, if a is a 4-bit reg and *aa* is a 6-bit **reg**, the following assignment places **1101** in *a* and **001001** in *aa*:

```
{a, aa} = 10'b1101001001
```

A concatenation operation on the right-hand side forms a vector of the size of all the variables that are being concatenated. This vector may be used in other operations or can be assigned to a left-hand side target.

A repetition multiplier can be used to form a vector. This vector may be used in another concatenation operation, or may be used as a vector in other operations, or as a right-hand side assignment. If the *a* and *aa* variables have the values assigned to them above, and *aaa* is a 16-bit **reg** data type, then the assignment,

```
aaa = {aa, {2{a}}, 2'b11}
```

puts **001001_1101_1101_11** in *aaa*. The leftmost 6 bits come from *aa*, the next 8 bits are two times repetition of *a*, and the least-significant 2 bits are the 2-bit constant **11**. Below are more examples of concatenation and replication operators:

```
{a, 2{b,c}, 3{d} is equivalent to: {a,b,c,b,c,d,d,d}
{2'b00, 3{2'01}, 2'b'11} results in: 10'b0001010111
```

3.2.9.6 Conditional operator.   The conditional operator in Verilog uses the **?:** notation, the general format of this operation is:

```
expression1 ? expression2 : expression3
```

If *expression1* is true, then *expression2* is selected as the result of the operation; otherwise *expression3* is selected. This operation provides a

TABLE 3.6   Conditional Operator

Example	Results in
1 ? 4'b1100 : 4'b|ZX0	4'b1100
0 ? 4'b1100 : 4'b1ZX0	4'b1ZX0
X ? 4'b1100 : 4'b1ZX0	4'b1XX0

compact if-then-else type of construct for in-line continuous assignment statements. If *expression1* is **X** or **Z**, both expressions 2 and 3 will be evaluated, and the result becomes the bit-by-bit combination of these two expressions. The bit-by-bit combination produces a **0** or a **1** when both expression bit positions are **0** or **1**, respectively, and produces **X** otherwise. Several examples of this operator are shown in Table 3.6.

3.2.9.7   Precedence of operators.   An expression involving several different operators discussed above is evaluated based on the precedence of operators, which determine which operation is performed first. Usually parenthesis override default precedences, and their use is recommended for clarity and unambiguity. All operators associate left to right with the exception of the conditional operator that associates right to left.

Figure 3.11 shows the precedence rules in Verilog. Precedence of operators is from top to the bottom of the list, and those on the same row have the same precedence. Figure 3.12 shows two examples for evaluating expressions based on precedence.

As shown in Fig. 3.12, because **+** has a higher precedence than **&**, adding is performed and the result is ANDed with *W*. In the other example in this

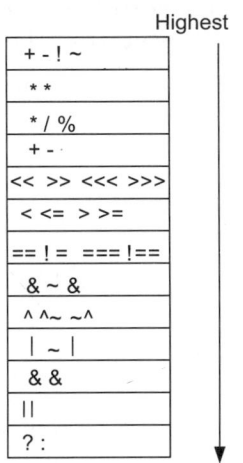

Highest

+ - ! ~
* *
* / %
+ -
<< >> <<< >>>
< <= > >=
== != === !==
& ~ &
^ ^~ ~^
| ~ |
& &
||
? :

Lowest

Figure 3.11   Operator Precedence

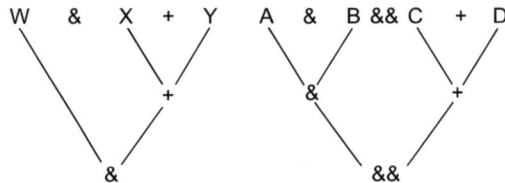

Figure 3.12   Precedence Examples

figure, because of left to right association, *A&B* is done, *C+D* is done, and the result of these two operations are logical ANDed to form a one bit result.

3.2.10   Verilog data types

Verilog has **net** and **reg** data types representing wires and variables, respectively. The **net** type represents data carriers such as interconnecting wires, gate outputs, and busses. The **reg** data type represents variables that hold the value they are assigned until they are overwritten. Additionally, a **net** or a **reg** can be declared as **signed**, which determines how they interpret data assigned to them. Declaring **net** and **reg** types and their significance in hardware modeling are discussed here.

3.2.10.1   net declarations.   A net represents a hardware wire driven by one or more gates or other types of signal sources. The simplest form of a **net** declaration begins with a keyword specifying the type of **net** followed by a list of identifiers.

Allowed **net** types are shown in Fig. 3.13. Types **wire** and **tri**, **wand** and **triand**, and **wor** and **trior** are equivalent. Types **supply0** and **supply1** are used for declaring signal names for supply voltages. The

	NET TYPES	PROPERTIES	INITIAL
Supply	Supply 0	Driven: 0	0
	Supply 1	Driven: 1	1
Three-state	Wire (tri)	Tri-state wired logic	Driven: X Not Driven: Z
	Wand (triand)	Wired-and logic	Driven: X Not Driven: Z
	Wor (trior)	Wired-or logic	Driven: X Not Driven: Z
Capacitive	trireg	Hold old value	X

Figure 3.13   net types and Properties

trireg net type declares three-state capacitive signals. Other **net** types (**wire**, **wand**, and **wor** or their equivalents, **tri**, **triand**, and **trior**) declare state signals that allow multiple driving sources. The keyword indicating a **net** type determines how multiple driving source values are resolved to form a single **net** value. Shown below is a wire declaration declaring wires *w, n, m*, and *p*. This statement declares wires used between gates or boolean expressions representing logic structures (see Fig. 3.1).

```
wire w, n, m, p;
```

By default, ports of a module are **net** of **wire** type. In Fig. 3.3, *co* and *s* represent wires that are driven by logic functions shown.

The **wire** type declares three-state signals. The table in Fig. 3.14 resolves multiple drivers on such signals. In this **net** type, an **X** value on any driving source overrides values from all other sources. The **Z** value is the weakest and is overridden by non-**Z** values from other driving sources. Driving a **wire** with multiple **0** and **1** conflicting values resolves in the **X** value for the wire.

The **wand** and **wor** type **net**s signify wired-and and wired-or functions, respectively. Figure 3.15 shows notations for these functions and tabulates their resolved values. The **wand** type implements a wired-and logic. For this type, a **0** value on a driving source overrides all other source values. Value **Z** is treated as null and is overridden by any other value driving a **wand net**. The **wor**, or wired-or, type performs the logical OR operation on all its driving sources. In this operation, logic value **1** on one source overrides all other source values. As in **wand**, the **Z** value is the weakest and is overridden by **0**, **1**, and **X** values.

Net types **tri0** and **tri1** are similar to **wire** (**tri**) in their resolved values except when the resolved value is to become **Z** in a **wire net**. While **Z** is generated in a **wire net**, **0** is generated in **tri0** and **1** is generated in **tri1**.

The **trireg** type net behaves as a capacitive wire and holds its old value when a new resolved value is to become **Z**. As long as there is at least one driver with **0**, **1**, or **X** value, **trireg** behaves the same as **wire**. When all

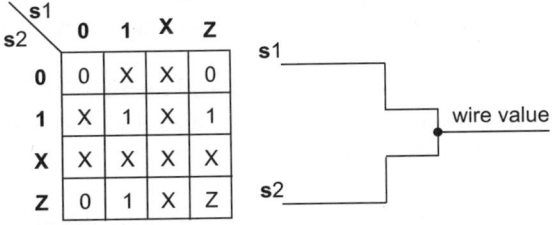

Figure 3.14 wire net Types

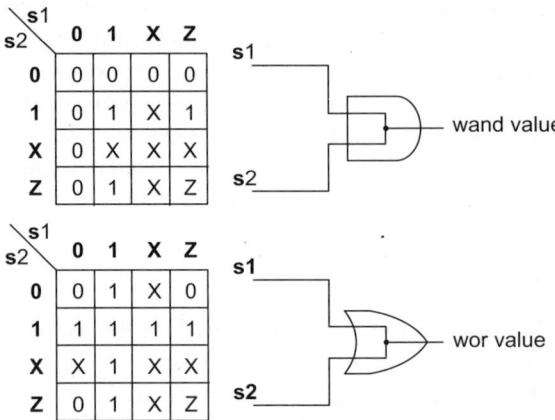

Figure 3.15 (a) **wand net** Types, (b) **wor** net Type

drivers are turned off (**Z**), a **trireg net** retains its previous value. The amount of time a **trireg net** holds a value is specified by a delay parameter in its declaration. Delay parameters will be discussed next. Chapter 7 shows examples of using **trireg** for CMOS flip-flop modeling.

In the above discussion we presented **net** declarations by **net** types. In addition to the type of the **net**, which specifies the resolution of driving-value conflicts, a **net** declaration may also include **net** delay values. Three delay values for **net** switching to 1, to 0, and to **Z** are specified in a set of parenthesis that are followed by a **#** sign after the **net** type keyword. A simpler format contains a single delay value. For example, wires for the gate outputs of Fig. 3.1 may be declared by:

```
wire #5 w, n, m, p;
```

This declaration specifies five time units of delay for all transitions of *w*, *n*, *m*, and *p* signals. This delay will be added to those of continuous assignments or gate outputs driving these **net**s. **Trireg net** types may also be declared with three delay parameters. Unlike the case with other **net**s, in this case the third timing parameter is not delay for the **Z** transition. Instead, this specifies the time that a declared **trireg net** holds an old value when driven by **Z**.

The initial value for all **net** types except **supply0** and **supply1** with at least one driver is **X**. A **net** with no driver assumes the **Z** value, except for **trireg**, which has the initial value **X**.

3.2.10.2   reg declarations.   In addition to the **net** type variable declarations, Verilog also supports the **reg** data type. Unlike a **net**, which models an interconnection or a gate output, a **reg** is a variable for holding intermediate signal values or nonhardware parameters and function values. The **reg** declaration shown below declares a, b, and ci as **reg** types with **0** initial values.

```
reg a = 0, b = 0, ci = 0;
```

This declaration is used in the *FulladderTester* module of Fig. 3.4 for assigning test values to the ports of the instantiated *Full_adder* module.

Because **reg** variables are not used exclusively for hardware modeling, other **reg** type declarations for more convenient forms of model parameters are provided in Verilog. These **reg** types are **integer** and **time**. An **integer** declaration declares a signed 2s-complement number, and a **time** declaration declares an unsigned **reg** variable of at least 64 bits. Verilog also allows declaration of **real** and **realtime** variables. These variables are similar in use to **integer** and **time** variables, but do not have direct bit-to-bit correspondence with **reg** type registers. The default initial value of a declared **reg** is (**X**). As shown in Fig. 3.4, this default can be overwritten when a **reg** is being declared.

3.2.10.3 Signed data. Verilog **net** and **reg** types can be declared as **signed**. Integer declaration is always considered **signed**. A signed 16-bit **reg** declaration is done as shown here:

```
reg signed [15:0] areg;
```

An expression using variables, **net**s, and constants as operand can become signed or unsigned depending on its operands. The left-hand side of an expression does not have any effect on the sign of the expression. A signed **reg** that is shifted right by the **>>>** operator is sign filled, whereas an unsigned **reg** shifted by this operator is zerofilled.

Integers are *signed*, and *based* numbers with the **s** notation are also signed. Other vector constants are considered unsigned. Concatenation and vector slicing always result in unsigned results. For other operators to result in signed results, all operands must be signed.

If the right-hand side of an assignment is determined as signed, it is sign extended to the size of its left-hand side and is placed on the left-hand side **reg** or **net**. As mentioned before, the extension of the sign only depends on the right-hand side and not on the left-hand side **reg** or **net**.

3.2.10.4 Parameters. Parameters in Verilog do not belong to either the variable or the **net** group. Parameters are constants and cannot be changed at runtime. Parameters can be declared as **signed**, **real**, **integer**, **time**, or **realtime**. Shown in Table 3.7 are several parameter declarations.

TABLE 3.7   Parameter Examples

Example	Explanation
parameter p1=5, p2=6;	32 bit parameters
parameter [4:0] p1=5, p2=6;	5 bit parameters
parameter integer p1=5;	32 bit parameter
parameter signed [4:0] p1=5;	5 bit signed parameter

3.2.11   Array indexing

Bit-select and part-select operators are used for extracting a bit or a group of bits from a declared array. Such addressing only applies to contiguous bits of an array. We use arrays of Sec. 3.2.6 (Fig. 3.8) to demonstrate bit-select and part-select operations. These arrays are shown here for reference.

```
reg [7:0] Areg;
reg Amem [7:0];
reg Dmem [7:0][0:3];
reg [7:0] Cmem [0:3];
reg [2:0] Dmem [0:3][0:4];
```

3.2.11.1   Bit selection.   Bit-selection is done by using the addressed bit number in a set of square brackets. For example *Areg[5]* selects bit 5 of the *Areg* array.

3.2.11.2   Part selection.   Verilog allows constant and indexed part-select. A constant part-select specifies range of bits to be selected. For example, *Areg[7:3]* selects the upper five bits of *Areg*. On the other hand an indexed part-select specifies starting index and the number of bits to be selected. For example, *Areg[3+:5]* selects the same five bits as *Areg[7:3]* does. Several examples follow:

```
Areg [5:3] selects bits 5, 4, and 3
Areg [5-:4] selects bits 5, 4, 3, and 2
Areg [2+:4] selects bits 5, 4, 3, and 2
```

3.2.11.3   Standard memory.   The standard format for declaring a memory in Verilog is to declare it as an array of a vector. For example *Cmem*, as shown above, is a 4-word memory of 8-bit words. The address space of this memory is 4. *Emem* as a byte-oriented memory of with a 10 bit address (1024 address space) is declared as:

```
reg [7:0] Emem [0:1023];
```

An expression can be used for addressing a memory. For example two bits of *Areg* can be used to extract an 8-bit word of *Cmem*. This is done as shown below:

```
Cmem [Areg[7:6]]  // addresses Cmem by Areg[7:6]
```

A memory word can be used as an address for itself. The following example uses the 8-bit word at location 0 of *Emem* to address this memory:

```
Emem [Emem[0]]  // addresses Emem by Emem[0]
```

Verilog allows selection rules for accessing part of an addressed word of a memory. For this purpose a second set of square brackets to the right of those used for memory addressing are used for bit- or part-select of the accessed memory word. For example the four least significant bits of the word at location 355 of *Emem* are accessed by:

```
Emem [355][3:0]  // 4 LSB of location 355
```

This operation is equivalent to:

```
Emem [355][3-:4]  // 4 bits starting from 3, down
```

Specifying a range of addresses in Verilog is not allowed. For example, *Emem* locations 355 to 358 cannot be addressed as shown below.

```
Emem [355:358]  // Illegal. Does not address a 4-word block
```

3.2.11.4 Indexing multi-dimensional memories. As discussed, declaring multi-dimensional memories is allowed in Verilog, e.g., *Dmem* above. For accessing such memories, simple indexings are allowed for specifying a word in the memory, and bit-select and part-select are allowed for accessing bit or bits of the addressed word. Figure 3.16 shows examples using arrays of Fig. 3.8.

3.3 Verilog Simulation Model

The previous section discussed **net** and **reg** declarations as well as operators, arrays, names, numbers, and other utilities that are used for representation and simulation of hardware entities. This section discusses how these utilities are used for correct modeling and simulation of hardware at various levels of abstraction. Data types **net** and **reg** represent very different entities in a hardware model, and the Verilog HDL

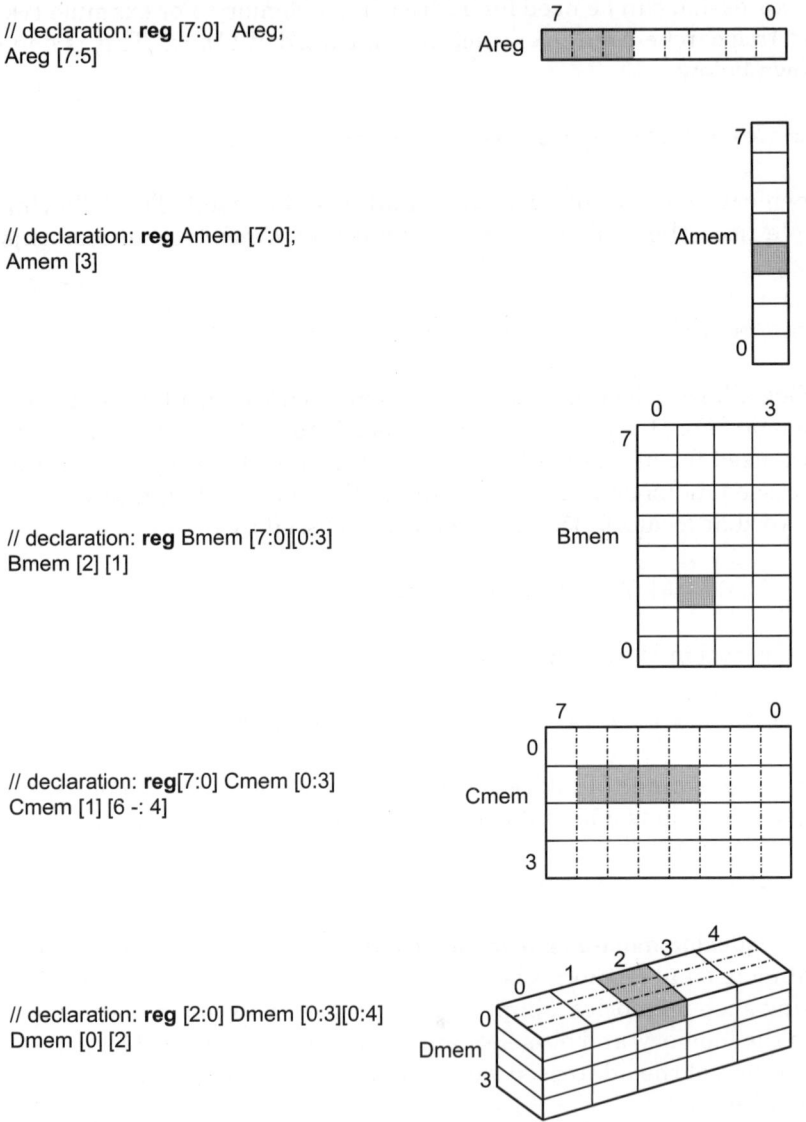

// declaration: **reg** [7:0] Areg;
Areg [7:5]

// declaration: **reg** Amem [7:0];
Amem [3]

// declaration: **reg** Bmem [7:0][0:3]
Bmem [2] [1]

// declaration: **reg**[7:0] Cmem [0:3]
Cmem [1] [6 -: 4]

// declaration: **reg** [2:0] Dmem [0:3][0:4]
Dmem [0] [2]

Figure 3.16   Array Addressing and Selection

is very specific as to the way they are used and the language constructs they are used in. We will show semantics of assignments to **reg** and **net** data types.

Variables declared as **net** are assigned values in Verilog concurrent bodies using continuous assignment statements. On the other hand, **reg** variables are assigned values in procedural bodies. The following discusses details of these constructs.

3.3.1   Continuous assignments

In this section we will discuss simple continuous assignments, assignments with delay, strength specification, **net** assignments, and assignments with multiple drivers.

3.3.1.1   Simple assignments.   A continuous assignment in Verilog is used only in concurrent Verilog bodies. This assignment represents a **net** driven by a gate output or a logic function. In its simplest form, a continuous assignment begins with the **assign** keyword, followed by the left-hand side **net** type variable, an equal sign, and a right-hand side expression. The example shown below models the OR gate at the output of Fig. 3.1.

```
assign w = m | p;
```

Like the OR gate, the above assignment becomes active only if an input of the gate (m or p), which represent **net**s on the right-hand side of the assignment, changes value.

3.3.1.2   Delay specification.   Continuous assignments may also include delay parameters. Using this format a better correspondence to the OR gate of Fig. 3.1 is:

```
assign #2 w = m | p;
```

This assignment becomes active when m or p changes. At this time, the new value of the $m \mid p$ expression is evaluated, and after a wait time of two time units, this new value is assigned to w.

The order in which continuous assignments appear in a Verilog concurrent body is not significant. Figure 3.17 shows four concurrent assignments corresponding to the gates of Fig. 3.1. Like the gates of this figure, regardless of its position in the code, each assignment waits for a right-hand side variable to change for it to execute.

```
`timescale 1ns/100ps

module Mux2to1 (input a, b, c, output w);
    wire n, m, p;
    assign #3 m = a & b;
    assign #3 p = n & c;
    assign #6 n = ~b;
    assign #2 w = m | p;
endmodule
```

Figure 3.17   Concurrent Continuous Assignments

Figure 3.18 shows simulation of the Verilog code of Fig. 3.17. Starting with $a = b = c = 1$, b changes to **0** at time 100 ns. This causes the first and the third **assign** statements to execute which causes m to become **0** at 103 ns and n to become **1** at 106 ns. The change on m at 103 ns causes the execution of the last **assign** statement causing w to go to **0** after 2 ns at 105 ns (103 ns + 2 ns). Meanwhile, the change on n at 106 ns causes the second **assign** statement to execute which causes p to become **1** after 3 ns at 109 ns. Note that p is used on the right-hand side of the assignment to w (the last **assign** statement). Because of this, at time 109 ns the right-hand side of w is evaluated again, and the calculated **1** value is assigned to w after 2 ns at time 111 ns.

The simulation of the above circuit results in a glitch due to a 1-hazard on w. The event driven simulation of concurrent statements makes this simulation to correspond to events in the actual circuit.

3.3.1.3 Strength specification. In addition to a logical value in the four-value system (**0**, **1**, **X**, and **Z**), Verilog allows **net**s to have strength values. Strength adds another degree of accuracy in modeling signal when the basic four values do not suffice. Strengths for **net**s are specified when assignments are done. As signals combine, new strength values are formed in the resulting signals.

Net strengths are specified by a pair of strength values bracketed by a set of parenthesis, as shown below.

```
assign (strong0, strong1) w = m | p;
```

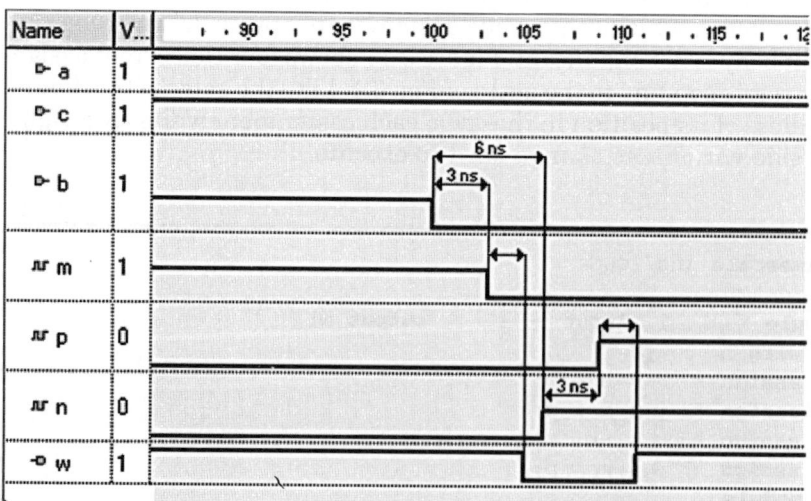

Figure 3.18 Simulation Run Showing a Glitch on w

One strength value is for logic **1** and one is for logic **0**, and the order in which the strength values appear in the set of parenthesis is not important. Strength value names for logic **1** end with a **1** (**supply1**, **strong1**, **pull1**, **weak1**, ...) and those for logic **0** end with a **0** (**supply0**, **strong0**, **pull0**, **weak0**, ...).

For **wire** and **tri** type **net**s, drive strength values are used, and for storage **net**s charge strength is used. Figure 3.19 shows strengths, their values, and the **net** types they apply to.

The strength value set for **wire** and **tri** type **net**s is referred to as *drive_strength* and is specified when an assignment to **net** takes place or when **net** is declared. Default values for these **net**s are **strong0** and **strong1** for logic **0** and logic **1**, respectively. The strength for **trireg** type **net**s specifies how weak or strong the charge storage capability of the **net** is. Three strength values, **large**, **medium**, and **small**, are used for these **net** types, and the default is **medium**. Examples in Chap. 7 illustrate how different strength values are resolved on gate outputs.

3.3.1.4   Net declaration assignments.   Using the **net** declaration assignment, a **net** assignment can be done at the same time that it is being declared. An assignment made as such, provides one driver for the

Wires (tri) wand (triand), Wor (trior), tri0, tri1		trireg	
Strength value	Level	Strength values	Strength 0
Supply 0	7		
Strong 0	6		
Pull 0	5		
	4	Large	
Weak 0	3		
	2	Medium (0)	
	1	Small (0)	
Highz 0	0		Weak values
Highz 1	0		
	1	Small (1)	
	2	Medium (1)	
Weak 1	3		
	4	Large (1)	
Pull 1	5		
Strong1	6		
Supply 1	7		Strength 1

Figure 3.19   net Types and Their Strengths

declared **net**. More drivers can be assigned to the same **net** using continuous assignment statements. As with continuous assignments and **net** declarations, strengths and timing may also be specified in a **net** declaration assignment.

The use of *net_declaration_assignment* construct is illustrated in Fig. 3.20. In this code, all the continuous assignments of Fig. 3.17 are replaced by a list of **net** declaration assignments providing drivers for *w, n, m*, and *p* signals. Drive strengths and/or delay values specified in a **net** declaration assignment apply to all **net** drivers. The syntax used here consists of a list of **net** declaration assignments as part of a **net** declaration construct. The same syntax can also be used with continuous assignments. In order to be able to specify delay values for individual wires, use of separate continuous assignments, as in Fig. 3.17, or separate **net** declaration assignments is recommended.

3.3.1.5 Multiple drivers. A situation in hardware in which several gate output are connected to the same wire is modeled with continuous assignments by having multiple assignments to the same left-hand side **net**. In this case, the **net** value is said to be driven with multiple sources simultaneously. The resulting **net** value is determined by the resolution of multiple driving values depending on the **net** types, as discussed in Sec. 3.2.10.

An example for multiple drivers using wired-or resolution is shown in Fig. 3.21. The code shown here models the circuit of Fig. 3.1 and is equivalent to the Verilog code shown in Fig. 3.17. The OR operation of the right-hand side of *w* in Fig. 3.17 is replaced by multiple assignments to the *w* **net** of **wor** type. This **net** has a delay of 2 ns, which is added to the individual driver delay values. A value assigned to *w* is first delayed by continuous assignment delay. Before this value appears on *w*, it is further delayed by 2 ns specified in **wor** declaration.

Figure 3.22 compares the simulation results of the Verilog codes of Fig. 3.17, Fig. 3.20, and Fig. 3.21. All simulations are event based. Waveforms of Fig. 3.17 and Fig. 3.21 are identical. The reason the timing

```
`timescale 1ns/100ps

module Mux2to1Net (input a, b, c, output w);
    wire #3
        m = a & b,
        p = n & c,
        n = ~b,
        w = m | p;
endmodule
```

Figure 3.20 Using *net_declaration_assignment*

```
`timescale 1ns/100ps

module Mux2to1Multiple (input a, b, c, output w);
    wire n;
    wor #2 w;
    assign #3 w = a & b;
    assign #3 w = n & c;
    assign #6 n = ~b;
endmodule
```

Figure 3.21   A net with Multiple Drivers

of the simulation of the Verilog code of Fig. 3.20 is different is that we have used a single delay value (#3 ns in *Mux2to1Net* **module**) for all wire delays.

3.3.2   Procedural assignments

Procedural assignments in Verilog take place in the **initial** and **always** procedural constructs, which are regarded as procedural bodies as discussed in Sec. 3.3.1. Primarily, assignments to **reg** data types take place in procedural bodies. This section discusses procedural flow control, blocking assignments, nonblocking assignments, and two forms of procedural continuous assignments.

3.3.2.1   Procedural flow control.   Statements in a procedural body are executed when program flow reaches them. Several flow control statements are available to control procedural flow in a procedural body. These language constructs are classified as delay control and event control. In

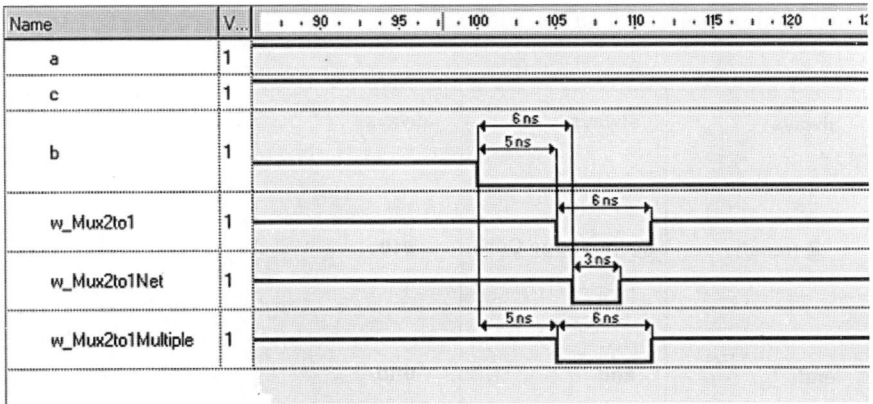

Figure 3.22   Simulation Run of Assignment Statements

the examples of Chaps. 4 to 6 we will show examples of these constructs and discuss their details. However, because these constructs affect the way procedural assignments are done, a brief discussion is included here.

Within a module, several procedural bodies may be used. Flows into these bodies concurrently begin at the same time at the start of simulation. An event or delay control statement in a procedural body causes program flow to be put on hold temporarily. The flow continues after an event occurs or a delay expires. Figure 3.23 shows three procedural flow control statements. In Fig. 3.23a, program flow stops when it reaches the @ *(reset)* statement and resumes when the value of *reset* changes. In Fig. 3.23b, program flow resumes after the positive edge of the *clk*, and in Fig 3.23c, program flow resumes after being put on hold for 10 time units.

3.3.2.2 Procedural blocking assignments.   A blocking assignment uses a **reg** data type on the left-hand side and an expression on the right-hand side of an equal sign. An event or delay control statement may delay execution of this statement. In addition to statements for control of program flow, procedural assignments may also contain intra-assignment delay or event control. The syntax of intra-assignment control constructs is similar to that of procedural flow control statements, discussed above, but these constructs appear on the right-hand side of an equal sign in a procedural assignment. Shown below is a procedural assignment that is delayed by 200 time units by a delay control statement and by 100 time units by an intra-assignment delay control.

        . . ._; #200 a = #100 b;

In this example, after 200 time units, when program flow reaches the procedural assignment, *b* is evaluated, and its value is assigned to *a* after 100 time units. The equal sign is used here for blocking assignment. In this case, the 100-time-unit delay blocks the procedural program flow until assignment to *a* takes place.

always	always	always
.	.	.
.	.	.
@ (reset)	@ (posedge CLK)	#10
.	.	.
.	.	.
end	end	end

Figure 3.23 Procedural Flow Control

```
initial begin : Blocking_Assignment_to_b
   b = 1;
   #100
   b = #80 0; b = #120 1;
   #100
   $display ("Initial Block with Blocking Assignment to b
             Ends at:", $time);
end
```

Figure 3.24   Blocking Procedural Assignments

Figure 3.24 shows an **initial** block that uses blocking assignments for assigning values to *b*. Signal *b* is a **reg** that is initialized to **1** at time 0 ns, set to **0** at time 180 ns, and set back to **1** at time 300 ns. The intra-assignment timing control statement blocks procedural flow and stops it from reaching assignment of **1** to *b* until *b* is assigned the value **0**. The **$display** statement shown in this figure displays 400.

3.3.2.2 Procedural nonblocking assignments.

Within a procedural block, nonblocking assignment to **reg** data types may be done. A nonblocking assignment uses the left arrow notation <= (left angular bracket followed by the equal sign) instead of the equal sign used in blocking assignments.

A nonblocking assignment is different from a blocking assignment only in the way in which intra-assignment control constructs are treated. Unlike a blocking assignment, a nonblocking assignment does not block the program flow in a procedural construct. When flow reaches a non-blocking assignment, the right-hand side of the assignment is evaluated and will be scheduled for the left-hand side **reg** to take place when the intra-assignment control is satisfied.

Figure 3.25 shows an **initial** block that uses nonblocking assignments for assignment of values to *a*. Signal *a* is a **reg** that is initialized

```
initial begin : Nonblocking_Assignment_to_a
   a <= 1;
   #100
   a <= #80 0; a <= #120 1;
   #100
   $display ("Initial Block with Nonblocking Assignment
             to a Ends at:", $time);
end
```

Figure 3.25   Nonblocking Procedural Assignments

to **1** at time 0 ns, set to **0** at time 180 ns, and set back to **1** at time 220 ns. The intra-assignment timing control statements used here, delay the assignment of values to a, but they do not block the flow of the program, while delays to a are being processed. In other words, an intra-assignment delay causes scheduling of a value into its left-hand side signal, and proceeds immediately to the next statement. The **$display** statement shown in Fig. 3.25 displays 200.

Figure 3.26 shows waveforms generated on b and a as a result of **initial** blocks shown in Fig. 3.24 and Fig. 3.25 respectively. We are also showing statements displayed in the simulation console. Note that when intra-assignment delays are used with nonblocking assignments, these delays do not affect the flow of program in the procedural blocks they are used in.

3.3.2.4 Multiple assignments.

Another important issue regarding procedural **reg** type assignments is the way multiple assignments from multiple procedural constructs interact. Because of the sequential flow in procedural bodies, an assignment takes place only when the program flow reaches it. If several assignments appear at the same real time in a procedural body, the last assignment overrides all others. However, if program flow in two procedural bodies reaches assignments to the same **reg** at exactly the same time, the outcome of the value assigned to the left-hand side of the assignment will not be known.

Consider, for example, the partial code shown in Fig. 3.27 for clock generation for testing a flip-flop. The intended operation of this code may have been for *clk* to get initialized to **0**, and after that get complemented every 17 time units. However, this is not necessarily the way the code simulates.

At time **0**, flows into the **initial** and **always** blocks begin simultaneously. In the **initial** block shown, *clk* is being set to **0**, and in the **always** block, at exactly the same time, *clk* is being complemented. Because the initial value of the **reg** type variable *clk* is **X**, the complement operation in the **always** block tries to complement **X** if *clk* is not initialized to **0** in the **initial** block. Whether *clk* is first set to **0** and then complemented, is first complemented, or is set to **X** because of the conflict is not known.

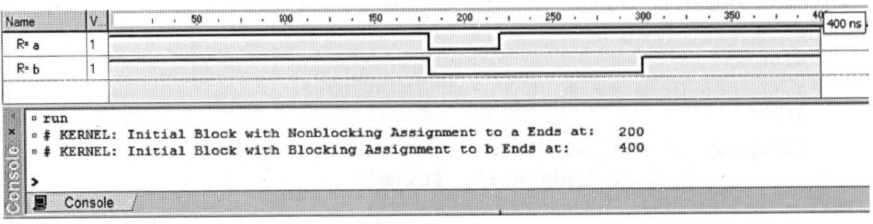

Figure 3.26 Comparing Blocking and Nonblocking Procedural Assignments

```
initial begin
   clk = 0;
end

always begin
   clk = ~clk;
   #17;
end
```

Figure 3.27 Multiple **reg** Assignments

This code works properly only if complementing of the *clk* is delayed until the *clk* is initialized to **0** in the **initial** block.

One way to correct this problem is to delay complementing the clock by one simulation cycle. This can be done by inserting *#0;* in the **always** block before complementing the *clk*, as shown in Fig. 3.28. This way, in the first simulation cycle, *clk* is set to **0** at time 0, and immediately after that, in the next simulation cycle, still at time 0, *clk* is set to the complement of **0**. Every time *clk* is complemented, flow in the **always** block is suspended for 17 time units, after which *clk* is complemented again for as long as the simulation run continues.

3.3.2.5 Procedural continuous assignment. Using a procedural continuous assignment construct, an assignment to a **reg** type variable can be made to stop all other assignments to this variable from taking place. The procedural continuous assignment to variable *qout* is done by the following procedural statement:

```
assign qout <= 0;
```

```
initial begin
   clk = 0;
end

always begin
   #0;
   clk = ~clk;
   #17;
end
```

Figure 3.28 Multiple **reg** Assignments; Delay Used for Deterministic Results

While *qout* is not deassigned, no other assignments affect its value. Deassigning a **reg** type variable *qout* is done by the following statement:

deassign qout;

Figure 3.29 shows an example of the use of procedural continuous assignments. Two **always** blocks are used in this example. The first block becomes active any time *reset* changes. In this block, if *reset* is **1**, the *qout* output of flip-flop is assigned a **0**. This value is forced on *qout* and cannot be overridden until *qout* is deassigned. In the first **always** block, if *reset* is not active (it is **0**), the flip-flop output *qout* is deassigned. Only after *qout* is deassigned can other assignments to *qout* change its value.

The second **always** block in Fig. 3.29 assigns *din* to *qout* on the positive edge of the *clk*. This assignment takes place only when reset is **0**. While an **assign** is in effect, another **assign** to the same variable, deassigns the one that is in effect and then assigns a new value to the variable.

3.3.2.6 Force and release.   Another form of procedural continuous assignments is provided by **force** and **release** statements. Unlike **assign** and **deassign**, which apply to **reg** type variables, **force**, and **release** constructs apply to **net** and **reg** types. Forcing a value on a **net** overrides all values assigned to the **net** through continuous assignments or connected to it through gate outputs.

Various forms of assignments to **net** or **reg** data types were presented in this section. Simulation semantics of each assignment was presented and examples were given. More examples of these assignments and their role in modeling hardware and writing testbenches will be given in chapters on combinational and sequential circuit modeling.

```
`timescale 1ns/100ps

module FlipflopAssign (input reset, din, clk, output qout);
    reg qout;
    always @ (reset) begin
        if (reset) assign qout <= 0;
        else deassign qout;
    end
    always @ (posedge clk) begin
        qout <= din;
    end
endmodule
```

Figure 3.29   Procedural Continuous Assignments

3.4 Compiler Directives

Certain hardware modeling requirements in Verilog are provided as compiler directives instead of being incorporated in the main syntax of the language. Compiler directives provide facilities for file inclusion, timing specification, default settings, and string substitution. This section provides a brief description of these language facilities. Directives are presented in the order of their importance in hardware modeling.

3.4.1 `timescale

The `timescale directive sets the time unit and time precision in a module. Including the

```
`timescale 1ns/100ps
```

directive before a module header causes all time-related numbers to be interpreted as having a 1ns time unit. When expressions manipulating timing values are performed, 100 ps precision will be used.

3.4.2 `default_nettype

Undeclared **net**s are implicitly declared as **wire** type **net**s. The default **wire** type can be changed by the `default_nettype`. For example,

```
`default_nettype wor
```

at the beginning of a module causes undeclared **nets** in constructs such as the terminal list of a module instance to be assumed to be **wor** type **nets**.

   This default setting stays in effect for all future compilations until it is set to another value or reset by use of the `resetall directive.

3.4.3 `include

To include a parameter definition file or a section of shared code in a module, the `include directive may be used. Because Verilog does not provide a common library of parts and utilities, a shared code must be explicitly inserted in modules that use the code. This can be achieved using the `include directive.

3.4.4 `define

For better code readability, a meaningful string (referred to as text macro) can be defined to represent a number or an expression. The `define` directives shown below define *word_length* of 32 and assign the **101** binary code to state *begin_fetch_state* of a state machine.

```
`define word_length 32
`define begin_fetch_state 3'b101
```

Defined strings may be used in Verilog code text by preceding their defined names by a back quote. For example, if *begin_fetch_state* is used anywhere in a Verilog code, it is replaced by 3'b101. The `undef` directive undefines a previously defined text macro.

3.4.5 `ifdef, `else, `endif

The `ifdef` directive tests whether a text macro that immediately follows the `ifdef` keyword has been defined using the `define` directive. If the next macro has been defined, the group of lines bracketed between `ifdef` and `else` is compiled. If the text macro has not been defined, the group of lines bracketed between `else` and `endif` is compiled.

These directives can appear anywhere in a Verilog source code. The two groups of lines bracketed in an `ifdef`, `else`, `endif` structure must independently have correct syntax.

3.4.6 `unconnected_drive

A port value left open in the connection list of a module instantiation assumes the default **net** value. To change this and to force **pull0** or **pull1** values on unconnected ports, the `unconnected_drive` directive can be used. The only arguments allowed with this directive are **pull0** or **pull1** for unconnected values **0** and **1**, respectively. The effect of this directive may be turned off by `nounconnected_drive` direction.

3.4.7 `celldefine, `endcelldefine

The `celldefine` and `endcelldefine` directives bracket modules that are to be considered as cells. The Verilog programming language interface (PLI) uses cell modules.

3.4.8 `resetall

The `resetall` directive turns off the effect of all compiler directives. Using this directive at the beginning of every module guarantees that no previous setting affects compilation of modules and that all defaults are set.

3.5 System Tasks and Functions

For testbench generation, data input and output, timing check, simulation flow control, data conversion, and memory initialization and specification, Verilog provides a number of system tasks and functions categorized into ten groups. The names of system tasks and functions begin with a dollar sign (**$**), followed by a task specifier. The name of the task or function usually contains characters and names that describe its functionality. A brief description of these language utilities will be given here. Application of several of these tasks has already been presented in this and the previous chapter, and more tasks and functions will be presented in the examples that use them in the chapters that follow.

3.5.1 Display tasks

Display tasks are used for outputting to the standard output device. The most basic display task is the **$display** task, which writes its string argument to the display device. Other tasks include those for monitoring and outputting variable values as they change (the **$monitor** group of tasks) and those for displaying variables at a selected time (the **$display** tasks). Display tasks can display in binary, hexadecimal, or octal formats. The character **b**, **h**, or **o** at the end of the task name specifies the data type a task handles. For all display tasks, a generic task can be used to display data with specified formats and data types. Chapter on testbenches presents examples that take advantage of many of these tasks for outputting test results.

3.5.2 File I/O tasks

File output tasks begin with a dollar sign followed by the letter **f** (for file) and then by the same task names as those of the display tasks. These tasks perform the same functionalities as their display task counterparts, except that their output is to a file instead of to the display terminal. The **$fopen** function opens a file and assigns an integer file description. The file descriptor will be used as an argument for all file I/O tasks. In addition, there are string write tasks (**$swrite**) that write their formatted outputs to a string.

Verilog also provides tasks for inputting data from files or strings. Such tasks allow reading characters, formatted data, or complete memory data from external data files or declared strings. Examples of these tasks are **$fgetc**, **$fscanf**, and **$sscanf** for getting a character from a file, reading formatted data from a file, and reading formatted data from a string, respectively. Other input tasks exist for reading memory data directly into a declared memory. Examples of such tasks are **$fread** and **$readmemh**.

File positioning tasks, **$fseek** and **$frewind** are also available for positioning file pointer for read or write.

Verilog I/O tasks are useful in developing complete hardware/software environments and developing testbenches. In the chapters that follow we will use these tasks for developing testbenches and in conjunction with such applications, we will describe utilization of tasks. Appendix B has a complete list of tasks and a simple example of each.

3.5.3   Timescale tasks

Timescale tasks are **$printtimescale** and **$timeformat**. The **$print-timescale** task displays the timescale and precision of the module whose hierarchical name is being passed to it as its argument. The **$timeformat** task formats time for display by file IO and display tasks.

3.5.4   Simulation control tasks

Simulation control tasks are **$finish** and **$stop**. The **$finish** task ends the simulation and exits. Usually, simulation environments require a confirmation before the action of exiting the environment is taken. The **$stop** task suspends the simulation and does not exit the simulation environment.

3.5.5   Timing check tasks

Timing check tasks are used for checking timings, such as pulse width duration and setup and hold times. In general, timing check tasks check the timing on one signal or the relative timing of several signals for certain conditions to hold. If a violation is detected, a message will be issued in the user simulation environment display area. For example, the statement shown below uses the **$nochange** timing check task to report a violation if *d_input* changes in the period of three time units before and five time units after the positive edge of the *clock*.

```
$nochange (posedge clock, d_input, 3, 5);
```

3.5.6   PLA modeling tasks

Programmable logical array (PLA) modeling tools use a declared memory as their first argument, configure it as synchronous or asynchronous, assign inputs and outputs to the PLA array, and configure the logical function of the PLA. Names used for PLA modeling tasks consist of three fields, *sync_async*, *and_or*, and *array_plane*. Tasks are named using these three fields separated by dollar signs. The general format that we use for describing these tasks is

```
$sync_async$and_or$array_plane
```

where *sync_async* can be either **sync** or **async**; *and_or* can be **and**, **or**, **nand**, or **nor**; and in place of *array_plane*, **array** or **plane** can be used.

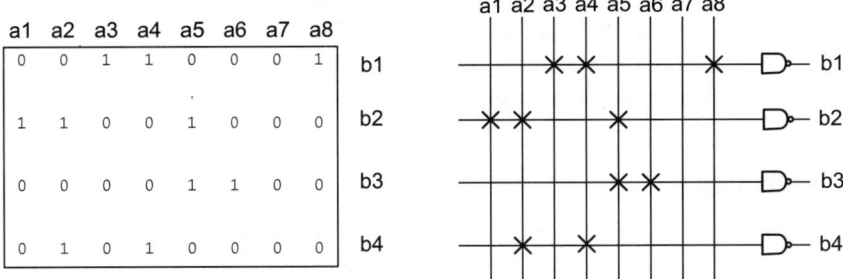

Figure 3.30 (*a*) Contents of *mem8by4*, (*b*) Corresponding PLA NAND Plane

Shown below is a PLA system task that defines an asynchronous PLA with a **nand** logical function, *a1* to *a7* inputs, and *b1* to *b4* outputs. PLA nand-plane fuses are determined by the contents of the *mem8by4* declared memory, as shown in Fig. 3.30.

```
$async$nand$array
    (mem8by4,
    {a1, a2, a3, a4, a5, a6, a7, a8},
    {b1, b2, b3, b4})
```

With this task and memory contents shown in Fig. 3.30, PLA outputs will be assigned values as shown in Fig. 3.31.

3.5.7 Conversion functions for reals

Verilog provides four system functions for converting from real to integer or bit, and for converting between bit or integer and real. These functions are **$bitstoreal**, **$realtobits**, **$itor**, and **$rtoi**.

3.5.8 Other tasks and functions

In addition to the above system utilities, Verilog provides several stochastic and probabilistic tasks and functions. Among these utilities,

---

b1 = ~(a3 & a4 & a8)

b2 = ~(a1 & a2 & a5)

b3 = ~(a5 & a6)

b4 = ~(a2 & a4)

Figure 3.31 PLA Output Equations

$random is a useful function for random data generation. We will use this task in examples in the chapters that follow.

There are also three time functions, **$realtime**, **$time**, and **$stime**, that return the simulation time in various formats.

3.6 Summary

The first part of this chapter presented general timing and concurrency concepts that are particular to hardware description languages. The second part, discussed utilities found in Verilog for describing hardware and hardware test environments. This part focused on general syntax of the language its operators, names, and data types. In the next section we discussed simulation of hardware described in Verilog using language constructs and utilities of this language. In the last part of this chapter tasks and compiler directives were discussed, that are part of the language utilities for hardware and testbench modeling, but are secondary to those discussed in Sec. 3.2. Overall this chapter presented most of Verilog without presenting a lot of examples and specific applications of the language constructs. The chapters that follow use this material to develop combinational and sequential circuit Verilog codes and their testbenches.

Problems

3.1 Starting at time 0 and assuming that no data is applied to the inputs of the circuit, show values on c, a, b, and y from time 0 until the circuit stabilizes. Include all **X** and **Z** initial values. List hexadecimal values in a tabular form.

```
`timescale 1 ns/1 ns
module quest1 (a, b, c, y);
   input [3:0] a, b, c;
   output [7:0] y;
   assign #3 a = 2'b2;
   assign #4 b = - 3'b3;
   assign #1 y = ~ c;
   assign #2 c = a ^ b;
endmodule
```

3.2 Starting at time 0 and assuming that a receives hex value A at time 10 ns, show values on a, b, w, and y from time 0 until the circuit stabilizes. Include all **X** and **Z** initial values. List values in a tabular form, use hexadecimal representation.

```
`timescale 1 ns/1 ns
module quest2 (a, b, w, y);
   input [3:0] a, b;
   output [7:0] y;
   output [11:0] w;
   assign #30 b = 1'hF;
   assign #40 y = {a, b};
   assign #50 w = {a, b, b[1:0], b[3:2]};
endmodule
```

3.3 Starting at time 0 and assuming that a receives the hexadecimal value A at time 10 ns and b receives hex value B at time 20, show values on a, b, w, and y from time 0 until the circuit stabilizes. Include all **X** and **Z** initial values. List values in a tabular form, use binary representation.

```
`timescale 1 ns/1 ns
module quest3 (a, b, w, y);
   input [3:0] a, b;
   output [7:0] y;
   wor [7:0] y;
   output [7:0] w;
   reg [7:0] w;
   assign #30 y = 8'h22;
   assign #20 y = {a, b};
   always @(a or b or y) #10w = y;
endmodule
```

3.4 In the following code, simulation begins at time 0; a receives hex value A at time 10 ns and it receives 0 at time 40; while b receives hex value 5 at time 20 and 0 at time 50. Show values on a, b, w, and y from time 0 until the circuit stabilizes. Include all **X** and **Z** initial values. List values in a tabular form, values must be in hexadecimal.

```
`timescale 1 ns/1 ns
module quest4 (a, b, w, y);
   input [3:0] a, b;
   output [7:0] w, y;
   reg [7:0] w, y;
   always @(a or b) #17 w = {a, b};
   always @(a or b) y = #17 {a, b};
endmodule
```

3.5 Rewrite the following code using nonblocking assignments such that timings of assignments remain the same.

```
initial
    begin
        a = #delay1 b;
        c = #delay2 d;
    end
```

3.6 Replace the three initial blocks shown below by only one.

```
initial
    a = #delay1 b;

initial
    c = #delay2 d;

initial
    begin
        e <= #delay3 f;
        k <= #delay4 g;
end
```

3.7 Write a positive edge sensitive 8-bit D-type register with synchronous set and reset inputs.

3.8 Given values of a, b, and c as shown, write the result of expressions shown below.

Assume: a is [3:0], b is [3:0], c is [5:0]

Assume: a = 4'b0010, b = 4'b1010, c = 6'b001101

```
a & b = ?              a || b = ?
a && b = ?             a | b = ?
a + b = ?              a = c , a = ?
a - b = ?              c = b , c = ?
&b = ?                 | a = ?
```

3.9 What does the following code do? Rewrite it using blocking assignments.

```
always @(posedge clock) begin
    a <= b;
    b <= a;
end
```

3.10 What does the following code do? Rewrite it using blocking assignments.

```
always @(posedge clock)
   #0 a <= b + c;
always @(posedge clock)
   b <= a;
```

3.11 Show waveform on *d* for the entire simulation time.

```
`timescale 1ns/100ps
module test;
reg b,c,d;
   initial begin
   b=1'b1;
   c=1'b0;
   #10 b=1'b0;
   end
   initial d = #25(b|c);
endmodule
```

3.12 Memory indexing. A) Declare a 4K memory with word size of 8, and a 1-bit flag. B) Select the 5th bit of the 12th word of the memory and put it in the flag. All these operations must be done as continuous **assign** statements.

3.13 Show waveforms for *a, b*, and *c* for the first 100 ns of simulation.

```
Module test;
   wire a, b;
   reg c;
   assign #60 a = 1 ;
   initial begin
      #20 c = b;
      #20 c = a;
      #20;
   end
 end
endmodule
```

3.14 Show values of *x*, *y*, and *z* in the first 100 ns of simulation.

```
module test;
   wire a;
   reg x, y, z;
   assign #25 a = 1 ;
   always begin
      #20;
      x = #10 a;
      #3 y = a;
      #3 z = a;
      #7;
   end
endmodule
```

Suggested Reading

Brown, S., and Z. Vranesic, *"Fundamentals of Digital Logic with Verilog Design"*, McGraw-Hill; New York, 2002, ISBN: 0-07-283878-7.

Navabi, Z., *"Verilog Computer-Based Training Course"*; CBT CD with hardcopy User's manual, McGraw-Hill, New York, 2002, ISBN 0-07-137473-6.

IEEE Std 1364-2001, *IEEE Standard Verilog Language Reference Manual*, SH94921-TBR (print) SS94921-TBR (electronic), ISBN 0-7381-2827-9 (print and electronic), 2001.

IEEE Std 1076-2002, *IEEE Standard VHDL Language Reference Manual*, SH94983-TBR (print) SS94983-TBR (electronic), ISBN 0-7381-3247-0 (print) 0-7381-3248-9 (electronic), 2002.

Combinational Circuit Description

This chapter focuses on combinational circuits. After discussing properties of ports and wires, we give a detailed presentation of combinational logic description in Verilog from gates to complex behavioral descriptions. Delay specifications in conjunction with gates, boolean expressions, and behavioral coding will be discussed. We will show how a complete description can be made by putting together components at various levels of abstraction. The last section in this chapter shows combinational logic coding for synthesis. In this section we highlight constructs that are synthesizable, and show styles of coding that are unambiguous in terms of the hardware that they synthesize to.

4.1 Module Wires

A **module** in Verilog defines a hardware component. Inside a module, various parts of a hardware component may be described by subcomponents or processes that define lower-level structures of a component. Wires or **net**s are used for interconnection of substructures together, and interconnection of module ports to appropriate ports of a module's substructures. By default, module ports are wires (**net**). Wires have delays, can take any of the four logic values (**0, 1, Z**, and **X**), and can be driven by multiple drivers.

4.1.1 Ports

Figure 4.1 shows the *Anding* module with *a*, *b*, and *y* ports. Verilog allows a port to be defined as **input**, **output**, or **inout**. An **input** port is always a **net** and can only be read. An **output** is a **net** by default, and can be declared as a **reg** if it is to be assigned a value inside a procedural block.

```
`timescale 1ns/100ps
module Anding (input a, b, output y);
    and (y, a, b);
endmodule
```

Figure 4.1 A Simple Module

An **inout** is a bidirectional port that can be written into or read from. An **inout** port is always a **net**.

In the schematics of this book, we make a correspondence between notations used in the schematics and their corresponding Verilog code. For the ports, we show an **input** port by a hollow box, and an **output** port by a solid box. An **inout** will be shown by a half-solid box. Figure 4.2 shows an example of this notation.

The default **net** type is **wire** for all module **net**s, including its ports. A **net** declaration within a module can change this.

4.1.2 Interconnections

Figure 4.1 shows that inside the *Anding* **module**, module port **net**s are connected to the ports of this module's substructure which is an **and** primitive. Basic gates are defined as Verilog primitives. The left-most argument of the **and** primitive is its output and the other arguments are its inputs.

Values put into the *a* and *b* inputs of *Anding* are carried through these wires to the inputs of **and**. The **and** primitive generates its output, which is carried through the *y* **net** to the output of *Anding*.

4.1.3 Wire values and timing

A **net** used for a module port or an internal interconnection can take any of the four Verilog logic values, i.e., **0**, **1**, **Z**, and **X**. Such a value assigned to a **net** can have a delay, which may be specified by the assignment to the **net** or as part of its declaration. Multiple simultaneous assignments

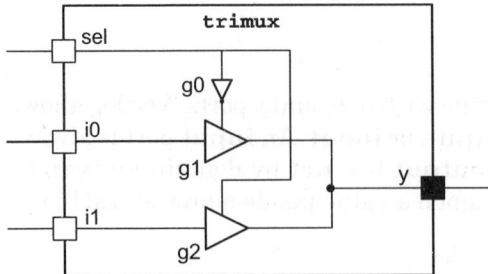

Figure 4.2 Multiplexer Using Tri-State Buffers

```
`timescale 1ns/100ps

module TriMux (input i0, i1, sel, output y);
    wire sel_ ;

    not #5 g0 (sel_, sel);
    bufif1 #4
        g1 (y, i0, sel_),
        g2 (y, i1, sel );
endmodule
```

Figure 4.3   Verilog for a Multiplexer with Tri-State Buffers

to a **net**, or driving a **net** by multiple sources is legal and the result is defined by the type of the **net**.

Figure 4.2 shows a multiplexer built by tri-state buffers. As shown, the *y* output is driven by two tri-state buffers. Furthermore, the upper buffer, which is labeled *g1*, receives the complement of *sel*, which may delay it because of the inverter.

The Verilog code corresponding to the multiplexer of Fig. 4.2 is shown in Fig. 4.3. This description uses **not** and **bufif1** Verilog primitives. The **not** gate uses the *g0* instance name (which is optional for primitive instantiations), takes *sel* as input, and puts its complement on the declared **net** *sel_* (we used underscore to read as logic BAR). This **not** gate has a delay of 5 ns, which is specified after the sharp sign after the name of the primitive.

Two instances of **bufif1** are combined into one gate instantiation construct by separating the two port connection lists by commas, and ending the construct by a semicolon.

In a **bufif1** port list the output comes first and is followed by the input, and then followed by the tri-state control input. In our example Verilog code, *g1* selects *i0* when *sel* is **0**, and *g2* selects *i1* when *sel* is **1**.

Figure 4.4 shows the simulation of **bufif1** for several changes on *i0*, *i1*, and *sel* between time 0 and 70 ns. Note that signal *s* in this waveform drives the *sel* input of the *TriMux* module. Test data inputs of this

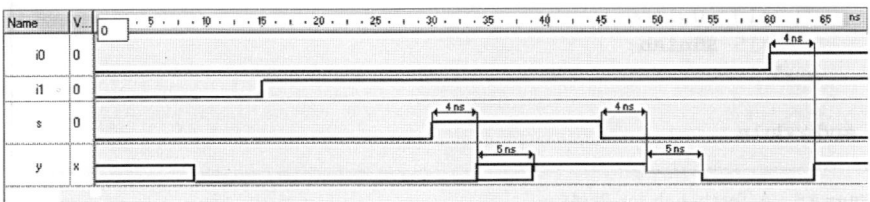

Figure 4.4   Simulation Results of **bufif1**

circuit are **0** at time 0. Note that initially y is **X** until the value of $i0$ propagates to this output at time 9 ns. The initial value of a driven **net** is **X** and that of an undriven net is **Z**. The 9 ns delay is due to a **0** propagating to $sel\_$ after 5 ns, and then $i0$ propagating to y after 4 ns.

At time 30 ns, when s (which drives sel) changes to **1**, it causes $g2$ to conduct after a 4-ns delay. At time 34 ns, both $g1$ and $g2$ conduct. $g1$ is still conducting because it takes the **not** gate 5 ns to change the value of $sel\_$ and **bufif1** an extra 4 ns to stop $g1$ from conducting. This causes the value **X** to appear on y. Because of the delay of the **not** gate, this **X** value remains on y for 5 ns. The opposite of this situation happens at time 45 ns when sel becomes **0**. Because of this change, after a 4 ns delay, neither $g1$ nor $g2$ conduct, causing a **Z** (high impedance) to appear on y for a period of 5 ns (inverter delay).

In this example we used simple delay constructs to demonstrate wire values and delays. Verilog allows more detailed delay and strength specifications to model physical properties of wires and gates more closely. These topics will be covered in the sections that follow.

4.1.4 A simple testbench

Values assigned to inputs of a circuit for examining its operation are either specified within a simulation environment using a waveform editor, or by a Verilog testbench. Values shown in Fig. 4.4 are generated by the testbench of Fig. 4.5. In this description, *TriMux* is instantiated

```
`timescale 1ns/100ps

module TriMuxTest;
    reg i0=0, i1=0, s=0;
    wire y;

    TriMux MUT (i0, i1, s, y);

    initial begin
        #15 i1=1'b1;
        #15 s=1'b1;
        #15 s=1'b0;
        #15 i0=1'b1;
        #15 i0=1'b0;
        #15 $finish;
    end

endmodule
```

Figure 4.5 A Testbench for *TriMux*

and *module-under-test (MUT)* is used for its instance name. Using an instance name for module instantiation is mandatory. Variables local to this testbench, *i0*, *i1*, *s*, and *y* are connected to the ports of the *MUT*. Because *i0*, *i1*, and *s* must be assigned values in this testbench, they are declared as **reg** and initialized to **0**. Wire *y* that connects to the output of *MUT* is declared as a **wire** and is driven by this module.

The **initial** statement is a procedural construct and uses delay control statements to delay the program flow in this procedural block. After each such delay, a value is assigned to *i0*, *i1*, or *s*. At the end of this block, after a 15-ns delay, the **$finish** simulation control task finishes the simulation run. The delay before **$finish** allows the last input change to have a chance to affect the circuit output. The delay values (15 ns) used in this example are chosen so that inputs remain stable while a change is propagating through the circuit. More elaborate testbenches will be shown in the sections that follow. The chapter on testbenches shows testbench writing techniques for combinational and sequential circuits.

4.2 Gate Level Logic

The previous section discussed the role of wires and basics of generating Verilog modules for simulation. Building upon that material, this section presents generation of Verilog modules using predefined gate primitives of this language. We will also discuss delay issues related to these gates and ways of defining them and the way they affect timing of an entire system.

4.2.1 Gate primitives

Verilog gate level primitives are shown in Fig. 4.6. This list includes standard *n_input*, *n_output*, and tri-state gates. Verilog instantiation of these gates are also shown in this figure. In addition, Verilog has switch level and transistor primitives that will be discussed in a later chapter.

Gates categorized as *n_input* gates are **and**, **nand**, **or**, **nor**, **xor**, and **xnor**. An *n_input* gate has one output, which is its left-most argument, and can have any number of inputs that may be listed as its argument separated by commas. These gates can have up to two delay parameters that can appear after the name of the gate in a set of parenthesis followed by a sharp sign. An example instantiation of a 4-input **nand** is shown here.

```
nand #(3, 5) gate1 (w, i1, i2, i3, i4);
```

In this example, t_{PLH} (low to high propagation) and t_{PHL} (high to low propagation) times are 3 and 5, respectively. Gate delays are optional,

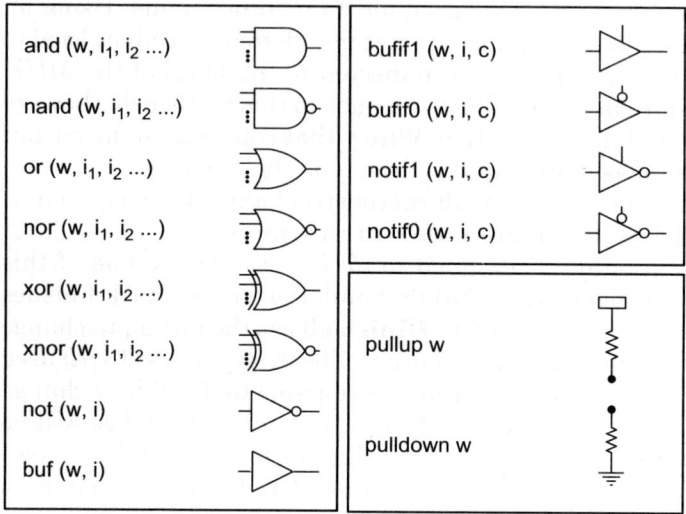

Figure 4.6   Basic Gate Primitives

and if not included, 0 delay values are assumed. If only one delay parameter is used, e.g., #3 or #(3), that delay parameter applies to all gate output transitions. Output transitions to **X** use the minimum of the delays specified. The above example uses *gate1* for the instance name. This parameter is optional and can be eliminated.

Figure 4.7 shows a majority circuit (*maj3*) with *a*, *b*, *c* input and *y* output. The circuit's primary inputs are directly connected to the **and** gate inputs. For connecting **and** gate outputs to the output **or** gate, intermediate wires *im1*, *im2*, and *im3* are used.

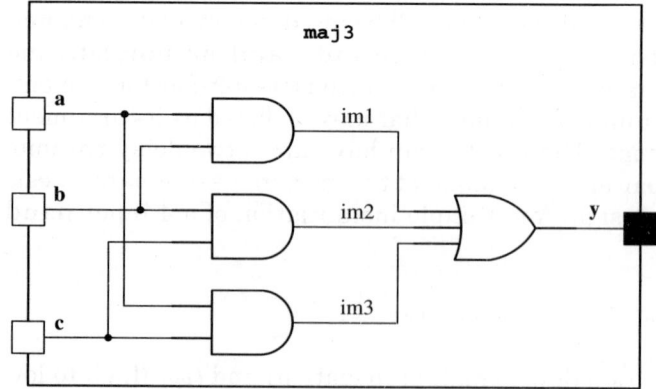

Figure 4.7   Majority Circuit

```
`timescale 1ns/100ps

module maj3 (a, b, c, y);
    input a, b, c;
    output y;
    wire im1, im2, im3;

    and #(2, 4)
        ( im1, a, b ),
        ( im2, b, c ),
        ( im3, c, a );
    or #(3, 5) (y, im1, im2, im3);

endmodule
```

Figure 4.8 Majority Verilog Code

Figure 4.8 shows the Verilog code of the majority circuit. As shown, the **and** primitive name and its delay specification are shared between the three instances of this gate. The **or** gate uses 3 and 5 ns for its delay parameters and drives the *y* output of the circuit.

4.2.2 User defined primitives

Basically, Verilog primitives are simple functions, and more complex functions can be formed based on these functions. Alternatively, for simple functions for which Verilog does not provide a primitive, a user-defined primitive (UDP) can be formed.

A combinational UDP can form a combinational function of up to 10 inputs and one output. The definition of a UDP can only include a table in the form of a logical Truth Table. A UDP output can only be specified as a **0** or a **1**. The **Z** value cannot be specified, and an **X** happens for unspecified input combinations. A UDP definition cannot include delay values, but when instantiating it, rise and fall (t_{PLH} and t_{PHL}) delay values can be specified in the same way as in Verilog primitives discussed in Sec. 4.2.1. The initial value of the output of a UDP is **X**, which changes to a **0** or a **1** after the specified delays.

Figure 4.9 shows a UDP defined for the majority function, the same function implemented by the module of Fig. 4.8. A UDP definition begins with **primitive** keyword, and after its name, its ports are defined. The output of a UDP is always first in its port list.

The UDP table shown in Fig. 4.9 has all eight combinations for the three inputs of this primitive. The ordering of table input columns is according to the declaration of inputs, i.e., *a*, *b*, and *c*. In the table, a **?**

```
primitive maj3 (y, a, b, c);
    output y;
    input a, b, c;
    table
        0 0 ? : 0;
        0 ? 0 : 0;
        ? 0 0 : 0;
        1 1 ? : 1;
        1 ? 1 : 1;
        ? 1 1 : 1;
    endtable
endprimitive
```

Figure 4.9   Majority UDP

stands for a **0**, **1** and an **X**. Therefore, **00?** expands to **000**, **001**, and **00X**. For all of such combinations *maj3* produces the same output.

The UDP table must contain all input combinations for **0** and **1** output values. Otherwise, an unspecified input combination produces an **X** output. A UDP is instantiated like a system primitive. Associated with a UDP instantiation, zero, one, or two delay values may be specified.

4.2.3   Delay formats

A two-value gate such as an **and** and an **xor** uses a *delay2* construct for its delay specification. For a tri-state gate a *delay3* language construct is used that specifies its delay to **1**, to **0**, and to **Z**. For these gates, zero, one, two, or three delay values may be specified. In the absence of the third value, minimum of the first two will be used for transitions to **Z**. Likely, transitions to **X** always take the minimum of the specified values.

Instead of a fixed delay value, **min:typ:max** delay may be used. This delay specifies the minimum, typical, and maximum delay values. This is called a delay expression and by default **typ** is used in simulation. Overriding this default and using the other specified values can be done by a simulation switch.

Figure 4.10 shows the schematic of a three-input XOR function. This circuit uses **not** and **nand** gates. The Verilog code corresponding to this diagram is shown in Fig. 4.11.

This Verilog code uses **min:typ:max** expressions for rise and fall delays of **not** and **nand** gates. By default, simulation will be done with **typ** delay values, i.e., 3 and 4 for **not** and 4 and 5 for **nand** gates. Delay control in procedural statements that will be seen later in this chapter can also use this delay format.

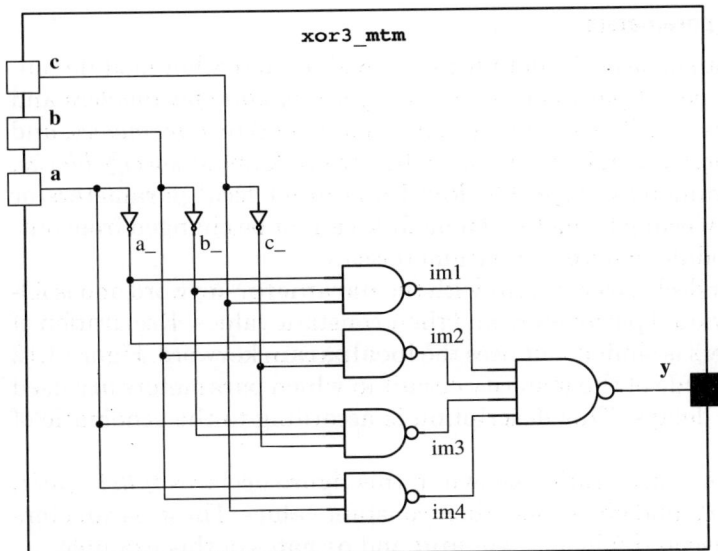

Figure 4.10   Three-Input XOR

```verilog
`timescale 1ns/100ps

module xor3_mtm (input a, b, c, output y);
    wire a_, b_, c_;
    wire im1, im2, im3, im4;

    not #(1:3:5, 2:4:6)
        ( a_, a ),
        ( b_, b ),
        ( c_, c );
    nand #(2:4:6, 3:5:7)
        ( im1, a_, b_, c ),
        ( im2, a_, b, c_ ),
        ( im3, a, b_, c_ ),
        ( im4, a, b, c );
    nand #(2:4:6, 3:5:7) (y, im1, im2, im3, im4);

endmodule
```

Figure 4.11   Verilog Code using **min:typ:max** Delay

4.2.4 Module parameters

Parameters can be used for defining delay values and other module constants. Two types of parameters in Verilog are *module parameters* and *specify parameters*. This section is dedicated to *module parameters*, and *specify parameters* will be discussed when we talk about *specify blocks*.

Module parameters are either **localparam** for local parameters or **parameter**. A local parameter of a module cannot be changed from outside of the module, whereas a parameter can.

A parameter declaration begins with the **parameter** keyword and is followed by individual parameters and their constant values. Declaration of local parameters is similar, but uses the **localparam** keyword. Figure 4.12 shows Verilog code of the majority circuit in which parameters are used for its timing delays. This description is according to the schematic of Fig. 4.7.

The parameter declaration shown in this figure declares *tplh1*, *tphl1*, *tplh2*, and *tphl2*, and gives them their constant values. These parameters are used for rise and fall delays of **and** and **or** gates of this example.

Figure 4.13 shows several ways of instantiating the *maj3_p* module and specifying some or all of its parameters. The *MUT1* instance in which no parameters are specified uses parameter values defined in the *maj3_p* module by **parameter** declaration.

The *MUT2* instance uses the ordered parameter assignment to override all parameters specified in the *maj3_p* module. The next format,

```
`timescale 1ns/100ps

module maj3_p (input a, b, c, output y);
    wire im1, im2, im3;

    parameter
        tplh1=2, tphl1=4,
        tplh2=3, tphl2=5;

    and #(tplh1, tphl1)
        ( im1, a, b ),
        ( im2, b, c ),
        ( im3, c, a );
    or #(tplh2, tphl2) (y, im1, im2, im3);

endmodule
```

Figure 4.12 A Parameterized Majority Circuit

```
maj3_p MUT1 (aa, bb, cc, y1);
  // Parameters are left as defined in module

maj3_p #(6, 8, 7, 9) MUT2 (aa, bb, cc, y2);
  // Parameters are overwritten by 6,8,7, and 9

maj3_p #(6, 8) MUT3 (aa, bb, cc, y3);
  // tplh1 and tphl1 are overwritten by 6 and 8,
  // and tplh2 and tphl2 are left as defined in the module

maj3_p #(.tplh2(7), .tplh1(6)) MUT4 (aa, bb, cc, y4);
  // tplh1 and tplh2 are overwritten by 6 and 7,
  // and tphl1 and tphl2 are left as defined in the module,
  // i.e, 4 and 5.

defparam MUT5.tplh2 = 7;
maj3_p MUT5 (aa, bb, cc, y5);
  //tplh2 is overwritten with 7 and all other parameter values
  // are left as defined in the module
```

Figure 4.13 Parameterized Module Instantiation

MUT3 also uses the ordered parameter assignment, but only overrides the first two *maj3_p* module parameters.

The *MUT4* instance of *maj3_p* of Fig. 4.13 uses named parameter assignment to override only those selected parameters. In this format parameter names and their corresponding values can appear in any order and are identified by the actual parameter names used in the module.

The last alternative shown in Fig. 4.13 uses *MUT5* for the instance name of *maj3_p* module. In this case, a **defparam** construct along with hierarchical naming are used to point to **parameter** *tplh2* of *maj3_p* and set its value to 7. This construct is referred to as *parameter override*, and can be used with hierarchical naming to reach parameters at any lower level of hierarchy.

Figure 4.14 shows output delays of *maj3_p* when this module is instantiated by alternative methods shown in Fig. 4.13. For each case rise and fall of the majority circuit output are shown. Because this circuit does not have any negations, and the distance of all inputs to the output is the same, rise (fall) delay of output *y* is calculated by simply adding rise (fall) delays of all gates in the input to output path. For example rise of *MUT4* is calculated by adding 6 and 7 (new values) and its fall is calculated by adding 4 and 5 that are the original module parameter

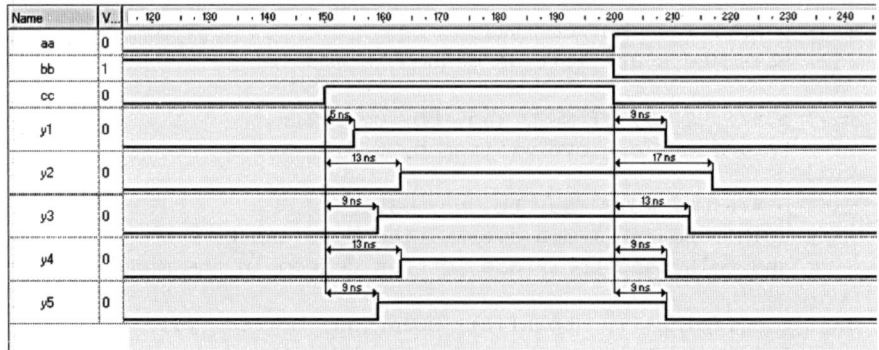

Figure 4.14   Overriding Parameters of *maj3_p*

values. Values shown in the waveforms of Fig. 4.4 verify the parameter override discussion presented earlier.

As described in Chap. 3, the `timescale directive defines a module's delay format. With 1ns/100ps, we can use delay values with one fractional digit. This format is illustrated in the parameterized three-input XOR of Fig. 4.15. Gate level diagram of this code is shown in Fig. 4.10.

```
`timescale 1ns/100ps

module xor3_p (input a, b, c, output y);
    wire a_, b_, c_;
    wire im1, im2, im3, im4;
    parameter
        tplh1=0.6, tphl1=1.1,
        tplh2=0.3, tphl2=0.9,
        tplh3=0.8, tphl3=1.3;
    not #(tplh1, tphl1)
        ( a_, a ),
        ( b_, b ),
        ( c_, c );
    nand #(tplh2, tphl2)
        ( im1, a_, b_, c ),
        ( im2, a_, b, c_ ),
        ( im3, a, b_, c_ ),
        ( im4, a, b, c );
    nand #(tplh3, tphl3) (y, im1, im2, im3, im4);
endmodule
```

Figure 4.15   A Parameterized XOR with Real Delay Values

4.3 Hierarchical Structures

Designs based on primitives or lower-level descriptions can be used in higher-level structures to form complete designs. There is no limit on the number of hierarchies in a design, but it is important to remember that simulation of such a design is done at the lowest level (gates or even switches). This causes slow simulations, but produces detailed timing results. A hierarchical primitive based design is often the output of a synthesis tool. However, if such a design is used as an input for synthesis, use of specific primitives is implied.

Verilog provides language constructs for easy description of large iterative hardware modules or array based regular structures. Furthermore, delay constructs, such as pin-to-pin delay specification, provide ways of fine tuning timings of upper-level modules independent of their lower-level details. This section uses examples of the previous section for formation and timing specification of higher-level structures.

In this book we use graphical notations for hardware components to correspond to the way these components are coded in Verilog. In these notations we use a rectangular box to represent a top-level component and rectangular boxes in the outer box to represent lower-level components of the upper-level hierarchy. Graphical notations for ports and port connections follow the conventions discussed in Sec. 4.1. Figure 4.16 is an example for this notation.

4.3.1 Simple hierarchies

Just as primitives can be wired for generating upper-level structures, existing modules can also be used as subcomponents of an upper-level

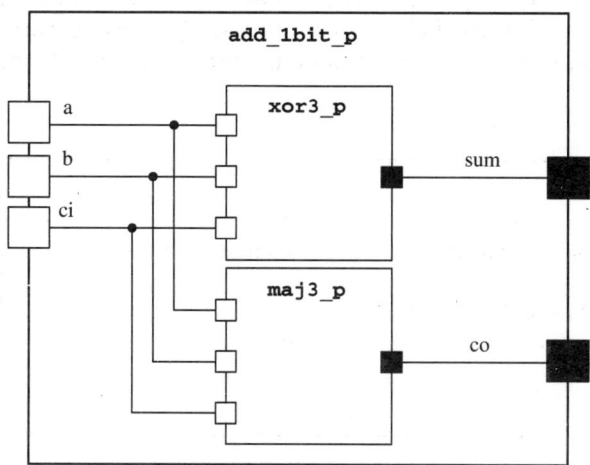

Figure 4.16   Full Adder Using *xor3_p* and *maj3_p*

```
`timescale 1ns/100ps

module add_1bit_p (input a, b, ci, output sum, co);

    xor3_p xr1 (a, b, ci, sum);
    maj3_p mj1 (a, b, ci, co);

endmodule
```

Figure 4.17   Full Adder Verilog Code Using *xor* and *maj*

module. Figure 4.16 shows a full adder that uses the *maj3_p* and *xor3_p* modules of Fig. 4.12 and Fig. 4.15 respectively. The circuit shown, *add_1bit_p*, uses *xor3_p* to generate *sum* and *maj3_p* to generate *co* (carry out).

Figure 4.17 shows module description corresponding to the diagram of Fig. 4.16. Instance name for *xor3_p* is *xr1* and that of *maj3_p* is *mj1*. Unlike primitive instantiations, instance names for module instantiations are not optional.

Connections of the ports of the *add_1bit_p* module to those of *xor3_p* and *maj3_p* are done according to the order of ports of these subcomponents, i.e., *a*, *b*, *ci*, and *sum* connect to *a*, *b*, *c*, and *y* of *xor3_p*, and *a*, *b*, *ci*, and *co* connect to *a*, *b*, *c*, and *y* of *maj3_p*. This kind of connection is called an *ordered connection*. An alternative connection format is to name each connection explicitly, in which case we are not required to list every port of an instantiated module according to its declared ports. Figure 4.18 shows another description for the full-adder of Fig. 4.6 using the *named connection* list. This explicit format reduces chance of errors, and also allows some connections to be left open by not specifying them.

```
`timescale 1ns/100ps

module add_1bit_p_named (input a, b, ci, output sum, co);

    xor3_p xr1 (.a(a), .b(b), .c(ci), .y(sum));
    maj3_p mj1 (.a(a), .b(b), .c(ci), .y(co));

endmodule
```

Figure 4.18   Named Connection List

4.3.2   Vector declarations

A module formed as described above can still be used in an upper-level structure. For example, four instances of *add_1bit_p* can be used to form the 4-bit adder of Fig. 4.19.

The Verilog code of Fig. 4.20 corresponds to the diagram of Fig. 4.19. This description uses four separate instantiations of *add_1bit_p* of Fig. 4.17. The timing of this structure is determined by those of its substructures, i.e., those of **and**, **or**, **nand**, and **not** gates of *xor3_p* and *maj3_p*.

Figure 4.20 shows declaration of *a*, *b*, and *s* ports of *add_4bit* as 4-bit vectors. The left-most bits of these vectors is bit 3 and their right-most bits are bit 0. In Verilog the left-most bit is always the most significant. For wiring 4-bit adder input and output vectors to the scalar

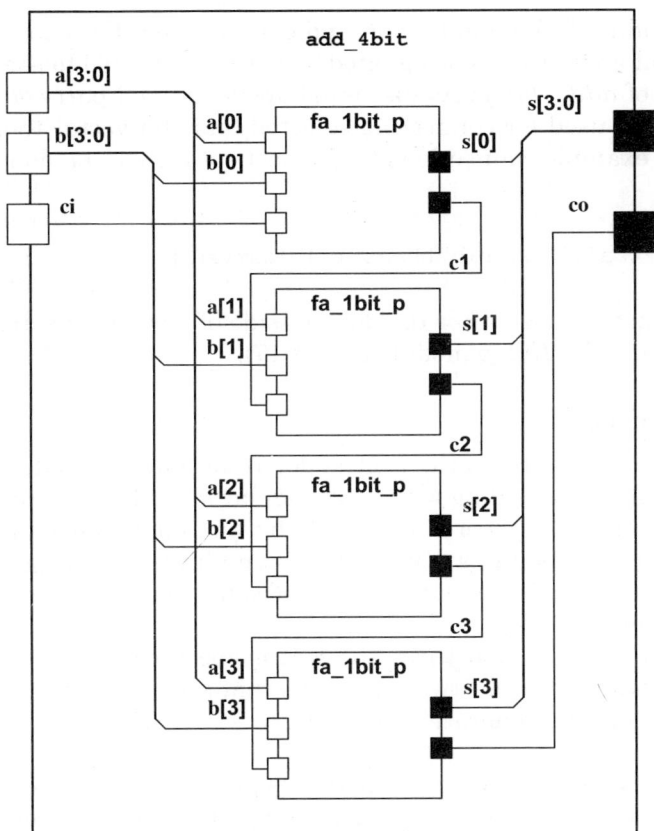

Figure 4.19   A 4-bit Adder by Wiring Four Full Adders

```
`timescale 1ns/100ps

module add_4bit (input [3:0] a, b, input ci,
                 output [3:0] s, output co);
   wire c1, c2, c3;

   add_1bit_p fa0 (a[0], b[0], ci, s[0], c1);
   add_1bit_p fa1 (a[1], b[1], c1, s[1], c2);
   add_1bit_p fa2 (a[2], b[2], c2, s[2], c3);
   add_1bit_p fa3 (a[3], b[3], c3, s[3], co);

endmodule
```

Figure 4.20   4-bit Adder Verilog Code

ports of the individual full adders, array indexing is used. For example bit 1 of the 4-bit adder that is designated by *a[1]* is connected to port *a* of *fa1* instance of *add_1bit_p*. As discussed above, named port connections can also be used for connecting selected bits of a vector to a scalar input. For example connections to the ports of *fa1* can be done as shown below:

```
add_1bit_p fa1 (.a(a[1]),.b(b[1],.ci(c1),.sum(s[1]),.co(c2));
```

In this format, names used after the dots outside of parenthesis are those of the ports of *add_1bit_p* module (Fig. 4.17).

4.3.3   Iterative structures

The format used for formation of a 4-bit adder from four full adders becomes impractical when dealing with large regular multidimensional structures. For example, if we were to wire a 64-bit combinational array multiplier, listing 64*64 instances of individual bit multipliers and specifying the wiring of each of these units would become a very time consuming and error-prone task.

For formation of iterative structures and wiring specification of regular structures, Verilog provides two constructs that are referred to as array of instances and the generate statement.

4.3.3.1   Array of instances.   An array of instances provides an easy way of instantiating similar modules in a single statement. Using this facility the 4-bit adder of Fig. 4.20 may be written as shown in Fig. 4.21. As shown, all four instances of Fig. 4.20 are merged into a vector notation including instance name of the module being instantiated.

```
`timescale 1ns/100ps

module add_4bit_vec (input [3:0] a, b, input ci,
                     output [3:0] s, output co);
    wire c1, c2, c3;
    add_1bit_p fa[3:0] (a, b, {c3, c2, c1, ci}, s,
                             {co, c3, c2, c1});

endmodule
```

Figure 4.21   4-bit Adder Using Array of Instances

In this format, for connection of vectored ports, the name of the vector should be used. For example, using s, that is a vector, in the port connection list of *fa[3:0]* connects bits 3 to 0 of this vector to the sum ports of instances *fa[3]*, *fa[2]*, *fa[1]*, and *fa[0]* of *add_1bit_p*. Where a direct vector connection cannot be made, a vector is created by the concatenation operation. For example, for connection of $c3$, $c2$, $c1$ and ci to ci ports of instances *fa[3]*, *fa[2]*, *fa[1]*, and *fa[0]*, a vector is formed by concatenation of {c3, c2, c1, ci}, and this vector is used in the port connection list of *fa[3:0]* to connect to the ci ports of four instances of *add_1bit_p* module.

4.3.3.2 Generating multiple instances.   A more flexible and powerful construct of Verilog for generating multiple similar instantiations is the *generate statement*. This construct is concurrent and with its generation scheme it multiplies any statement or group of statements that is encloses. A simple format of these constructs is discussed here.

Figure 4.22 shows another version of our 4-bit adder example that is easily expandable to an adder of any size. Declaration of ports and intermediate signals in this description are vectors whose sizes depend on the *SIZE* parameter.

An n-bit adder has a regular structure, but its right-most and left-most bits are exceptions. While the carry input of a full adder that is not on either side of an n-bit adder is taken from the carry output of its right-hand side full adder, the carry input of the right-most full adder is driven by the external carry input of the adder. A similar exception exists for the carry output of the left-most full adder of an n-bit adder. To put this irregularity into a regular structure, we have declared *carry* as a vector of *[SIZE:0]*. The right-most bit of this vector is driven by ci of the n-bit adder, and its left-most bit drives co of the adder. Using this vector, full adder i in an n-bit adder takes its carry input from *carry[i]* and puts its carry output on *carry[i+1]*. This regular structure is described by the **generate** statement shown in Fig. 4.22.

```
`timescale 1ns/100ps

module add_4bit_gen (a, b, ci, s, co);
    parameter SIZE = 4;
    input [SIZE-1:0] a, b;
    input ci;
    output [SIZE-1:0] s;
    output co;
    wire [SIZE:0] carry;
    genvar i;
    assign carry[0] = ci;
    assign co = carry[SIZE];

    generate
        for (i=0; i<SIZE; i=i+1) begin : full_adders
            add_1bit_p fa (a[i], b[i], carry[i], s[i], carry[i+1]);
        end
    endgenerate
endmodule
```

Figure 4.22   Adder Using Generate Statement

A **generate** statement requires a **genvar** variable, that needs to be declared. Using this variable the generate multiplication scheme can use **for**, **if**, **case**, or a combination or nesting of these statements. For multi-dimensional array generations, several **genvar** variables must be used. Our simple example here uses a *generate_loop_statement* to generate and interconnect four copies of the *add_1bit_p* full-adder.

The **generate-for** construct shown, loops for *i* values 0 to 3. In each iteration it generates and wires a full-adder from bit 0 to 3. The identifier for the **for** statement used in this example is *full_adders*. This identifier is mandatory and is used for referencing individual full-adders or their internal signals. For example, the *sum* output of *add_1bit_p* in position 3 is referenced by *full_adders[3].fa.sum*. the Verilog code of the *add_4bit_gen* module shows the use of the *SIZE* **parameter**. Depending on this parameter, that can be overridden when *add_4bit_gen* is instantiated, the **generate** statement shown expands to *SIZE-1* number of instances of *add_1bit_p*.

Instead of using the 5-bit *carry* intermediate signal and the **assign** statement of Fig. 4.22, a **generate-if** can be used as shown in Fig. 4.23. This statement separates 0 and SIZE-1 instances of *add_1bit_p* from the rest of the instances of this module.

```
`timescale 1ns/100ps

module add_4bit_genif (a, b, ci, s, co);
    parameter SIZE = 4;
    input [SIZE-1:0] a, b;
    input ci;
    output [SIZE-1:0] s;
    output co;
    wire [3:0] carry;
    genvar i;

    generate
        for (i=0; i<SIZE; i=i+1) begin : full_adders
            if (i==0) add_1bit_p fa (a[0], b[0], ci, s[0], carry[0]);
            else if (i==SIZE-1)
                    add_1bit_p fa (a[i], b[i], carry[i-1], s[i]co);
            else add_1bit_p fa (a[i], b[i],
                                carry[i-1], s[i], carry[i]);
        end
    endgenerate

endmodule
```

Figure 4.23   Using **generate-if** to From an Adder

4.3.4   Module path delay

Verilog delay specifications discussed so far are referred to as *distributed delay*. This means that the overall timing of an upper-level component, e.g., our 4-bit adder, depends on the time it takes events to propagate through gates and **net**s inside this module. In other words, delays are distributed through out various components of this design. A *module path delay specification*, however, describes the time it takes an event at a source (input or inout port) to propagate to a destination, e.g., an output. This form of delay specification is also referred to as pin-to-pin delay.

Module path delays (pin-to-pin) are specified in a **specify** block inside a module. A **specify** block begins with the **specify** keyword and ends with **endspecify**. Within a **specify** block, input to output path delays can be specified. Figure 4.24 shows our full adder example with pin-to-pin delay specifications.

In the specify block shown in this figure, a *simple path delay* construct lists *a*, *b*, and *ci* as delay sources and after the ***>** symbol its lists the delay destinations that are *co* in one case, and *sum* in another. Any time

```
`timescale 1ns/100ps

module add_1bit_p2p (input a, b, ci, output sum, co);
    specify
        (a, b, ci *> co) = 12;
        (a, b, ci *> sum) = 15;
    endspecify

    xor3_p xr1 (a, b, ci, sum);
    maj3_p mj1 (a, b, ci, co);

endmodule
```

Figure 4.24   Module Path Delay Specifications

a or *b* or *ci* changes if, as a result, *co* changes, it is delayed by 12 ns and if *sum* changes, it is delayed by 15 ns. Figure 4.25 shows input to output delays of the full adder of Fig. 4.24.

Module path delays can coexist with gate level distributed delays. However, they affect output timing only when they are larger than the sum of all internal distributed delays. Module paths not specified in a **specify** block are considered with 0 pin-to-pin delays, in which case, distributed delays will be used. To use a different rise than fall, two delay values enclosed by a set of parenthesis should be used. As with other delay parameters, **min:typ:max** delay format may be used in path delays.

As another example of path delays consider the **specify** block of *add_4bit_p2p* module of Fig. 4.26. Path delays shown for *a* to *s* and *b* to *s* are full-path, which means an event on any bit of *a* to a resulting change on any bit of *s* is delayed by 31 ns. The same for bits of *b* to *s* is 32 ns. In addition, we are specifying the delay from any bit of *a*, or *b* or *ci* to *co* as 37 ns.

The **specify** block of this figure does not specify a path delay for the situation that an event on *ci* results in a change on *s*. Therefore, for this situation, gate level distributed delays will be used.

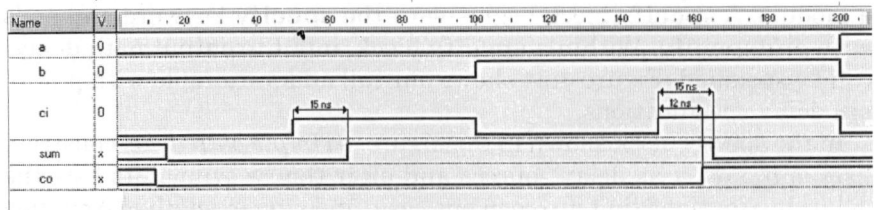

Figure 4.25   Path Delays Simulation Run

```
`timescale 1ns/100ps

module add_4bit_p2p (input [3:0] a, b, input ci,
output [3:0] s, output co);
    wire c1, c2, c3;

    specify                        .
        (a *> s) = 31;
        (b *> s) = 32;
        (a, b, ci *> co) = 37;
    endspecify

    add_1bit_p fa0 (a[0], b[0], ci, s[0], c1);
    add_1bit_p fa1 (a[1], b[1], c1, s[1], c2);
    add_1bit_p fa2 (a[2], b[2], c2, s[2], c3);
    add_1bit_p fa3 (a[3], b[3], c3, s[3], co);

endmodule
```

Figure 4.26   Full Path Delay for a 4-bit Adder

Figure 4.27 shows timing comparison of distributed delays of Fig. 4.20 versus path delays of Fig. 4.26. Signals $s1$ and $co1$ are outputs of *add_4bit* of Fig. 4.20, and $s2$ and $co2$ are outputs of *add_4bit_p2p*. After *b* changes, the distributed delay output goes through several transitions before it reaches its stable value of **F**. On the other hand, the path delay output, $s2$, changes after 32 ns to its stable value. On the other hand, when *ci* changes, both $s1$ and $s2$ go through transitional distributed delay values. This is because the **specify** block of Fig. 4.26 does not specify a path delay from *ci* to *s*. The path delay from *ci* to *co* of this code is 37 ns as shown in Fig. 4.27.

Another path delay specification format in Verilog is what is referred to as parallel path. This format uses => instead of *> for vector source

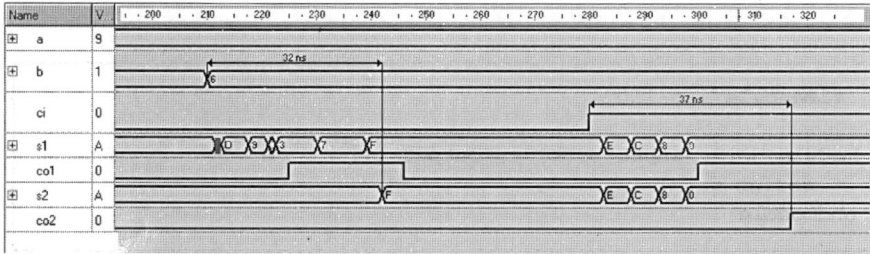

Figure 4.27   Comparing Distributed and Path Delays

to destinations. The *(a=>s) = 12* path delay specification defines events on bits of *a* to similar bits of *s* to be 12 ns. This means that if an event on *a[1]* causes *s[1]* and *s[3]* to change, the 12 ns delay only applies to *s[1]* and distributed delays will be used for changes on *s[3]*.

4.4   Describing Expressions with Assign Statements

At a more abstract level than gates or complex gate structures as described in the previous sections, is the level of describing hardware using boolean expressions. Verilog uses **assign** statements for assigning a boolean expression to a wire or a vector of wires (**net**s). As with the gate primitives, **assign** statements allow the use of delays and tristate specification. However, the higher abstraction level results in a less detailed timing specification. This section shows the use of the **assign** statement in coding components in Verilog.

Corresponding to a Verilog module, we use a graphical notation to represent the way the module is coded. Some of the conventions we are using are discussed in Sec. 4.1 and Sec 4.3. Another convention we use is for representation of an **assign** statement. A logic block described by an **assign** statement is represented by a dotted rectangular box (an example is Fig. 4.62). The right-hand side signals of the **assign** statement are considered the inputs of the box and its left-hand side **net** is the output of the box.

4.4.1   Bitwise operators

Instead of instantiating individual gates of a hardware structure like the three-input XOR of the previous sections, a simple boolean expression using a *continuous assign statement* can express module functionality. The *xor3* module of Fig. 4.28 takes advantage of this language utility.

As with gate structure outputs, the left-hand side of an **assign** statement must be a **net** or a vector of **net**s. On the right-hand side of an **assign** statement, any scalar or vector expression using Verilog operators can be used. Our *xor3* example uses Verilog bitwise ^ operation.

```
`timescale 1ns/100ps

module xor3 (input a, b, c, output y);
   assign y = a ^ b ^ c;
endmodule
```

Figure 4.28   *xor3* with **assign** Statement

```
`timescale 1ns/100ps

module maj3 (input a, b, c, output y);
    assign #(4) y = ( a & b ) | ( b & c ) | ( a & c );
endmodule
```

Figure 4.29   *maj3* with **assign** Statement

As another example of using an **assign** statement consider the *maj3* module of Fig. 4.29. This example uses **&** and **!** bitwise operators and uses a 4 ns delay for assignments to the output of *maj3*.

The delay structure used with an **assign** statement only allows specification of rise and fall delays. In our example, the 4-ns delay applies to rise and fall delays. Verilog allows the use of **min:typ:max** delay format for delay values associated with an **assign** statement.

Multiple independent **assign** statements are also allowed in a module. Figure 4.30 shows another version of our *add_1bit* module of the previous sections. Such **assign** statements are regarded as concurrent statements, and the order of their execution does not depend on the order in which they appear in a module. An **assign** statement is said to be sensitive to its right-hand side events and begins evaluation of its right-hand side when an event occurs on any of its right-hand side variables. The result of the evaluation will be scheduled for the left-hand side **net** after the specified delay expires.

Real hardware components are regarded as concurrently active elements. However, true concurrency cannot exist in software execution of hardware description language (HDL) models of such hardware components. Verilog and other HDL simulators model (or, in a way "fake") this concurrency by applying event-based execution of statements, as described earlier.

```
`timescale 1ns/100ps

module add_1bit (input a, b, ci, output s, co);

    assign #(10) s = a ^ b ^ ci;
    assign #(8) co = ( a & b ) | ( b & ci ) | ( a & ci );

endmodule
```

Figure 4.30   *add_1bit* using **assign** Statements

```
`timescale 1ns/100ps

module add_1bit (input a, b, ci, output s, co);
  assign #(3, 4) {co, s} = {(a & b)|(b & ci)|(a & ci), a^b^ci};
endmodule
```

Figure 4.31   Full Adder Using Concatenation

4.4.2 Concatenation operators

We can use concatenation operators on the right or left-hand sides of an **assign** statement, for as long as we keep track of proper right-hand side values assigned to the left-hand side **net**s. Figure 4.31 shows another version of a full adder using concatenation operations on the right and left-hand sides of an **assign** statement. In the example shown, the XOR expression is assigned to s and the *and-or* expression is assigned to the *co* output. Delay values used here apply to both outputs, i.e., *co*, and *s*.

4.4.3 Vector operations

An example showing vector operations is shown in Fig. 4.32. In this example, bitwise XOR operation of bits of a and b is formed on *im* and a NOR reduction ($\sim|$) of *im* is generated on *eq*. The *eq* output becomes 1 if a and b are equal. In this example, delay values can be specified for individual **assign** statements. This circuit is a simple comparator with an equal output.

Another example of using vector operands is shown in Fig. 4.33. This odd-even parity circuit also demonstrates that an **assign** can be shared

```
`timescale 1ns/100ps

module comp_4bit (input [3:0] a, b, output eq );

    wire [3:0] im;
    assign im = a ^ b;
    assign eq = ~|im;

endmodule
```

Figure 4.32   Bitwise Operation with Vector Operands

```
`timescale 1ns/100ps

module parity_gen (input [7:0] a, output even, odd);
    assign #(3, 4)
        even = ^a,
        odd = ~^a;
endmodule
```

Figure 4.33   Reduction Operation with Shared Assign

between various assignments. Individual assignments are separated with commas. Operations used here are XOR and XNOR reduction. The specified delays apply to assignments to *even* and *odd* output.

A reduction operation applies to two bits at a time starting with the left-most bit. The description shown in Fig. 4.33 corresponds to two gate structures one driving *odd* and one driving the *even* output. Negating a reduction first reduces all bits and then negates the result.

4.4.4   Conditional operation

Conditional operation (**?:**) used with various relational operators provide convenient mechanisms for description of fairly complex logical functions.

Figure 4.34 shows Verilog code of a quad 2-to-1 multiplexer circuit. The **assign** statement shown, assigns the 4-bit *i0* vector to *y* if *s* is **0**, and if *s* is **1** then *i1* is assigned to *y*. If *s* is ambiguous, i.e., **X** or **Z**, then a vector consisting of all **X**s except in those positions that both *i1* and *i0* are equal (**0** or **1**) will be assigned to the left-hand side.

Ambiguous conditions also occur if *s* is multi-bit and any of its bits are **X** or **Z**. This case also results in a vector of **X**s with exception of bits of the same value. For example if *s* is ambiguous and *i1* is **0Z11** and *i2* is **1Z1X**, then *y* receives **XX1X**. For illustration of how ambiguous values

```
`timescale 1ns/100ps

module quad_mux2_1 (input [3:0] i0, i1, input s,
                        output [3:0] y);
    assign y = s ? i1 : i0;
endmodule
```

Figure 4.34   Conditional Operation Describing a Multiplexer

```
`timescale 1ns/100ps

module test_quad_mux2_1;
    reg [3:0] i0, i1;
    reg s;
    wire [3:0] y;
    quad_mux2_1 MUT (i0, i1, s, y);
    initial begin
        i0=4'b1010; i1=4'b0101; s=1'b0;
        #20 i0=4'b0000;
        #20 s=1'b1;
        #20 i1=4'b1111;
        #20 i0=4'b0z11; i1=4'b1z1x; s=1'bz;
        #20 $finish;
    end
endmodule
```

Figure 4.35   Testbench Generating an Ambiguous Value on *s*

affect a conditional assignment, the testbench of Fig. 4.35 generates an ambiguous value for the *s* input of *quad_mux2_1*. At time 80 ns, this testbench puts an ambiguous value on *s*. The resulting waveform is shown in Fig. 4.36.

Conditional operations (?:) can be nested to imply a nested *if-then-else* expression. The example of Fig. 4.37 describing a 2-to-4 decoder illustrates this. The left-hand side of the **assign** statement concatenates *d3*, *d2*, *d1*, and *d0* outputs of the decoder to form a 4-bit vector. This vector receives **0001**, **0010**, **0100**, or **1000** if *{a, b}* is 0, 1, 2, or 3, respectively. The *{a, b}* concatenation forms a 2-bit select vector in which *a* is the most and *b* is the least significant bit. The order of evaluation of nested conditional operators is from left to right.

The example of Fig. 4.37 uses the **==** equality operator to check *{a, b}* against its four possible values. This operator returns a **1** if bit-by-bit matching of its operands occurs, provided they are **0** or **1** (known values).

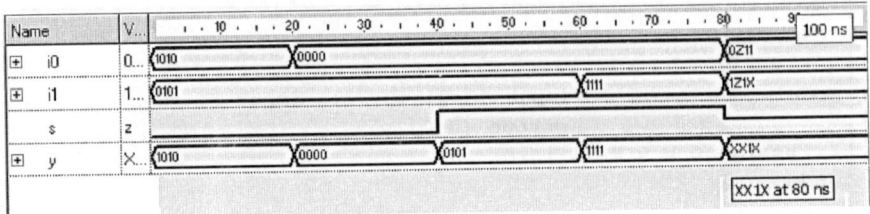

Figure 4.36   Ambiguous Values in Condition Operation

```
`timescale 1ns/100ps

module dcd2_4 (input a, b, output d0, d1, d2, d3 );
   assign
      {d3, d2, d1, d0} = ( {a, b} == 2'b00 ) ? 4'b0001 :
                         ( {a, b} == 2'b01 ) ? 4'b0010 :
                         ( {a, b} == 2'b10 ) ? 4'b0100 :
                         ( {a, b} == 2'b11 ) ? 4'b1000 :
                         4'b0000;
endmodule
```

Figure 4.37   Nested Condition Operations

An **X** or a **Z** in any bit position of the operands of an equality operator results in an **X**. In Fig. 4.37, the last value **4'b0000** is put on the outputs if none of the equality checks before it succeed.

Figure 4.38 shows another example of using relational and equality operators. $a>b$ returns a **1** if a is greater than b, the same applies to $a==b$ and $a<b$. The 1-bit values obtained from these operations are assigned to the comparator outputs of the *comp_4bit* module of this figure.

In this circuit, an **X** or a **Z** in any bit position of the inputs a or b, causes an **X** to be assigned to all three outputs of the circuit.

As our last example of use of condition operators in conjunction with relational and equality operators consider the cascadable comparator of Fig. 4.39. This comparator has two 4-bit data inputs, three compare outputs, and three control inputs (*gt, eq,* and *lt*) for cascading purposes.

The *a_gt_b* output becomes **1** if $a>b$ or if $a==b$ and the *gt* input is **1**. The *a_lt_b* output is similar. The *a_eq_b* output becomes **1** if a is equal

```
`timescale 1ns/100ps

module comp_4bit ( input [3:0] a, b,
                   output a_gt_b, a_eq_b, a_lt_b);
   assign
      a_gt_b = (a>b),
      a_eq_b = (a==b),
      a_lt_b = (a<b);
endmodule
```

Figure 4.38   4-bit Comparator Using Relational and Equality Operators

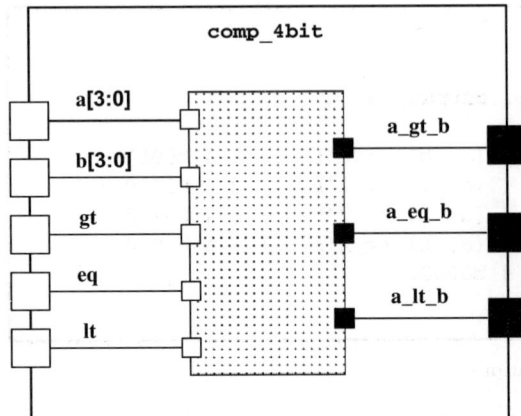

Figure 4.39   Cascadable 4-bit Comparator

to *b* and the *eq* input is **1**. This procedure is described by three assignments to *a_gt_b*, *a_eq_b*, and *a_lt_b* outputs in the Verilog code of Fig. 4.40.

4.4.5   Arithmetic expressions in assignments

Use of arithmetic expressions for description of hardware functions provides a more abstract form of hardware representation than expressions dealing with boolean functions or bit-level details. Verilog allows the use of arithmetic expressions in forming expressions describing hardware. Arithmetic operators are **+, -, *, /, ****, and **%**. Generally, these operators are used in behavioral and RT level hardware descriptions.

As an example consider a 4-bit adder with a carry-in and a carry-out. In the Verilog code of Fig. 4.41, *a, b* and *ci* are added together on the

```
`timescale 1ns/100ps

module comp_4bit (input [3:0] a, b, input gt, eq, lt,
                  output a_gt_b, a_eq_b, a_lt_b );
   assign
      a_gt_b = (a==b) ? gt : (a>b),
      a_eq_b = (a==b) ? eq : 1'b0,
      a_lt_b = (a==b) ? lt : (a<b);

endmodule
```

Figure 4.40   Cascadable Comparator Verilog Code

```
`timescale 1ns/100ps

module add_4bit (input [3:0] a, b, input ci,
                 output [3:0] s, output co);
   assign { co, s } = a + b + ci;
endmodule
```

Figure 4.41   A 4-bit Adder

right-hand side of an **assign** statement. Adding ci that is a 1-bit unsigned operand of **+** causes it to be padded with **0**s to turn it into a 4-bit vector. The result of the add operation generates a 5-bit result in which the left-most bit is the carry and the other four bits constitute the sum. On the left-hand side of the **assign** statement, concatenation of co and s captures the 5-bit result.

Verilog is very forgiving in size mismatches in assignments and operations. As seen here vectors of different bits are added together and the operands are expanded according to the type of expression being signed or unsigned. Assignments to left-hand sides will be truncated if the left-hand side has fewer bits than the right-hand side.

4.4.6   Functions in expressions

Functions help design modularity and reuse. Verilog functions can be used for test data, management, response analysis, or for describing hardware blocks or boolean functions. Figure 4.42 demonstrates the use of functions for describing boolean functions.

```
`timescale 1ns/100ps

module add_1bit_f (input a, b, ci, output s, co );

   function [1:0] adder (input a, b, c);
   begin
     adder = {(a & b)|(b & c)|(a & c), a^b^c };
   end
   endfunction

   assign #(3, 4) {co, s} = adder ( a, b, ci );

endmodule
```

Figure 4.42   Using Functions

The *adder* function of the above example has a 2-bit output vector. As shown, the inputs are *a*, *b*, and *c*. In the body of this function the function name is given the return value of the function. In the body of the *add_1bit_f* module, this function is used on the right-hand side of an **assign** statement. On the left-hand side of this statement concatenation of *co* and *s* capture the two bits of this function. The time delays #(3, 4) apply to rise and fall of *co* and *s* when receiving values calculated by the *adder* function.

The way the *adder* function is used in this example, only the *add_1bit_f* module can use it. If other modules are to take advantage of a function, the function must be defined in a separate file, and included where needed by the `include directive.

In writing and using functions, several guidelines should be used. The following paragraph discusses function inputs and output, function body, function calls, and timing.

The function name is a variable via which a function returns its value. This variable, and thus, the return value can be a scalar or a vector. A function requires at least one input, which can be a scalar or a vector. In the body of a function, procedural statements are used to define a mapping of function inputs to the output of the function. Timing statements are not allowed inside a function. Nonblocking procedural assignments are not allowed in a function. When calling a function, the order of the arguments must be according to the declaration of function inputs. Calling is done by using the name of the function on the right-hand side of an expression. Calling several functions can be nested.

4.4.7   Bus structures

In an RT level description, a bus is referred to as a shared vector that is driven by logic units, registers, memories, circuit inputs, or other busses. For every source, a bus has a select input that selects its corresponding source to drive the bus. Selecting multiple sources causes bus contention which can produce different results depending on the bus type. Bus destinations are all those hardware units that read the bus. Usually, multiple simultaneous readings of a bus do not cause any problems.

Figure 4.43 shows a bus structure with several sources and destinations. Select lines are single bit control lines, and the width of the bus and its sources and destinations is *n*. A *Seli* select line selects *Sourcei* for driving the bus.

What happens when several *Sel* lines are simultaneously active depends on the type of the bus. In such cases, the information on the bus is useless, but may be different from one bus type to another. Verilog allows the use of **wire, (tri), wand (triand), wor (trior), tri0,** and **tri1** for bus types. In a **tri** bus **(wire)**, the output of the bus when it is

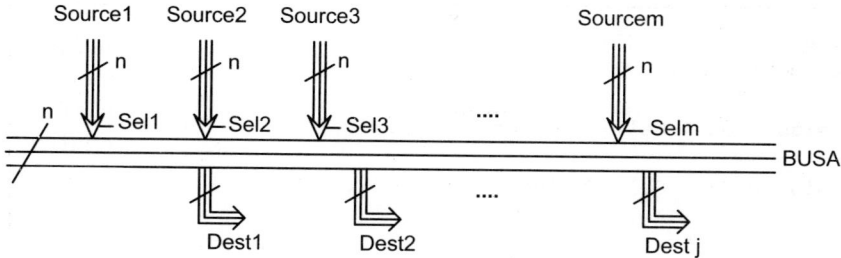

Figure 4.43   A Bus System

simultaneously driven by multiple sources is the bit-by-bit wiring result of all its active sources. In a **triand** bus (**wand**), this becomes bit-by-bit ANDing of all active sources, and in a **trior** bus (**wor**), the function of each bit of a bus is the OR result of all its active sources. Figure 4.44 shows various bus type declarations for the bus of Fig. 4.43. The inactive value for a **wire** bus is **Z**. This means that a source not driving such a bus puts **Z**'s on the bus. Inactive values for a **wand** and a **wor** bus are **1** and **0**, respectively.

An **assign** statement is the most common language construct for making source assignments to a bus. For a **wire** type bus, assignments shown in Fig. 4.45 specify sources of this bus. Assignments to **wand** and **wor** type busses are only different from those shown in Fig. 4.45 in the value of the *inactive* **parameter**. For **wand** this parameter is *n'b11...1*, and for **wor** it is *n'b00...0*.

4.4.8   Net declaration assignment

Verilog allows combining **net** declarations and **assign** statements. As delay associated with **assign** statements correspond to gate level delays, those associated with **net** declarations correspond to wire delays. A simple example demonstrating the **net** declaration assignment is shown in Fig. 4.46. This example shows an AND-OR multiplexer in which ANDing s and $\sim s$ with $i0$ and $i1$ data inputs is done with **net** declaration assignments and ORing the results is done by an **assign** statement.

```
wire [n-1 : 0] BUSA;
   . . .
wand [n-1 : 0] BUSA;
wor  [n-1 : 0] BUSA;
```

Figure 4.44   Bus Type Declarations

```
wire [n-1 : 0] BUSA;
parameter [n-1 : 0] inactive = n'bZZ...Z;

assign BUSA = Sel1 ? Source1 : inactive;

assign BUSA = Sel2 ? Source2 : inactive;
        .

        .

        .
assign BUSA = Selm ? Sourcem : inactive;
```

Figure 4.45   wire Type Bus Assignment

4.5 Behavioral Combinational Descriptions

As discussed before, **assign** statements of Sec. 4.4 provided a higher-level of abstraction over primitive based designs of Sec. 4.3. In the same way, procedural statements provide a mechanism for describing hardware at still a higher level of abstraction than any of the formats discussed so far. This level of abstraction is often referred to as the behavioral level.

Verilog's procedural blocks are bodies within which statements are executed sequentially. This form of hardware description is, generally, easier for designers to describe their complex hardware. Procedural bodies do provide mechanisms for specification of timing, but, in general, a hardware described with procedural structures contains less timing details than a hardware described with **assign** statements or primitives.

In this section we first present basics of procedural blocks, we will then discuss timing and flow control in procedural blocks. Various types of statements and their simulation semantics and hardware implications

```
`timescale 1ns/100ps

module mux2_1 (input i0, i1, s, output y);

    wire #(0.6, 0.8)
         im0 = i0 & ~s,
         im1 = i1 & s;

    assign #(3, 4) y = im0 | im1;

endmodule
```

Figure 4.46   Using **net** Declaration Assignments

will be discussed next. After a thorough presentation of the mechanics of procedural blocks, we will show high-level-language constructs such as **if-else** and **case** statements that are used for a behavioral description of hardware.

The graphical notation that we started presenting in Sec. 4.1 completes in this section. Recall that, hollow boxes in a graphical notation of a Verilog code correspond to module instantiations, and dotted boxes are for **assign** statements. In the examples that follow, a gray shaded box used in a rectangular box of a module represents the use of a procedural block in the module. For such a box, inputs are signals in the sensitivity list of the procedural block and outputs of the gray box are variables and signals that appear on the left-hand side of assignments in the procedural block. Figure 4.58 is an example of this notation.

4.5.1 Simple procedural blocks

Any of the circuits of the previous section can be described with procedural blocks. We will use our *xor3* and *maj3* examples to demonstrate some of the main concepts of this Verilog construct.

Figure 4.47 shows the *xor3* module using an **always** statement. This statement is a procedural statement and within its body all statements are executed sequentially. An **always** statement begins with the **always** keyword and can enclose any number of procedural statements. Variables used on left-hand side of procedural statements must be of type **reg**. The **reg** type can be declared separately in the module body or if it is an output that is a **reg**, it can be declared along with the output declaration.

An **always** block has an implicit infinite loop, causing it to run forever unless halted by flow control statements. Examples of flow control statements are timing and event control statements.

4.5.2 Timing control

Following the **always** keyword in Fig. 4.47, the **always** block contains an event control statement. This statement that begins with an @ sign

```
`timescale 1ns/100ps

module xor3 (input a, b, c, output y);
    reg y;

    always @(a, b, c)
        y = a ^ b ^ c;
endmodule
```

Figure 4.47 A Simple Use of an **always** Block

halts the flow into the body of the **always** statement until an event occurs on any of the signals listed, i.e., *a* or *b*, or *c*. An event control (like that of our example) right at the beginning of an **always** block is referred to as the procedural block's sensitivity list. It is said that this block is sensitive to events on its sensitivity list. This is like using these same variables on the right-hand side of an **assign** statement. When an event occurs on a variable on the right-hand side of an assign statement, the right-hand side is calculated and the calculated value is assigned to the left-hand side **net**. In an **always** statement, an event on a variable in its sensitivity list causes it to wake up and execute the procedural statement(s) in its body. The sensitivity list of an always statement provides an added flexibility to choose variables instead of the default of all right-hand side variables as it is with the continuous **assign** statements.

In the example of Fig. 4.47, when an event occurs on *a, b*, or *c*, the right-hand side of the assignment to *y* **reg** is executed and the calculated value is assigned to *y*. This variable retains its value until the next time a new value is assigned to it.

Another example for a simple always statement is shown in Fig. 4.48. In this example the **reg** type for *y* is specified with its output declaration. Another difference between this example and the *xor3* module is the use of **begin** and **end** keywords.

Multiple procedural statements within the body of a procedural block must be bracketed by **begin** and **end** keywords. In our *maj3* example, the **begin-end** bracketing encloses only one procedural statement and is not mandatory. However, this bracketing helps readability of a code and allows insertion of new procedural statements for module debugging or modifications.

An event-control statement that begins with the @ sign is a regular procedural statement and can appear any where in a procedural block.

```
`timescale 1ns/100ps

module maj3 (input a, b, c, output reg y);

    always @(a, b, c)
    begin
        y = (a & b) | (b &c) | (a & c);
    end

endmodule
```

Figure 4.48   Majority Circuit Using an **always** Block

```
`timescale 1ns/100ps

module maj3 (input a, b, c, output reg y);

    always begin
       @(a, b, c)
       y = (a & b) | (b & c) | (a & c);
    end

endmodule
```

Figure 4.49   Bracketing Multiple Procedural Statements

As shown in the example of Fig. 4.49, the **begin-end** bracketing encloses two procedural statements separated by a semicolon.

A flow control statement, like the event control of Fig. 4.49, can attach to its succeeding statement to form a combined procedural statement by removing their separating semicolon. This means that the semicolon after the @*(a, b, c)* statement in Fig 4.49 is optional. When this mergence of the two procedural statements occurs, only one statement is formed, and then again, the **begin-end** bracketing becomes optional. Separating. signals and variables of an event control statement can be done by commas (as above) or use of the **or** operation. The @ *(a or b or c)* statement is equivalent to what is shown above.

Another type of a timing control statement is a delay control statement. This statement begins with a sharp sign (#) and in the set of parenthesis that follows it contains a delay expression that halts the flow of the procedural block until the specified delay expires. The parenthesis following # can be eliminated if a single delay value is used. Like an event control statement, a delay control statement merges with its proceeding procedural statement by removing their separating semicolon.

Figure 4.50 shows another version of the *maj3* circuit that uses an event and a delay control statement. The control statements and the procedural assignment are merged into a single statement by removing

```
`timescale 1ns/100ps

module maj3 (input a, b, c, output reg y);
    always @(a, b, c) #5 y = (a & b) | (b &c) | (a & c);
endmodule
```

Figure 4.50   *Maj3* with Delay and Event Control Statements

their separating semicolons. This mergence eliminates the need for **begin-end** bracketing needed for multiple procedural statements.

When an event occurs on a, b, or c, flow into this block begins. This is immediately halted by the #5 delay control statement. When this delay elapses, evaluation of the right-hand side of y begins. The calculated value is then assigned to the y **reg**. The flow into this block is blocked until y receives its new value. After this, the flow into this **always** block goes back to its sensitivity list, waiting for another event on a, b, or c.

4.5.3   Intra-assignment delay

Another form of delay specification in procedural statements is intra-assignment delay. Unlike the delay control statement that is by itself a separate procedural statement, an intra-assignment delay (or event) is considered as part of an assignment. Figure 4.51 shows our *maj3* example using a 5 ns intra-assignment statement.

The way the **always** statement of this figure is executed is different than that of Fig. 4.50, discussed in Sec. 4.5.2. The flow into the **always** block of Fig. 4.50 begins when an event occurs on a, b, or c. Following such an event, the right-hand side of y is immediately evaluated (instead of waiting for 5 ns as is done in the code of Fig. 4.50). The result of evaluation of the right-hand side of y will be held for 5 ns and after this time elapses this result will be assigned to y. While waiting for this time to elapse, the flow of the **always** block is blocked.

The difference in simulation of the intra-assignment delay versus the delay control of Fig. 4.50 is that, after an event on a, b, or c code of Fig. 4.51 waits 5 ns before reading new values of a, b, and c for evaluating the right-hand side of y. However, in Figure 4.51, the reading of a, b, and c is done immediately after an event occurs on any of these signals. If in the 5 ns delay of the intra-assignment delay a, b, or c changes, the new value does not affect the value of y.

4.5.4   Blocking and nonblocking assignments

Procedural assignments discussed so far in this chapter are all of the blocking type. This means that while the assignment is taking place, the flow

```
`timescale 1ns/100ps

module maj3 (input a, b, c, output reg y);
    always @(a, b, c) y = #5 (a & b) | (b &c) | (a & c);
endmodule
```

Figure 4.51   Using Intra-Assignment Delay in *maj3*

of the program into the procedural block is halted (or blocked). This is especially noticeable when using intra-assignment delays as discussed above. A different type of procedural assignment is a nonblocking assignment that uses **<=** instead of **=**. This type of assignment schedules its right hand side into the left-hand side **reg** and continues on to the next statement.

Figure 4.52 shows a full-adder circuit using two procedural blocking assignments with intra-assignment delays for generating *s* and *co* outputs. After a change on *a, b,* or *ci,* current values of these inputs are read and used for evaluating *s*. This new value will be assigned to the *s* output after 5 ns, while flow into the procedural block is blocked.

After this delay elapses, *a, b,* and *ci* are read again and the new value for *co* is evaluated. This value will be assigned to *co* after 8 ns (5+3 = 8 ns from the time of the first event on an input). This timing has the potential problem of reading a different set of input values for calculating *s* and *co*.

The use of nonblocking assignments resolves the above said problem and at the same time makes the timings of the two statements independent. Figure 4.53 shows the full-adder circuit using nonblocking assignments with intra-assignment delays. After an event on *a, b,* or *c,* the right-hand side of *s* is evaluated and the new value is scheduled into *s* for 5 ns later. At the exact same time, the right-hand side of *co* is evaluated and scheduled into this output for 3 ns later. Output *s* receives its new value after 5 ns, and *co* receives its new value 3 ns if an input changes. In this case, delay values do not accumulate.

Verilog allows multiple procedural blocks in a module. Such statements have independent timings and are executed concurrently. The Verilog code of Fig. 4.54 shows another version of our full-adder using two **always** statements. Because the sensitivity list of these statements contains the same set of inputs, an event on an input begins the flow of both **always** statements simultaneously. The result is that assignments

```
`timescale 1ns/100ps

module add_1bit (input a, b, ci, output s, co );
    reg s, co;
    always @(a, b, ci)
    begin
        s = #5 a ^ b ^ ci;
        co = #3 (a & b) | (b &ci) | (a & ci);
    end
endmodule
```

Figure 4.52   Blocking Assignments

```
`timescale 1ns/100ps

module add_1bit (input a, b, ci, output s, co );
   reg s, co;

   always @(a, b, ci) begin
      s <= #5 a ^ b ^ ci;
      co <= #3 (a & b) | (b &ci) | (a & ci);
   end

endmodule
```

Figure 4.53   Nonblocking Assignments

to *co* and *s* are executed independent of each other and each uses its own delay value.

4.5.5 Procedural if-else

Examples discussed thus far demonstrated the use of procedural block assignment types and timings. In this and the sections that follow, we will show the use of higher-level procedural statements in procedural blocks. These statements include **if-else, while, case,** and **for,** and are referred to as control flow statements.

Figure 4.55 shows an **if-else** statement used for describing a 2-to-1 multiplexer. The **always** block shown is sensitive to multiplexer inputs. The body of this statement contains an **if-else** statement. Since this entire statement is considered as a single procedural statement, the use of **begin-end** bracketing is optional.

If the condition of the **if** part of an **if-else** statement becomes **1,** the procedural statement in the **if** part is taken, otherwise if the condition expression is **0, Z,** or **X,** the **else** part is taken. The **if** part and the **else** part can contain any other procedural statement such as **if-else** or **case** statements.

```
`timescale 1ns/100ps

module add_1bit (input a, b, ci, output reg s, co );
   always @(a, b, ci) #3 co = (a & b) | (b &ci) | (a & ci);
   always @(a, b, ci) #5 s = a ^ b ^ ci;
endmodule
```

Figure 4.54   Full adder Using Two Independent always Blocks

```
`timescale 1ns/100ps

module mux2_1 (input i0, i1, s, output reg y);

    always @(i0, i1, s) begin
        if (s==1'b0)
            y = i0;
        else
            y = i1;
    end
endmodule
```

Figure 4.55 Multiplexer Using **if-else**

Another example of using an **if-else** statement, which also shows the use of arithmetic expressions and generation of carry, is shown in Fig. 4.56. This example is an adder/subtractor circuit that adds its vector operands if *m* is **1** and subtracts if *m* is **0**. This circuit is described by an **always** statement that is sensitive to *a* and *b* 4-bit vectors and *ci* and *m* input signals. When a vector appears in an event control statement, a change in any of the vector bits triggers the event.

When the flow begins in the **always** block, the value of *m* is checked. If it is **1**, then the add result of *a*, *b* and *ci* is assigned to concatenation of *co* and *s*. Otherwise, if *m* is **0**, **Z**, or **X**, subtraction is done. This causes the most significant bit of the add or subtraction operation result to be given to the *co* output.

```
`timescale 1ns/100ps

module add_sub_4bit (input [3:0] a, b, input ci, m,
                     output reg [3:0] s, output reg co );

    always @(a, b, ci, m)
        if ( m )
            { co, s } = a + b + ci;
        else
            { co, s } = a - b - ci;

endmodule
```

Figure 4.56 Adder/Subtractor using **if-else**

If the **if**-condition contained a vector, any non-ambiguous (i.e., not containing **X** or **Z**) non-zero value would cause the **if**-part to be taken. In coding combinational circuits, it is recommended that the last **else** of a group of nested **if-else** statements appears without an **if**. This gives a default case to the nested statements which will be taken if conditions of the preceding **if**-statements are not satisfied. Failure to do so, may cause certain variables to retain their old values, which will imply latches on these variables.

4.5.6 Procedural case statement

Another flow control statement of Verilog is a **case** statement. For larger number of selections, this statement is more convenient than nested **if-else** statements. A **case** statement has a *case expression* that is compared with all its *case alternatives*. A procedural statement that follows a *case alternative* is taken if the *case expression* value matches the value of the *case alternative*.

Figure 4.57 shows a 2-to-4 decoder that uses a **case** statement. The *a, b* inputs are regarded as its 2-bit encoded input (*a* is the most significant bit). The *{a, b}* vector becomes the *case expression* of this statement and is compared with **00, 01, 10, 11** that are the **case** statement's *case alternatives*. When a match is found, the corresponding encoded output appears on *{d3, d2, d1, d0}* outputs of the decoder.

The last *case alternative* in the *dcd2_4* module is **default**. If none of the other *case alternatives* match the *case expression*, the **default** is taken. This is necessary to assign a value to the *{d3, d2, d1, d0}* output

```
`timescale 1ns/100ps

module dcd2_4 (input a, b, output reg d0, d1, d2, d3 );

    always @(a, b) begin
      case ( { a, b } )
          2'b00 : { d3, d2, d1, d0 } = 4'b0001;
          2'b01 : { d3, d2, d1, d0 } = 4'b0010;
          2'b10 : { d3, d2, d1, d0 } = 4'b0100;
          2'b11 : { d3, d2, d1, d0 } = 4'b1000;
          default: { d3, d2, d1, d0 } = 4'b0000;
      endcase
    end

endmodule
```

Figure 4.57 Decoder Using **case** Statement

if an invalid value appears on the *{a, b}* input. If this is not done and an invalid or ambiguous (**X** or **Z** on *a* or *b*) value appears on *{a, b}*, then the output will retain its old value. Retaining the old value not only causes an output to have an erroneous value, it also implies a latched output. Since we are modeling a combinational circuit, latches on our outputs are not allowed. Using a **default** as the last *case alternative* is recommended in all Verilog models. This is equivalent to the last **else** without an **if** in a nesting of **if-else** statements, which is done for preventing latches.

We will use a different type of a **case** statement for describing the arithmetic logical unit (ALU) of Fig. 4.58. As shown, the ALU has a two-bit *f* input that selects its function.

The Verilog code of Fig. 4.59 describes this ALU. We are making this an n-bit ALU by declaring the *N* **parameter** and using this constant in input and output declarations. This parameter can be overridden when the *alu_n_bit* module is instantiated.

As shown, the function of the ALU is described by an **always** statement that is sensitive to all ALU inputs. A **casez** statement, instead of a **case** statement, is used to select ALU functions depending on its *f* input.

Unlike a **case** statement that treats all **X**s and **Z**s in the *case expression* as ambiguous, a **casez** treats **Z**s as don't cares. If the value of *f* contains a **Z**, that bit position matches a **0, 1**, or a **Z** in the same bit position of the *case alternatives*. For example *f* = **0Z** matches **00, 01**, and **0Z** *case alternatives*. In our example, because *case alternatives* are examined sequentially, *f* of **0Z** matches **00** first, which causes the ALU to perform the *a+b* operation.

This ALU example adds, subtracts, ANDs, or XORs its inputs depending on *f*. If *f* contains an **X** the **default** alternative is taken which puts a **0** on the output. Our ALU performs n-bit operations and puts the result in its n-bit output. Therefore, carry from add and subtract operations is ignored.

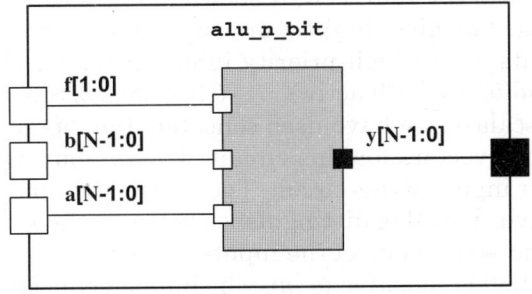

Figure 4.58   A Simple 4-bit ALU

```
`timescale 1ns/100ps

module alu_n_bit (a, b, f, y );
    parameter N=4;
    input [N-1:0] a, b;
    input [1:0] f;
    output [N-1:0] y;
    reg [N-1:0] y;
    always @ (a, b, f)
    begin
        casez ( f )
            2'b00 : y = a + b;
            2'b01 : y = a - b;
            2'b10 : y = a & b;
            2'b11 : y = a ^ b;
            default: y = 0;
        endcase
    end

endmodule
```

Figure 4.59   ALU Using casez

Another form of a case is **casex**. In this case all **X** and **Z** values in the *case alternative* or *case expression* are treated as don't cares.

4.5.7 Procedural for statement

A **for** statement is a procedural statement that is used for indexed looping. As with other procedural statements, this statement can be used inside an **always** or an **initial** statements, as well as in functions or user defined tasks. Furthermore, this statement can be nested or combined with other procedural statements. For example, a *case alternative* can contain a **for** statement. An example of a **for** statement is shown in Fig. 4.60.

The priority encoder of Fig. 4.60 gives highest priority to its *i3* input, and generates an index of its active high priority input on its *y1*, *y0* output. For example if *i2* and *i0* are both active *y1*, *y0* becomes 2. Since all ports of this circuit are scalars, we have used concatenation operators to form input and output vectors for this circuit. The *im* four-bit vector corresponds to the four inputs of this circuit. The *f* output becomes 1 if at least one input is active. This flag distinguishes between the situations when only *i0* is 1 and when none of the inputs are active.

The **always** block of Fig. 4.60 is sensitive to *im*, which represents circuit inputs. In this **always** block all circuit outputs are initially set to their

```
`timescale 1ns/100ps

module priority_encoder (input i0, i1, i2, i3,
                         output reg y1, y0, f);

   wire [3:0] im = { i3, i2, i1, i0 };
   reg [2:0] indx;

   always @(im) begin
      { y1, y0 } = 2'b00;
      f = 1'b0;
      for (indx=0; indx<4; indx=indx+1)
      begin
         if ( im[indx] )
         begin
            { y1, y0 } = indx;
            f = 1'b1;
         end
      end
   end
endmodule
```

Figure 4.60   Priority Encoder using **for**

inactive (zero) values. This is a good practice which guarantees that no output retains its value from a previous activation of the **always** block.

The **for** statement in this **always** block uses *indx* for its index. This index is initialized to 0, its final value is less than 4 (*indx* <4), and is incremented by 1 (*indx* = *indx* + 1). The *indx* variable is declared as a 3-bit **reg** to allow it to take values between 0 and 4. In the body of the **for** procedural statement an **if** statement sets the proper *indx* value corresponding to the high-priority active input to the *{y1, y0}* output of the circuit.

In this example, our **if** statement does not use a final **else**, which contradicts our recommendations for completeness of all cases. However, since our outputs are initially set to their inactive values at the beginning of the **always** block, they will never be able to retain their old values. Therefore, the last **else** part for making sure that the outputs do receive some values, and thus avoiding output latches, is not necessary.

4.5.8 Procedural while loop

Unlike a **for** loop that has a fixed looping procedure based on an index, a **while** loop is more flexible in terms of specification of looping conditions.

An example **while** loop is shown in the Verilog code of Fig. 4.61. The *parity_gen* module is an n-bit parity generator. The size of its input vector is specified by the *SIZE* parameter. The **while** loop that handles generation of the parity is in an **always** block that is sensitive to the input of the circuit. Each time through the **while** loop, a partial parity is calculated and an index is incremented. Looping continues for as long as the index is less than *SIZE*.

The code shown in Fig. 4.61 is only for the purpose of demonstrating the use of the **while** loop. Generating parity can be done much easier than what is done in this example. E.g., a reduction XOR operator is all that is needed to generate parity of a vector.

4.5.9 A multilevel description

All coding styles discussed so far, i.e., gate instantiations, continuous assignments, module instantiations, and various forms of behavioral descriptions can be combined into a complete design description.

Figure 4.62 shows the diagram of a 4-bit 4-function ALU with overflow, parity and compare outputs, and a tri-state data output. This ALU has

```
`timescale 1ns/100ps

module parity_gen (a, p);

    parameter SIZE = 8;
    input [SIZE-1:0] a;
    output reg p;

    reg im_p;
    integer indx;

    always @( a ) begin
        im_p = 0;
        indx = 0;
        while (indx < SIZE)
        begin
            im_p = im_p ^ a[indx];
            indx = indx + 1;
        end
        p = im_p;
    end

endmodule
```

Figure 4.61 Parity Circuit using **while**

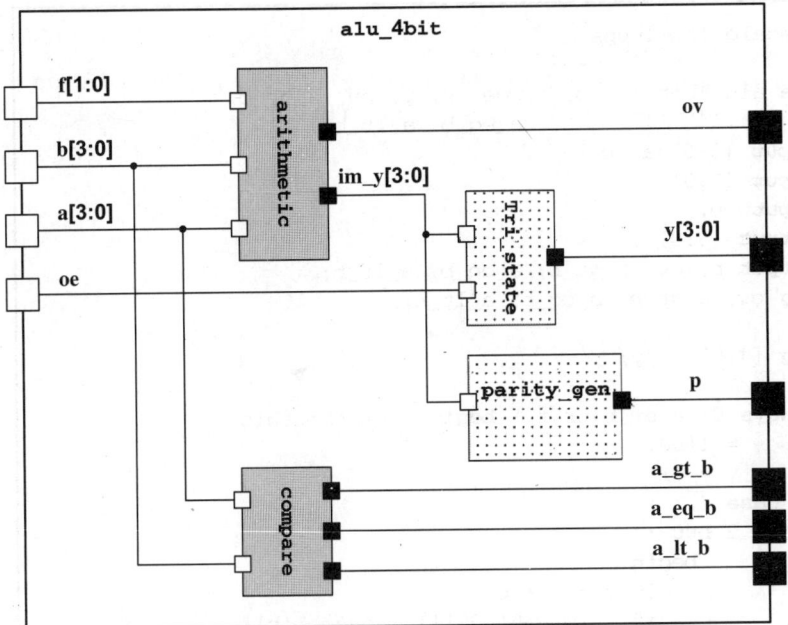

Figure 4.62   Block Diagram of a Multi-Function ALU

four subcomponents for performing arithmetic operations, generating compare outputs, generating parity, and making the output tri-state.

Figure 4.63 shows the Verilog code that corresponds to the block diagram of Fig. 4.62. The arithmetic part of the ALU is handled with an **always** block (labeled *arithmetic*) that uses a **case** statement for selecting the ALU operation based on *f*. The output of this part goes on *im_y*, which is the ALU's intermediate output before it goes through the output tri-state logic. The compare outputs are generated in an **always** block (labeled *compare*) that is sensitive to *a* and *b* that are the operands being compared. This **always** statement uses an **if-else** statement for assigning appropriate values to *a_gt_b, a_eq_b,* and *a_lt_b* outputs.

The parity output of the ALU performs XOR reduction on its internal *im_y* output of the *arithmetic* part. The last part of the *alu_4bit* module of Fig. 4.63 takes the *im_y* output and if *oe* (output enable) is active it puts it on the *y* output of the module. If *oe* is **0**, *y* becomes all **Z**'s.

4.6   Combinational Synthesis

Discussions so far in this chapter have concentrated on general coding styles for describing hardware. On the other hand, an important concern of any designer is translation of his or her code into hardware. As

```
`timescale 1ns/100ps

module alu_4bit (a, b, f, oe, y, p, ov, a_gt_b,
                           a_eq_b, a_lt_b);
    input [3:0] a, b;
    input [1:0] f;
    input oe;
    output [3:0] y;
    output p, ov, a_gt_b, a_eq_b, a_lt_b;
    reg ov, a_gt_b, a_eq_b, a_lt_b;

    reg [4:0] im_y;

    always @( a or b or f ) begin : arithmethic
       ov = 1'b0;
       im_y = 0;
       case ( f )
          2'b00 :
             begin
                im_y = a + b;
                if ( im_y>5'b01111 ) ov = 1'b1;
             end
          2'b01 :
             begin
                im_y = a - b;
                if ( im_y>5'b01111 ) ov = 1'b1;
             end
          2'b10 : im_y[3:0] = a & b;
          2'b11 : im_y[3:0] = a ^ b;
          default: im_y[3:0] = 4'b0000;
       endcase
    end

    always @( a or b ) begin : compare
       if ( a > b ) { a_gt_b, a_eq_b, a_lt_b } = 3'b100;
       else if ( a < b ) { a_gt_b, a_eq_b, a_lt_b } = 3'b001;
       else { a_gt_b, a_eq_b, a_lt_b } = 3'b010;
    end

    assign p = ^ im_y[3:0];

    assign y = oe ? im_y[3:0] : 4'bz;

endmodule
```

Figure 4.63   Verilog Code of a Multi-Function ALU

discussed in Chap. 1, a synthesis tool is used for automated translation of Verilog code of a design into hardware. It is important that the same code developed by a hardware designer and simulated for functional verification can be directly fed to a synthesis tool. This way, the designer is sure that what he or she gets after synthesis is what has been verified by simulation. Although, synthesis and synthesizability of codes presented were not mentioned directly, except for timing parameters all codes presented thus far in this chapter are synthesizable. Timing parameters specify detailed physical characteristics of a hardware component. Furthermore exact gate and wire delays cannot be known until a design is completely synthesized, mapped to its target library, and specific placement and routings are done. Therefore, timing parameters specified in a presynthesis description are either not allowed by a synthesis tool or they are ignored.

This section focuses on specific coding styles that are synthesizable. We discuss gates, continuous assignments and behavioral coding styles.

4.6.1   Gate level synthesis

Gate level designs are synthesizable; however most synthesis tools do not allow use of user defined primitives. Switch level primitives are also not allowed in a presynthesis description. Tri-state gates are allowed, but if a target technology does not support them, a warning message is issued by the synthesis tool. The message indicates the replacement of the tri-state structures by equivalent logic functions.

Figure 4.64 shows a full-adder circuit with *s* and *co* outputs. The Verilog code that corresponds to this structure is shown in Fig. 4.65. This code synthesizes to a logical circuit with the same functionality as that

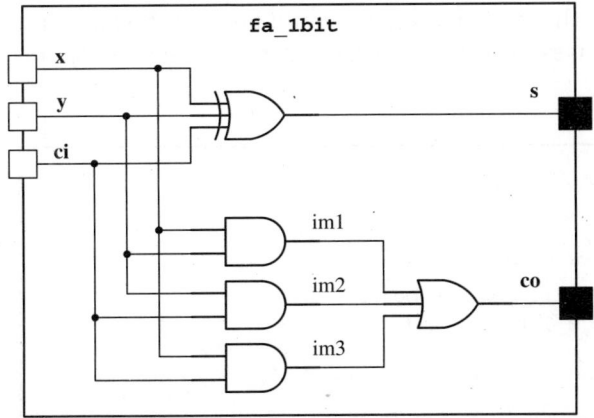

Figure 4.64   A Gate level Full Adder

```
module fa_1bit ( x, y, ci, s, co );
   input x, y, ci;
   output s, co;

   wire im1, im2, im3;

   xor ( s, x, y, ci );
   and ( im1, x, y ),
       ( im2, y, ci ),
       ( im3, ci, x );
   or  ( co, im1, im2, im3 );

endmodule
```

Figure 4.65   Synthesizable Gate Level Verilog Code

of Fig. 4.64. However, the exact postsynthesis hardware of this circuit may be very different than this diagram.

4.6.2   Synthesizing continuous assignments

Continuous assignments in any of the forms discussed in Sec. 4.4 are synthesizable. Multiple assignments are also allowed in synthesis. Generally, a synthesis tool performs logic optimizations, which enables designers to focus more on design functionality than having to get into the details of gate or structure level optimizations.

Figure 4.66 shows a 4-bit comparator with *gt, eq,* and *lt* outputs. We have used **assign** statements with conditional statements for ease of expressing functionality of the outputs of this circuit. After optimizations, the hardware that this synthesizes to would be no different than a presynthesis code that uses logic gates for generating compare outputs.

```
module compartor ( a, b, gt, eq, lt );
   input [3:0] a, b;
   output gt, eq, lt;
   assign gt = (a>b) ? 1'b1 : 1'b0;
   assign eq = (a==b) ? 1'b1 : 1'b0;
   assign lt = (a<b) ? 1'b1 : 1'b0;

endmodule
```

Figure 4.66   Synthesizable Code Using Assign Statements

Bussing and tri-state assignments like those of Fig. 4.45 are synthesizable, but like the tri-state gates, if a synthesis target library does not support tri-state structures it replaces tri-state busses with AND-OR busses and issues proper warning messages.

4.6.3 Behavioral synthesis

An **always** block with procedural statements using coding styles described in Sec. 4.5 is synthesizable. Because procedural blocks offer more flexibility in coding than gates or concurrent assignments, certain guidelines must be followed to make sure an **always** statement intended to synthesize a combinational block actually corresponds to such hardware.

4.6.3.1 Input Sensitivity.

A combinational circuit continuously monitors its input and an event on **any** of its inputs causes circuit outputs to be evaluated. In other words, a combinational circuit is sensitive to **all** its inputs. A Verilog **always** block that is to correspond to a combinational block of hardware must have a similar behavior. For this correspondence, such an **always** block must include all its inputs in its sensitivity list. We define inputs of an **always** block as all variables or signals that are being read. For example if a variable is defined outside of the **always** block that is not necessarily an input of the circuit being synthesized and participates in a relational operation inside the **always** block, this variable must be included in the sensitivity list of the **always** block. Verilog descriptions of Sec. 4.5 synthesize to combinational circuits, and in all **always** blocks used (see for example Fig. 4.63) all inputs are included in their sensitivity lists.

Verilog makes this synthesis rule easier to follow by allowing an asterisk (*) in the sensitivity list of **always** blocks to imply all inputs. For example, the headings of the *arithmetic* and *compare* **always** statements of Fig. 4.63 can be replaced with:

```
always @ (*) begin: arithmetic
```

and

```
always @ (*) begin: compare
```

4.6.3.2 Output assignments.

Another property of a combinational circuit is that its outputs are always affected by a change in its inputs. Another way of saying this is that outputs of a combinational circuit never retain their old values, i.e., no latching occurs on the outputs. As with input sensitivity described above, an **always** statement that corresponds to a combinational block must have provisions for preventing output latches.

There are two ways we can make sure outputs of a combinational **always** statement are always affected when an event occurs on its input. One way is for the designer to trace through all conditions of **case**, **if-else**, and other procedural statements of the **always** block, and make sure **all** outputs receive some value, no matter what the input conditions are. For large designs with many nested procedural statements, this may become a formidable task. If the designer misses an output, and under certain input conditions an output does not receive a value, the synthesis tool generates a latch on this output, which is contrary to what we expect from outputs of combinational blocks.

Figure 4.67 shows the block diagram of an ALU. As shown in the Verilog code of Fig. 4.68, the *y* and *co* outputs of this ALU are given values inside an **always** block. Notice in this block that regardless of input conditions both outputs receive some values. In this example the **case default** helps assigning values to the outputs under conditions that are not explicitly specified in other *case alternatives*.

To make this output assignment requirement easier to implement, a designer can place procedural assignments at the beginning of an **always** block to assign all outputs of the **always** block to their inactive values. Outputs of an **always** block are regarded as all those variables

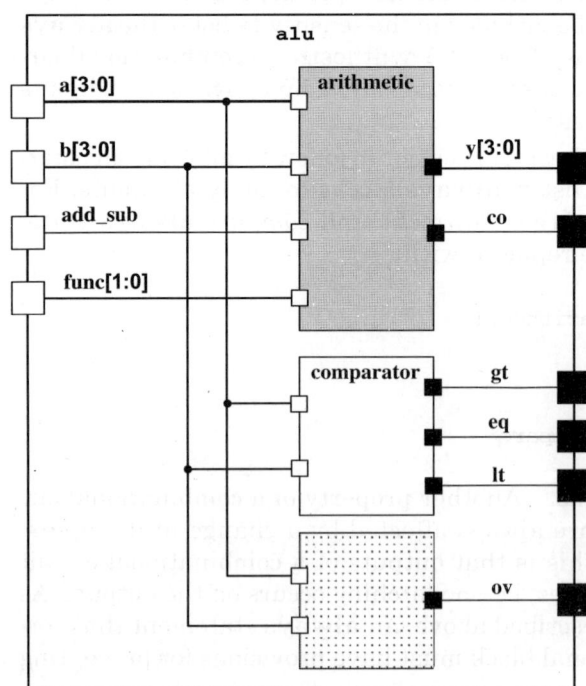

Figure 4.67   ALU Block Diagram

```
module alu ( a, b, add_sub, func, y, co, gt, eq, lt, ov );
    input [3:0] a, b;
    input add_sub;
    input [1:0] func;
    output [3:0] y;
    reg [3:0] y;
    output co, gt, eq, lt, ov;
    reg co;

    always @( a or b or add_sub or func ) : arithmetic
        case (func)
            2'b00 :
                if (add_sub) { co, y } = a - b;
                else { co, y } = a + b;
            2'b01 : { co, y } = { 1'b0, a };
            2'b10 : { co, y } = { 1'b0, a & b };
            2'b11 : { co, y } = { 1'b0, ~a };
            default: { co, y } = , 5'b00000 ;
        endcase

    compartor cmp ( a, b, gt, eq, lt );

    assign ov = (func==2'b00)
            ? ((a[3] & b[3] & ~y[3]) | (~a[3] & ~b[3] & y[3]))
            : 1'b0;
endmodule
```

Figure 4.68   ALU Synthesizable Verilog Code

that appear somewhere in the block on the left-hand side of a procedural assignment. Consider, for example, the *arithmetic* **always** block of the example of Fig. 4.63. In this code *ov* and *im_y* are set to **0** right at the beginning of the **always** block. In the **case** statement that follows these assignments, the *ov* output receives values when add or subtraction operations ($f = $ **00**, or $f = $ **01**) are taking place. However, this output is left unaffected by this **case** statement if f has any value but **00** and **01**. The assignment of *1'b0* to *ov* at the beginning of the **always** statement guarantees that no latches are put on this output.

The situation with the *im_y* output of the *arithmetic* block of this code is somewhat different. In the **case** statement, this output receives some value regardless of input conditions, and therefore no latches are implied on the bits of this vector. Therefore, the *im_y = 0;* assignment at the beginning of this block is not necessary, but does not hurt the synthesis process. The effect of the redundant assignment of **0** to this output is eliminated in the synthesis optimization process.

4.6.4 Mixed synthesis

Gate level, module instantiations, **assign** statements and **always** blocks can all coexist in a synthesizable description. For example the code of Fig. 4.68 has an **always** block that assigns values to *co* and *y* outputs of the ALU. Instantiation of the *comparator* module of Fig. 4.66 provides hardware for driving the compare outputs of the ALU. The last statement in the synthesizable code of the *alu* module is an **assign** statement that generates logic for driving the *ov* output. Note that overflow only occurs if the ALU is performing an add operation. Therefore assignment to *ov* is conditioned by *func* == *2'b00*, where *2'b00* is the add opcode.

4.7 Summary

Although the focus of this chapter was primarily on combinational circuits, most language constructs used for combinational and sequential circuits were presented. Therefore this chapter can be regarded as our main chapter presenting the Verilog language in an example oriented fashion. In addition to constructs that directly correspond to hardware structures, this chapter also presented constructs for specification of timing of a hardware unit. Obviously, such constructs cannot be used in designs for synthesis.

Problems

4.1 Write a Verilog description for the following function.

$$f (A, B, C, D) = \Sigma_m(0, 2, 4, 5, 6, 7, 9, 10, 11), d(1, 13)$$

4.2 Write Verilog code for a 4-to-1 multiplexer with a tri-state output and an active low *OutputEnable* input. The multiplexer has four data inputs (*d0, d1, d2, d3*) and two select inputs (*s1, s0*). Use **case** and **if** statements.

4.3 Use two of the multiplexers of Prob. 4.2 to build an 8-to-1 multiplexer.

4.4 Use the 8-to-1 multiplexer of Prob. 4.3 to implement the function of Problem 4.1.

4.5 Show Verilog code for a cascadable 4-to-2 priority encoder. Your circuit should have an enable input, four data inputs, an enable output, an interrupt output, and two source id outputs. All inputs and outputs must be active high. For cascading purposes and to be able to use wired-OR logic, use tri-state for your id outputs. Adjust the details of your design for a better cascading capability.

4.6 Write a Verilog description for a 4-bit adder-subtractor that adds when *as* is **1** and subtracts when *as* is **0**.

4.7 Write a Verilog function to implement a 4-bit BCD to seven-segment display (BDC_to_7Seg) converter. Use vector inputs and outputs. Number your output segments from 0 to 6 in clock-wise direction starting with the top segment and ending with the middle segment.

4.8 Show gate level details of the circuit described by the following Verilog code.

```
module infer (q, d, e, c);
input d, c;
output q;
reg q;
   always @(c, d, e)
   if (c == 1 && e == 0) q = d; else q = 1;
endmodule
```

4.9 Consider two 4-bit binary numbers A and B. Bits of A are $a3$, $a2$, $a1$, and $a0$, and bits of B are $b3$, $b2$, $b1$, and $b0$. A is greater than B if $a3$ is 1 and $b3$ is 0, but if $a3$ and $b3$ are the same, then if $a2$ is 1 and $b2$ is 0, we can determine that A is greater than B. This evaluation continues until $a0$ and $b0$ are considered. If $a0$ and $b0$ are equal then A and B are equal. Using discrete gates and Verilog gate primitives build a 4-bit comparator that generates a **1** on its GT output when its 4-bit input A is greater than its B input, and a **1** on its EQ output when A and B are equal. Use the **generate** statement and other Verilog iterative logic support structures.

4.10 Using the comparator of Prob. 4.9 and discrete logic gates, build a MIN circuit that takes two 4-bit inputs and puts the smaller of its two inputs on its output. Write the Verilog description of this circuit.

4.11 Write a Verilog description for a multiplier bit. Using this multiplier bit write a Verilog code for a 4×4 array multiplier. Then, using the 4×4 multiplier design an ALU that performs eight functions according to the following table. The data inputs of the ALU are 8 bits. Use **always** blocks and **case** statements in your Verilog description.

s_2	s_1	S_0	Function
0	0	0	A add B
0	0	1	A sub B
0	1	0	A add 2*B
0	1	1	A sub 2*B
1	0	0	A * B (4-bit)
1	0	1	min (A, B)
1	1	0	Abs (A)
1	1	1	B

Suggested Reading

Bhasker, J., *Verilog HDL Synthesis, A Practical Primer*, Star Galaxy Pub, 1998, ISBN: 0965039153.

Brown S., and Z. Vranesic, *Fundamentals of Digital Logic with Verilog Design*, McGraw-Hill, New York, 2002, ISBN: 0-07-283878-7.

IEEE Std 1364-2001, *IEEE Standard Verilog Language Reference Manual*, SH94921-TBR (print) SS94921-TBR (electronic), ISBN 0-7381-2827-9 (print and electronic), 2001.

Navabi, Z., *Digital Design and Implementation with Field Programmable Devices*, Kluwer Academic Publishers, Boston, 2005, ISBN: 1-4020-8011-5.

Navabi, Z., *Verilog Computer-Based Training Course*, CBT CD with hardcopy User's manual, McGraw-Hill, New York, 2002, ISBN 0-07-137473-6.

Palnitkar, S., *Verilog HDL*, 2d ed, Prentice Hall PTR, New Jersey, 2003, ISBN: 0130449113.

Thomas, P. R., and P. Moorby, *The Verilog® Hardware Description Language*, Springer, Boston, 2002, ISBN: 1402070896.

Sequential Circuit Description

Based on the material of Chap. 3, the previous chapter discussed ways of modeling timing and functionality of combinational circuits. This chapter follows a similar flow as that of Chap. 4, but concentrates on using Verilog constructs for description of sequential circuits. For the sake of completeness we discuss use of gate level and assignments for describing memory elements; however, most of our attention will be given to modeling such parts with procedural statements. Because most of Verilog language constructs have already been discussed in conjunction with the combinational circuits, this chapter focuses less on language aspects and more on hardware modeling. Furthermore, timing-related issues of the language, as discussed in Chap. 4, apply as well to sequential circuits and will not be discussed here.

The next section in this chapter discusses hardware properties that make a logic circuit store data. Section 5.2 shows ways of describing basic memory components. Based on models discussed, Sec. 5.3 shows Verilog description of functions with storage capability such as counters and shift registers. This will be followed by state machine and controller modeling, and finally as in Chap. 4, we will have a discussion of modeling for synthesis.

5.1 Sequential Models

In digital circuits, storage of data is done either by feedback, or by gate capacitances that are refreshed frequently. Verilog provides language constructs for building memory elements using both these schemes. However, more abstract models also exist and are used in most sequential circuit models.

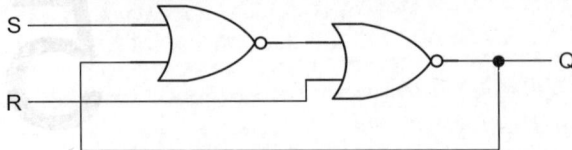

Figure 5.1 Basic Feedback

5.1.1 Feedback model

The construct shown in Fig. 5.1 is the most basic feedback circuit that has data storage capability.

This circuit has one feedback line that makes it a two-state (feedback = **0** and feedback = **1**), or a 1-bit, memory element. Many Verilog constructs can be used for proper modeling of this circuit.

5.1.2 Capacitive model

Another hardware structure with storage capability is shown in Fig. 5.2. When C becomes **1** the value of D is saved in the input gate of the inverter and when C becomes 0, this value will be saved until the next time that c becomes **1** again. The output of the inverter is equal to the complement of the stored data.

Because of powerful switch level capabilities of Verilog, the circuit of Fig. 5.2 can be very closely modeled in Verilog. Chapter 7 of this book discusses switch level modeling for combinational and sequential circuits in great detail.

5.1.3 Implicit model

Feedback and capacitive models discussed above are technology dependent, and they have the problem of being too detailed and thus too slow to simulate. Of course, where such details are needed, this level of modeling is possible in Verilog.

Verilog also offers language constructs that model storage elements at more abstract levels than the previous models. Such modelings are technology independent and allow much more efficient simulation of circuits

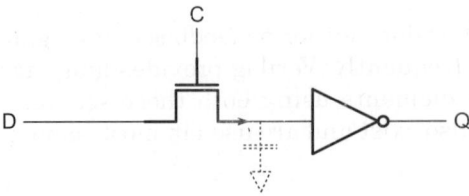

Figure 5.2 Capacitive Storage

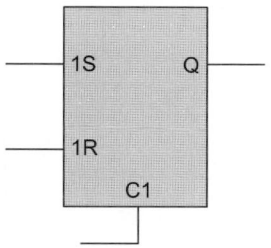

Figure 5.3 An SR-Latch Notation

with a large number of storage elements. Figure 5.3 shows an SR-latch model without gate level details.

Because gate and transistor details of models at the block diagram level are not known, Verilog provides timing check constructs for ensuring correct operation of this level of modeling. The sections that follow present language constructs for feedback modeling of storage elements, but concentrate on the more abstract models in which storage is implied by the Verilog code.

5.2 Basic Memory Components

This section shows modeling memory components in Verilog. We start with latches and discuss 1-bit and multidimensional memories. In the use of Verilog constructs, we show how gates, primitives, assignments, and procedural blocks are used for memory modeling.

5.2.1 Gate level primitives

Figure 5.4 shows a cross-coupled NOR structure that forms a 1-bit storage element. This circuit is no different than that of Fig. 5.1, and its storage is due to the feedback from q back to $g1$.

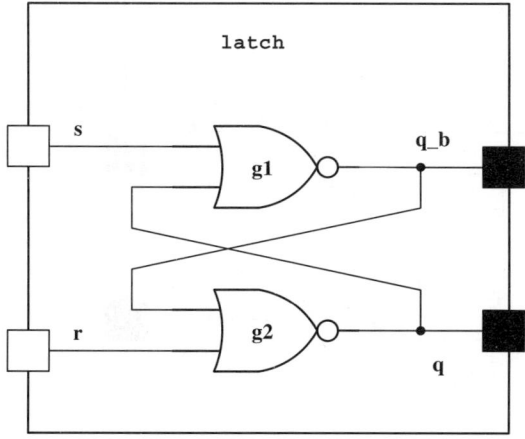

Figure 5.4 Cross-Coupled NOR Latch

```
`timescale 1ns/100ps

module latch (input s, r, output q, q_b );
    nor #(4)
        g1 ( q_b, s, q ),
        g2 ( q, r, q_b );
endmodule
```

Figure 5.5   SR-Latch Verilog Code

The Verilog code of this diagram is shown in Fig. 5.5. The q and q_b outputs are driven by two NOR gates, and are therefore initially **X**. The outputs remain at this ambiguous state for as long as s and r remain **0**. After a delay of 4 ns after s becomes **1**, q becomes **1** and after another 4 ns delay, q_b becomes **0**. Simultaneous assertion of both inputs results in loss of memory.

This memory element is the base of most static memory components. Adding control gates and a clock input results in a clocked SR-latch. Figure 5.6 shows an all-NAND version of a clocked SR-latch.

The Verilog code that corresponds to the diagram of Fig. 5.6 is shown in Fig. 5.7. As shown, the circuit is parameterized so that delay values can be controlled when the latch is instantiated. We have declared our parameters in the module header along with module name and ports. Wire names $\_s$ and $\_r$ are used for the set and reset inputs to the cross-coupled core of this memory element.

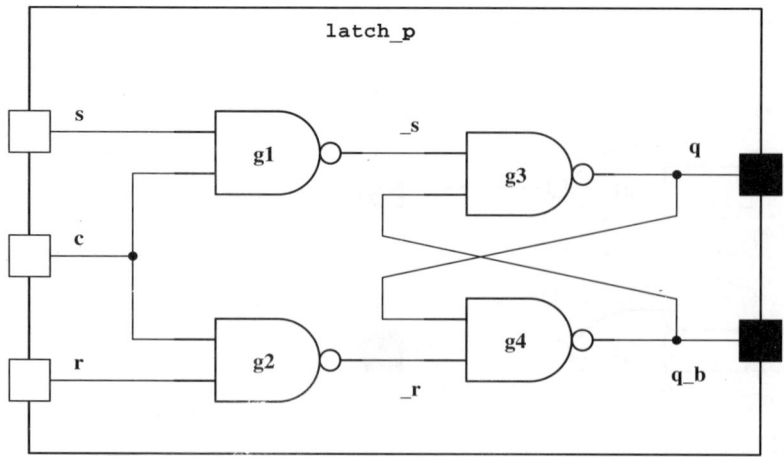

Figure 5.6   All NAND Clocked SR-Latch

```
`timescale 1ns/100ps

module latch_p #(parameter tplh=3, tphl=5) (input s, r, c,
                    output q, q_b );
    wire _s, _r;
    nand #(tplh,tphl)
          g1 ( _s, s, c ),
          g2 ( _r, r, c ),
          g3 ( q, _s, q_b ),
          g4 ( q_b, _r, q );
endmodule
```

Figure 5.7 All NAND Clocked Latch

The simulation of Fig. 5.7 is shown in Fig. 5.8. Initially both q and q_b are **X**, and 8 ns after the c clock becomes **1** while s is **1** the q output is set to **1**. This delay is due to a fall of 3 ns and a rise of 5 ns in the NAND gates of the circuit. While c is **1** when r becomes **1** at time 50 ns, the q output becomes **0** after 13 ns. This output becomes **1** after 8 ns of s becoming **1** at time 90 ns.

Using the latch of Fig. 5.6, Fig. 5.9 shows the formation of a master-slave D flip-flop.

The Verilog code of Fig. 5.10 corresponds to the diagram of Fig. 5.9. In this code, hierarchical naming is used for overriding parameters of the *master* and *slave* instantiations of *latch*_p module.

5.2.2   User defined sequential primitives

For faster simulation of memory elements and for correspondence with specific component libraries, Verilog provides language constructs for defining sequential user-defined primitives (UDPs). A sequential UDP has the format of the combinational UDP of Chap. 4, except that in addition to the inputs, and output of the circuit, its present state is also specified.

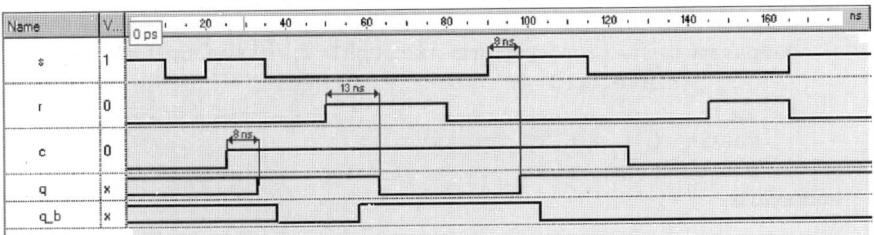

Figure 5.8 SR Latch Simulation

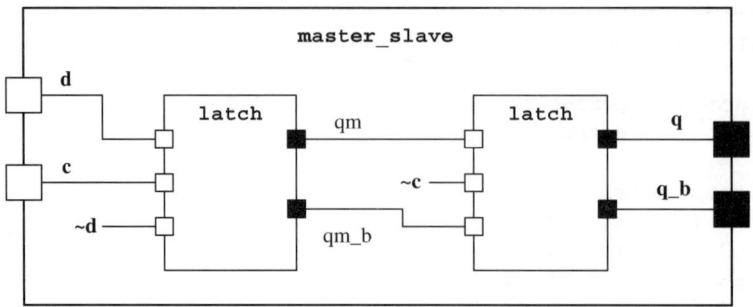

Figure 5.9   Master-Slave D Flip-Flop

Figure 5.11 shows a sequential UDP defining an SR-clocked latch. The behavior described here is the same as that of circuit of Fig. 5.4 or the Verilog code of Fig. 5.5. As shown, the table defining the latch output has a column for specifying its present state. This column comes before the output specification column and is separated from inputs and output by colons. Question marks (**?**) in the table signify "any value" and dashes (-) are for "no change." For example the first row of the table of **primitive** *latch* reads as: any value on *s*, *r* and the present sate (*q*), for as long as *c* is **0**, keeps the next state of the machine (*q*+) unchanged.

When instantiated, rise and fall delay values can be specified for a sequential UDP in the same way as specifying delays for other language primitives.

5.2.3   Memory elements using assignments

As discussed in Chap. 3, a continuous assignment is equivalent to a gate structure driving the left-hand side of the assignment. We can use these

```
`timescale 1ns/100ps

module master_slave (input d, c, output q, q_b );
   wire qm, qm_b;
   defparam master.tplh=4, master.tphl=4, slave.tplh=4,
            slave.tphl=4;
   latch_p
      master ( d, ~d, c, qm, qm_b ),
      slave  ( qm, qm_b, ~c, q, q_b );
endmodule
```

Figure 5.10   Master-Slave D Flip-Flop Verilog Code

```
primitive latch( q, s, r, c );
   output q;
   reg q;
   input s, r, c;
   initial q=1'b0;
   table
      // s r c    q    q+;
      // ————:——:——;
         ? ? 0 : ? : - ;
         0 0 1 : ? : - ;
         0 1 1 : ? : 0 ;
         1 0 1 : ? : 1 ;
   endtable
endmodule
```

Figure 5.11   Sequential UDP Defining a Latch

statements for specifying specific gates of a hardware module for a latch, or for specifying a feedback circuit.

Figure 5.12 shows two feedback blocks forming a master-slave flip-flop. Each block has three inputs and one output. When a block's clock input is **0**, it puts its output back to itself (feedback), and when its clock is **1** it puts its data input into its output.

Figure 5.13 shows a master-slave flip-flop that uses **assign** statements to implement feedbacks shown in Fig. 5.12. In the first assignment the use of *qm* on the right and left of the assign statement corresponds to the feedback of this output back to its input. Each **assign**

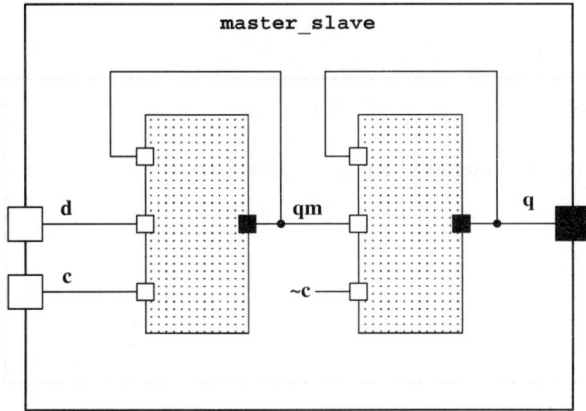

Figure 5.12   Master-Slave Using Two Feedback Blocks

```
`timescale 1ns/100ps

module master_slave_p #(parameter delay=3) (input d,c, output q);
    wire qm;
    assign #(delay) qm = c ? d : qm;
    assign #(delay) q = ~c ? qm : q;
endmodule
```

Figure 5.13   Assign Statements Implementing Logic Feedback

statement implements a latch, and the module that uses two latches with complementary clocks implements a master-slave flip-flop.

5.2.4   Behavioral memory elements

The previous sections showed Verilog models for latches and flip-flops by explicit use of feedback or present and next states. Such a model corresponds to the actual hardware implementing a memory element, and has the potential of having all gate level delays specified.

A more abstract and easier way of writing Verilog code for a latch or flip-flop is by behavioral coding. This way, the storage of data and its sensitivity to its clock and other control inputs will be implied in the way model is written.

5.2.4.1   Latch modeling.   As our first behavioral model of a memory element consider the Verilog code of the D latch of Fig. 5.14. This code reads as: when c or d changes, if c is **1**, q gets d. This means that if c or d changes and c is not **1**, then q does not change and it retains its old value.

```
`timescale 1ns/100ps

module latch (input d, c, output reg q, q_b );
    always @( c or d )
        if ( c ) begin
            #4 q = d;
            #3 q_b = ~d;
        end
endmodule
```

Figure 5.14   A D-Type Latch Verilog Code

```
`timescale 1ns/100ps

module latch (input d, c, output reg q, q_b );
    always @( c or d )
        if ( c ) begin
            q <= #4 d;
            q_b <= #3 ~d;
        end
endmodule
```

Figure 5.15   Latch Model Using Nonblocking Assignments

It can also be read from the Verilog code that while c is **1**, changes on d directly affect q and q_b outputs. This behavior implies a storage unit that is level sensitive to c and is thus a latch.

In the body of the **always** block used in this latch, timing control statements are used for delaying assignments to q and q_b. When the flow begins in this procedural statement, if c is **1**, then after 4 ns the d input is read and assigned to q. After another wait of 3 ns, d is read again and its complement ($\sim d$) is assigned to q_b. If d changes between the time it is read for q and q_b (in the 3 ns time), erroneous results happen. To avoid this problem, nonblocking assignments, with intra-statement delay controls should be used. Figure 5.15 shows another version of the D latch that corrects this timing problem. Figure 5.16 shows the timing behavior of this model for storing a **1** at time 30, and a **0** at time 50 into our latch.

Because of timing problems such as those described above, use of nonblocking assignments for describing sequential circuits is recommended. However, it is important to understand the semantics of the language constructs and their timing implications. With this understanding and detailed analysis of the behavior of a model, decisions as to the use of proper constructs should be made.

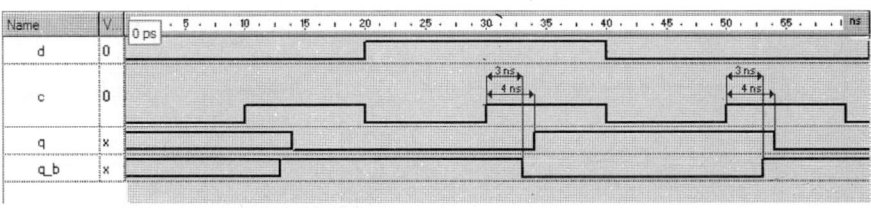

Figure 5.16   Testing Latch with Nonblocking Assignments

```
`timescale 1ns/100ps

module d_ff (input d, clk, output reg q, q_b );
    always @( posedge clk ) begin
        q <= #4 d;
        q_b <= #3 ~d;
    end
endmodule
```

Figure 5.17   Positive Edge Trigger Flip-Flop

5.2.4.2   Flip-flop modeling.   A basic edge trigger flip-flop model at the behavioral level is shown in Fig. 5.17. This model is sensitive to the positive edge of the clock, and uses nonblocking assignments for assignments to q and q_b.

Flow into the procedural block of Fig. 5.17 is controlled by the *event control statement* that has **posedge** *clk* as its *event expression*. Assignments to q and q_b are reached immediately after the flow in the **always** block begins. As shown, the actual assignment of values to q and q_b are delayed by 4 and 3 ns, respectively. With each clock edge, the entire procedural block is executed once from **begin** to **end**.

Figure 5.18 shows a partial waveform of a simulation run of our D flip-flop. At 60 ns when we have the positive edge of the clock, the value of d is read and scheduled into q and q_b for times 64 and 63 ns, respectively. The sensitivity to the positive edge of the clock in this example is illustrated by the fact that during the time that *clk* is **1** (from 60 to 80 ns, exclusive of 60 ns, and inclusive of 80 ns), changes on d do not affect the state of the flip-flop.

5.2.4.3   Flip-flop with set-reset control.   The style presented in *d_ff* of Fig. 5.17 can be expanded to cover flip-flops with synchronous and asynchronous set and reset control inputs. The Verilog code of Fig. 5.19 is a D-type flip-flop with synchronous set and reset (s and r) inputs.

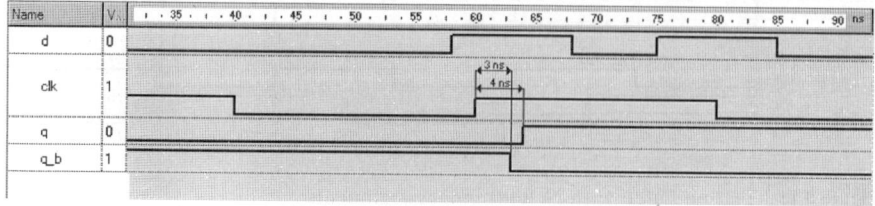

Figure 5.18   Simulation of a Positive Edge Flip-Flop

```
`timescale 1ns/100ps

module d_ff_sr_Synch (input d, s, r, clk, output reg q, q_b );

    always @(posedge clk) begin
        if( s ) begin
            q <= #4 1'b1;
            q_b <= #3 1'b0;
        end else if( r ) begin
            q <= #4 1'b0;
            q_b <= #3 1'b1;
        end else begin
            q <= #4 d;
            q_b <= #3 ~d;
        end
    end
endmodule
```

Figure 5.19   D Type Flip-Flop with Synchronous Control

As shown in this figure, a single **always** statement is used for describing the *d_ff_sr_Synch* module. The flow into the **always** block is only initiated by the **posedge** of *clk*. Therefore, the **if**-statements with *s* and *r* conditions are only examined after the positive edge of the clock. This behavior is in accordance with synchronicity of *s* and *r* control inputs.

The Verilog code of Fig. 5.20 is a D-type flip-flop with asynchronous set and reset inputs. Unlike the code of Fig. 5.19, the sensitivity list of the **always** block in the *d_ff_sr_Asynch* module includes ***posedge s*** and ***posedge r*** as well as ***posedge clk***. Inclusion of ***posedge s*** and ***posedge r*** enables flow into the **always** block when clock changes to 1 or when *s* or *r* become active. The arrangement of **if** conditions and this sensitivity makes this model a positive edge trigger with asynchronous set and reset control inputs.

Although **posedge** is used for *clk*, *s*, and *r*, the *d_ff_sr_Asynch* is sensitive to the edge of the clock, but to the levels of *s* and *r*. This is because the **if** statements examine *s* and *r* and the default **else** is used for *clk*; being last in the condition of the **if** statements makes assignment of *d* to *q* sensitive to the clock edge. Examining this description with various values of *s*, *r*, and *clk* proves correctness of this model.

Figure 5.21 shows output waveforms of *d_ff_sr_Synch* and *d_ff_sr_Asynch* for data applied to *d*, *s*, and *r*. In the first half of this waveform (before 120 ns), changes to *q* are triggered by the clock and *q_Synch* and *q_Asynch* are exactly the same. In the second half of this

```
`timescale 1ns/100ps

module d_ff_sr_Asynch (input d, s, r, clk, output reg q, q_b );

    always @( posedge clk, posedge s, posedge r ) begin
        if( s ) begin
            q <= #4 1'b1;
            q_b <= #3 1'b0;
        end else if( r ) begin
            q <= #4 1'b0;
            q_b <= #3 1'b1;
        end else begin
            q <= #4 d;
            q_b <= #3 ~d;
        end
    end
endmodule
```

Figure 5.20   D-type Flip-Flop with Asynchronous Control

waveform, *s* and *r* become active and cause changes to the flip-flop output. Note that *q_Synch* still waits for the edge of the clock to set or reset, while *q_Asynch* changes occur independent of *clk* when *s* or *r* becomes active.

5.2.4.4 Other storage element modeling styles.   Models presented above are the most commonly used models for latches and flip-flops. Verilog provides other language constructs that can be used for this purpose, and for the sake of completeness and presentation of such constructs, we discuss storage elements utilizing them.

Figure 5.22 shows a latch using a **wait** statement instead of an event control statement. The code shown models a positive level sensitive

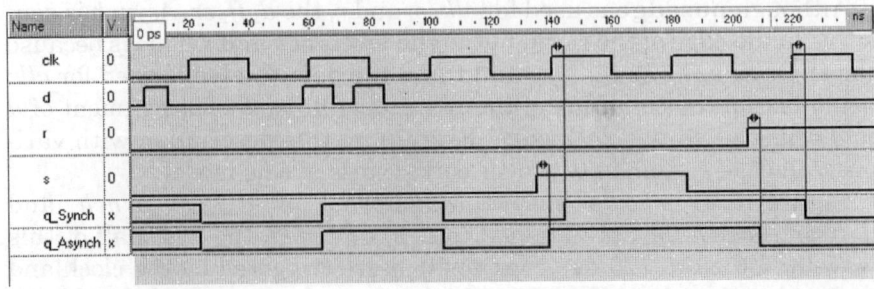

Figure 5.21   Comparing Synchronous and Asynchronous Flip-Flop Controls

```
`timescale 1ns/100ps

module latch (input d, c, output reg q, q_b );

    always begin
        wait ( c );
            #4 q <= d;
            #3 q_b <= ~d;
    end
endmodule
```

Figure 5.22   Latch Using **wait**, a Potentially Dangerous Model

D-type latch. The **wait** statement shown is a procedural statement that blocks the flow of the procedural block when c is **0**. When c becomes **1**, the **wait** statement allows the program flow to pass it and reach assignments to q and q_b. Because an **always** statement repeats itself forever, if c becomes **1** and remains at this value, the body of the **always** statement repeats itself every 7 ns due to the delay control statements. If these delays are omitted, then the looping of the body of the **always** statement happens in zero time causing an infinite loop in simulation.

Figure 5.23 shows a D-type flip-flop using a **fork-join** construct. A **fork-join** bracketing instead of **begin-end** causes all sequential statements or blocks of sequential statements that are immediately within this bracketing to be executed in parallel. In the Verilog code of d_ff shown in Fig. 5.23, delay control statements and their following assignments are executed in parallel. Therefore assignment to q is delayed by 4 ns and to q_b by 3 ns. Unlike **begin-end** bracketing, accumulation of delays does not occur.

```
`timescale 1ns/100ps

module d_ff (input d, clk, output reg q, q_b );

    always @( posedge clk )
        fork
            #4 q <= d;
            #3 q_b <= ~d;
        join

endmodule
```

Figure 5.23   Flip-Flop using **fork-join**, no Delay Accumulation

Another mechanism for modeling storage elements is by the use of sequential **assign** and **deassign** statements. The D-type flip-flop of Fig. 5.24 with asynchronous *s* and *r* control inputs uses sequential **assign** statements to force set (**1**) and reset (**0**) values onto the flip-flop *q* output.

A sequential **assign** statement forces a value into a **reg** type variable, and a sequential **deassign** removes it. A value forced into a variable by **assign** can only be removed by **deassign**. A sequential **assign** statement implies de-assigning any previously assigned values. While a sequential **assign** is in effect, all non-**assign** statements will be ineffective.

Our example of Fig. 5.24 uses the *force_a_1* block to force a **1** into *q* when *s* is **1**, and uses *force_a_0* block to force a **0** into *q* when *r* is **1**. The third block (i.e., *clocked*) cannot affect *q* or *q_b* when any of the forcing

```verilog
`timescale 1ns/100ps

module d_ff (input d, clk, s, r, output reg q, q_b );
    always @( s ) begin : force_a_1
        if ( s )
            begin
                #6 assign q = 1'b1;
                #4 assign q_b = 1'b0;
            end else begin
                deassign q;
                deassign q_b;
            end
    end

    always @( r ) begin : force_a_0
        if( r )
            begin
                #6 assign q = 1'b0;
                #4 assign q_b = 1'b1;
            end else begin
                deassign q;
                deassign q_b;
            end
    end

    always @( posedge clk ) begin : clocked
        #4 q = d;
        #3 q_b = ~d;
    end

endmodule
```

Figure 5.24 D-type Flip-Flop Sequential **assign** and **deassign**

blocks are in effect. The **deassign** statements of any of the forcing blocks are able to release the flip-flop outputs, allowing the *clocked* block to clock *d* into the flip-flop with the positive edge of the clock.

5.2.5 Flip-Flop timing

As discussed above, behavioral modeling of flip-flops only allows a limited timing specification. Furthermore, the body of a procedural statement, where a flip-flop behavior is described, is not an appropriate place for checking for inappropriate timings of flip-flop inputs and clock. This section discusses Verilog language constructs for detecting and reporting timing violations. Such constructs include system tasks for checking *setup, hold, period,* and *width* parameters.

5.2.5.1 Setup time. Setup time is defined as the minimum necessary time that a data input requires to setup before it is clocked into a flip-flop. Verilog construct for checking the setup time is **$setup**, which takes the flip-flop data input, active clock edge and the setup time as its parameters. The **$setup** task is used within a **specify** block.

Figure 5.25 shows a D-type flip-flop with a positive edge trigger flip-flop and asynchronous set and reset controls. In the *d_ff* module shown,

```
`timescale 1ns/100ps

module d_ff ( input d, clk, s, r, output reg q, q_b );

    specify
        $setup ( d, posedge clk, 5 );
    endspecify

    always @( posedge clk or posedge s or posedge r ) begin
        if( s ) begin
            q <= #4 1'b1;
            q_b <= #3 1'b0;
        end else if( r ) begin
            q <= #4 1'b0;
            q_b <= #3 1'b1;
        end else begin
            q <= #4 d;
            q_b <= #3 ~d;
        end
    end

endmodule
```

Figure 5.25 Flip-Flop with Setup Time

the **$setup** task that is used within the **specify** block continuously checks the timing distance between changes on *d* and the positive edge of the *clk* clock. If this distance is less than 5 ns, a violation message will be issued.

Figure 5.26 shows the simulation of the flip-flop of Fig. 5.25. The *d* input changes at 57 ns, and before this change is allowed the setup time of 5 ns, the data is clocked into the flip-flop at 60 ns, only 3 ns after *d*. The simulation run reports the violation as shown in Fig. 5.26.

5.2.5.2 Hold time. Hold time is defined as the minimum necessary time a flip-flop data input must stay stable (hold its value) after it is clocked. The Verilog construct for checking the hold time is **$hold**, which takes the flip-flop active edge of the clock, its data input, and the required hold time as its parameters. The **$hold** task must be used inside a **specify** block.

Figure 5.27 shows a flip-flop with hold time check and Fig. 5.28 shows the input signals that cause the hold time violation of 3 ns. As shown, *clk* samples the *d* value of **1** at 20 ns. However, at 22 ns, *d* changes. This violates the minimum required hold time of 3 ns, and the message shown in Fig. 5.28 is displayed.

The Verilog **$setuphold** task combines setup and hold timing checks. The following replaces both tasks used in Figs. 5.25 and 5.27.

```
$setuphold (posedge clk, d, 5, 3);
```

5.2.5.3 Width and period. Verilog **$width** and **$period** check for minimum pulse width and period. Pulse width checks the time from a specified edge of a reference signal to its opposite edge. Period checks the time from a specified edge of a reference signal to the same edge. Figure 5.29 shows a D-type flip-flop with **$setuphold**, **$width**, and **$period**.

This flip-flop behaves the same as those of Figs. 5.25 and 5.27. The **always** block shown and assignment to *q_b* are done differently, just to demonstrate other coding styles possible for flip-flops. Simulation waveform demonstrating width, setup, and period violations is shown in Fig. 5.30.

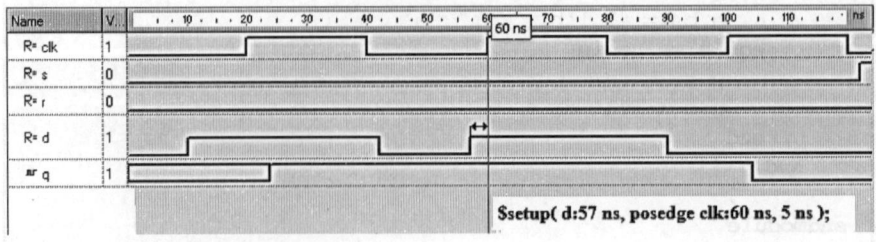

Figure 5.26 Setup Time Violation

```
`timescale 1ns/100ps

module d_ff ( input d, clk, s, r, output reg q, q_b );

    specify
        $hold ( posedge clk, d, 3 );
    endspecify
    always @( posedge clk or posedge s or posedge r ) begin
        if( s ) begin
            q <= #4 1'b1;
            q_b <= #3 1'b0;
        end else if( r ) begin
            q <= #4 1'b0;
            q_b <= #3 1'b1;
        end else begin
            q <= #4 d;
            q_b <= #3 ~d;
        end
    end

endmodule
```

Figure 5.27   Flip-Flop with Hold Time

5.2.6   Memory vectors and arrays

Coding styles and timings discussed in the previous sections apply to arrays and vectors as well. The only difference is that when one-dimensional vectors or multi-dimensional arrays are being considered, their input output ports and their memory structures should be declared accordingly.

5.2.6.1   Vectors.   Figure 5.31 shows an 8-bit transparent D-latch. Data input and latch output are declared as 8-bit vercors. The **always** block shown is sensitive to *c* and all eight bits of *d*.

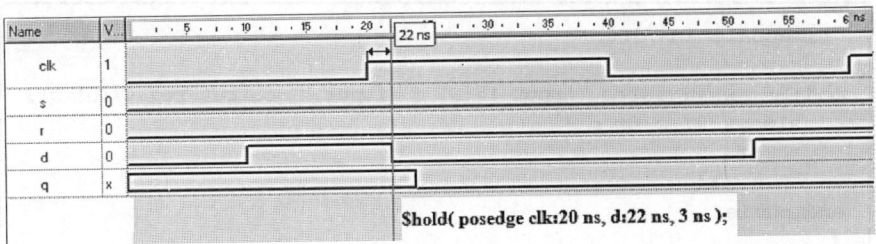

Figure 5.28   Hold Time Violation

```
`timescale 1ns/100ps

module d_ff ( input d, clk, s, r, output reg q, output q_b );

    specify
        $setuphold ( posedge clk, d, 5, 3 );
        $width (posedge r, 4);
        $width (posedge s, 4);
        $period (negedge clk, 43);
    endspecify

    always @( posedge clk or posedge s or posedge r )
        if( s ) q <= #4 1'b1;
        else if( r ) q <= #4 1'b0;
        else q <= #4 d;
    assign #2 q_b = ~q;

endmodule
```

Figure 5.29   Setup, Hold, Width, and Period Checks

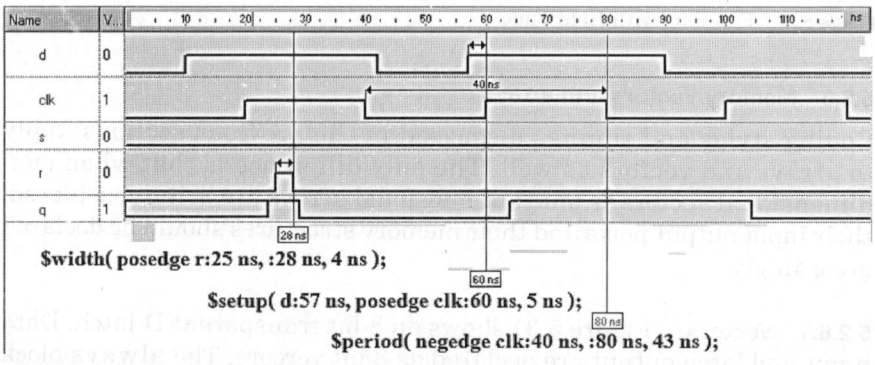

$width(posedge r:25 ns, :28 ns, 4 ns);

$setup(d:57 ns, posedge clk:60 ns, 5 ns);

$period(negedge clk:40 ns, :80 ns, 43 ns);

Figure 5.30   Setup, Width, and Period Violation

```
`timescale 1ns/100ps

module vector_latch (input [7:0] d, input c, output reg [7:0] q);
    always @( c or d )
        if( c )
            #4 q = d;
endmodule
```

Figure 5.31   8-bit Transparent D-Latch

Figure 5.32 shows an 8-bit register with a synchronous *rst* input, and a tri-state output controlled by *oe*. The structure of the **always** block is the same as those of the flip-flops of the previous section. This block assigns eight **0**s (i.e., *8'b0000_0000*) or the input *d* to the *internal_q*. A separate **assign** statement puts *internal_q* or eight **Z**s on the *q* output of the register. While *oe* is **0**, the *q* output will always be at the high-impedance state.

Another example of a register is shown in Fig. 5.33. This Verilog code corresponds to a sizable register whose size can be specified when instantiated. This register acts as an asynchronous active low reset input.

5.2.6.2   Arrays.   Figure 5.34 shows the block diagram of an unclocked memory with an address space of 8 and word length of 4. This memory has a *rd* read and a *wr* write inputs.

Figure 5.35 shows a sizable Verilog code that corresponds to this block diagram. Parameters used in this code are *M* and *N* for the length of the address and data, respectively. When written into, *mem* holds the data, and when being read, the addressed location drives *data*. When *rd* is not active, *data* is float.

5.2.6.3   Memory initialization.   Verilog has the **$readmemh** and **$readmemb** tasks for reading external data files and using them for initialization of memory blocks. The code of Figure 5.35, that describes a memory with bidirectional data lines, also shows memory initialization from an external file. Data from *mem.dat* external data file is read and the memory is initialized with this data.

As shown in this example, an **initial** statement (last statement in Fig 5.35) invokes **$readmemh** at time 0. At this time, data words from

```
`timescale 1ns/100ps

module vector_ff (input [7:0] d, input clk, rst, oe,
                  output [7:0] q);
    reg [7:0] internal_q;

    always @( posedge clk )
        if( rst )
            #4 internal_q <= 8'b0000_0000;
        else
            #4 internal_q <= d;

    assign q = oe ? internal_q : 8'bZ;
endmodule
```

Figure 5.32   An 8-bit Register with Tri-State Output

```
`timescale 1ns/100ps
module sizable_reg #(size) ( input [size-1:0] d, input clk, rst,
                                output reg [size-1:0] q );
    always @( posedge clk, negedge rst )
       begin
          if( ~rst )
             #4 q <= 0;
          else
             #4 q <= d;
       end

endmodule
```

Figure 5.33   A Sizable Register

mem.dat are read and are sequentially placed in the words of *mem*. Each line in *mem.dat* corresponds to a word of *mem* (starting from location 0), and data words are expected to be in hexadecimal. Optionally, invocation of **$readmemh** task may contain a range of the memory words to fill. The **$readmemb** task is similar to **$readmemh**, except that each line of the file read is expected to contain binary data equivalent to the word length of the memory.

5.2.6.4 Bidirectional memory.   Another feature of the memory of Fig. 5.35 is its bidirectionality. As shown, *data* is declared as **inout**. An **inout** bus is only considered as a **net**, and cannot be declared as a **reg**. Therefore, if such a port is to receive values in a procedural block, a temporary **reg** must be declared to contain data that is being put on the bidirectional bus. In our example, we have declared **reg** *temp* to hold data being read

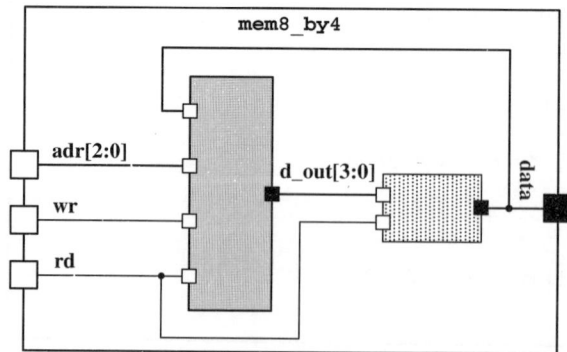

Figure 5.34   Memory Block Diagram

```
`timescale 1ns/100ps

module Memory_2Power_M_by_N #(parameter M=3, N=4)
        (input [M-1:0] adr, input rd, wr, inout [N-1:0] data);

    reg [N-1:0] mem [0:2**M-1];
    reg [N-1:0] temp;
    assign data = rd ? temp : `bZ;
    always @( data, adr, rd, wr )
       begin
           if ( wr )
               #4 mem[adr] = data;
           else if ( rd )
               #4 temp = mem[adr];
           else
               #4 temp = `bZ;
       end
    initial $readmemh("mem.dat", mem);

endmodule
```

Figure 5.35   Memory with **inout** and External File Initialization

from the memory and that is going out on the *data* bidirectional output. An **assign** statement puts *temp* on *data* if *rd* is **1**. Writing into this memory is more direct, and it is done by reading data directly from the bidirectional *data* lines of the memory.

In general, certain rules apply for declaring and using bidirectional (**inout**) lines. An **inout** port cannot be declared as a **reg**. For reading from an **inout** line, its name, as appears in its declaration, should be used in right-hand side expressions and/or conditional expressions. For assigning values to an **inout** line in a procedural block, a temporary **reg** of the same size must first be declared. Then by using an **assign** statement, the temporary **reg** must be conditionally assigned to the **inout** line. The condition should include all the read conditions, and if false, the **assign** statement should drive the **inout** line with all **Z**s. Procedural blocks wanting to output something through the **inout** line, must make assignments to the declared temporary **reg**. Figure 5.36 outlines these steps.

If a design does not require writing into an **inout** from a procedural block, the use of temporary **reg** is not necessary, but writing into the **inout** line must still be conditioned with all read conditions.

5.2.6.5 PLA modeling. Although programmable logic arrays (PLAs) can be described with combinational expressions, Verilog has PLA modeling

1. Declaring a bidirectional bus:
   inout [Size-1 : 0] *mybus*;
2. Reading the bidirectional bus:
   Some_LHS_Assigned = *mybus*;
3. Writing into the bidirectional bus by a procedural block:
   a. Declare a temporary **reg** as shown:
      reg [Size-1 : 0] *mytemp*;
   b. Put the temporary **reg** into the bidirectional bus:
      assign *mybus* = (*read_condition*) ? *mytemp* :'bZ;
   c. Write into the temporary **reg** representation of the bus:
      always . . . **begin**
          mytemp = *What_is_being_outputted*;
      end

Figure 5.36   Reading and Writing Bidirectional Lines

tasks that have a better correspondence with actual PLAs than using combinational modeling styles for describing them. Verilog PLA modeling tasks use a personality memory whose contents determine PLA fusing. The name of a PLA modeling task determines its array logic type.

Figure 5.37 shows a Verilog code that uses the contents of *mem* to define a PLA. The **$readmemb** task reads PLA personality data from *pla.dat* external file and loads it into *mem*. In this example, PLA configuration is determined by concatenation of system tasks **$asynch**, **$nand**, and **$array**.

```
`timescale 1ns/100ps

module pla ( in, out );
    input [7:0]in;
    output [3:0]out;
    reg [3:0]out;
    reg [1:8] mem[1:4];
    initial begin
        $readmemb ("pla.dat", mem);
        // Contents of pla.dat external file:
        //11000000
        //10110100
        //00001100
        //00000111
        $async$nand$array ( mem,
                  {in[7],in[6],in[5],in[4],in[3],in[2],in[1],in[0]},
                  {out[3],out[2],out[1],out[0]} );
    end
endmodule
```

Figure 5.37   PLA Verilog Example

A PLA task uses a task name that is a concatenation of the PLA type, its logic and its format. The arguments of a PLA task are the personalization memory, the input terms, and the output terms. Input and output terms must be concatenation of scalars, concatenated to form the size, the memory word-size, and the number of words respectively. The output terms must be declared as **reg**.

The type of a PLA can be synchronous or asynchronous (**$synch** or **$asynch**). The PLA logic is determined by its logic specification that can be **$and**, **$or**, **$nand**, or **$nor**. For a PLA with the **$nand** logic and **$array** format, each of its outputs is bit-by-bit NAND result of those input bits that have a corresponding **1** in the PLA personality data. With *pla.dat* contents as shown in comments of the Verilog code of Fig. 5.37, this PLA generates the following expressions on its outputs.

```
out3 = nand(in7, in6);
out2 = nand(in7, in5, in4, in2);
out1 = nand(in3, in2);
out0 = nand(in2, in1, in0);
```

5.3 Functional Registers

A register is defined as a group of flip-flops with a common clock. The term register also applies to a group of latches, but to differentiate them we will be very precise in using the correct terminology. We will refer to a group of latches by its size, e.g., octal latch, or nibble latch. We define a functional register as a group of flip-flops with a common clock and with some functionality, such as counting and shifting. Styles used for Verilog coding of functional registers are similar to those of flip-flops and registers described in the previous section. The difference is the added arithmetic or logical functionality to the code of functional registers.

5.3.1 Shift registers

Coding shift registers in Verilog is very similar to registers of the previous section. The addition of shift operations to these styles will be discussed here. Several shift register examples in this section take advantage of shift operators and concatenation.

5.3.1.1 Basic shifter.
Shown in Fig. 5.38 is a 4-bit shift register with load, reset, and shift capabilities. The l_r input controls left or right shifting. In either case, the vacated bit will be filled with the contents of s_in serial input.

Verilog code of Fig. 5.39 corresponds to this shift register. As shown, all shift register operations are synchronized with the positive edge of the circuit clock. The active high rst input causes $4'b0000$ to be loaded into the q output. The ld input performs parallel loading of d into q. If

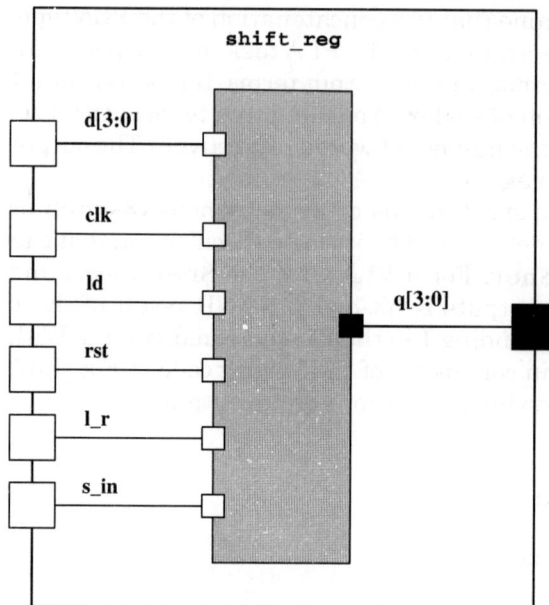

Figure 5.38   A Basic Shift Register

neither *rst* nor *ld* are active, *l_r* determines left or right shifting. Left shifting is performed by concatenating *s_in* to the right of *q[2:0]* forming a 4-bit vector that is clocked into *q[3:0]*. Similarly, for right shifting, a 4-bit vector is formed by concatenating *s_in* to the left of *q[3:1]*. In this case *s_in* goes into *q[3]*, and *q[3]*, *q[2]*, and *q[1]* go into *q[2]*, *q[1]*, and *q[0]*, respectively, causing the right shifting of *q*. All operations of this circuit are done in an **always** block that is sensitive to the positive edge of the circuit clock. A nesting of **if-else** statements handles assignments to the *q* output. The last **else** covers all conditions not mentioned in the previous **if** statements. This technique guarantees that all conditions are taken care of by the **if** statement, and leaves no room for ambiguities.

5.3.1.2   Universal shift register.   Figure 5.40 shows the Verilog code of a universal shift register with bidirectional *io*. The circuit has *s1*, *s0* inputs forming a 2-bit number ranging from 3 to 0. The shifter does nothing, shifts right, shifts left, or performs a parallel load depending on the value of *{s1, s0}*. The synchronous *rst* input resets the shifter.

Because this circuit has a bidirectional **inout** port, we have declared *q_int* to hold the shift register output at all times. Inside an **always** block that is sensitive to the positive edge of the clock, assignments to *q_int* take place. If *rst* is **1**, this variable is set to 0. Otherwise, a **case**

```
`timescale 1ns/100ps

module shift_reg (input [3:0] d, input clk, ld, rst, l_r, s_in,
                   output reg [3:0] q);
    always @( posedge clk ) begin
        if( rst )
            #5 q <= 4'b0000;
        else if( ld )
            #5 q <= d;
        else if( l_r )
            #5 q <= {q[2:0], s_in};
        else
            #5 q <= {s_in, q[3:1]};
    end

endmodule
```

Figure 5.39   Basic Shifter Verilog Code

```
`timescale 1ns/100ps

module shift_reg (input clk, rst, r_in, l_in, en, s1, s0,
                   inout [7:0] io);
    reg [7:0] q_int;
    assign io = (en) ? q_int : 8'bz;
    always @( posedge clk ) begin
        if( rst )
            #5 q_int = 8'b0;
        else
            case ( {s1,s0} )
                2'b01 : // Shift right
                q_int <= { r_in, q_int[7:1] };
                2'b10 : // Shift left
                q_int <= { q_int[6:0], l_in };
                2'b11 : // Parallel load
                q_int = io;
                default : // Do nothing
                q_int <= q_int;
            endcase
        end

endmodule
```

Figure 5.40   Universal Shift Register

statement uses *{s1, s0}* to decide value assigned to *q_int*. The **case** statement uses the **default** alternative to cover *{s1, s0}* of *2'b00* and all possible ambiguous values. This **default** alternative is like the **else** of the previous example, which guarantees that all conditions are accounted for.

The *io* **net** is the bidirectional port of this shift register. When *{s1, s0}* is *2'b11* (parallel loading the shift-register), *io* is read and put into *q_int*. For outputting through *io*, an **assign** statement assigns *q_int* or eight **Z**s to this bidirectional bus. If *en* is **1** *q_int* is put on *i0*, and if it is **0**, *8'bZ* drives *io*. An external device wanting to drive *io* from outside of this module can only do so when *en* is **0**.

5.3.1.3 Separate register and combinational blocks.

A style of coding that is often used for describing sequential circuits with complex functionalities is to use separate combinational and sequential blocks. We use this style of coding for a shift register circuit that can shift its contents a specified number of positions to the right or to the left. The block diagram of the shift register is shown in Fig. 5.41.

The shifter shown shifts right or left when *sr* or *sl* is active. The number of shifts is determined by *s_cnt*. The shifter uses **0**s for filler for vacated shift positions. The *ld* input loads *d_in* into the shift register. The *rst* input directly affects the register block and provides a synchronous reset.

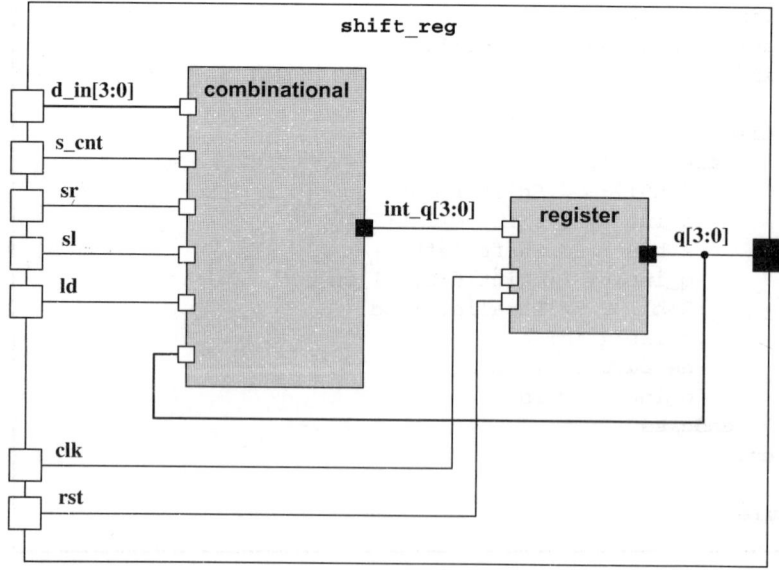

Figure 5.41   Multi-bit Shifter with Separate Register Block

Figure 5.42 shows the Verilog code of the shifter of Fig. 5.41. The *combinational* **always** block is sensitive to all inputs that affect the *int_q* output of this block. This includes *q* that is the output of our shift register and the output of the *register* block that is fed back into the *combinational* block. A nesting of **if-else** statements sets the *int_q* output of the *combinational* **always** block to *d_in*, to *q* shifted right *s_cnt* positions, to *q* shifted left *s_cnt* positions, or to *q*. As before, the last **else** guarantees that the output of the combinational block receives a value no matter what input combinations occur.

The *register* block of Fig. 5.42 takes *int_q*, which is the output of the *combinational* block, as input and clocks it into the shift register output. This block is also responsible for resetting the shift register. In this example, synchronous resetting is used. If asynchronous resetting were to be used, **posedge** *rst* would have to be included in the sensitivity list of the **always** block of the *register* block.

5.3.2 Counters

Coding styles for describing counters is like those for shift registers, except that arithmetic add (+) and subtract (−) operations must be used for count-up and count-down. We dedicate this section to describing counters as well as presenting coding techniques that are general and can be applied to other sequential circuits as well.

```
`timescale 1ns/100ps

module shift_reg( input [3:0] d_in, input clk, sr, sl, ld, rst,
                  input [1:0] s_cnt, output reg [3:0] q);
    reg [3:0] int_q;
    always @( d_in, q, s_cnt, sr, sl, ld ) begin: combinational
        if( ld )
            int_q = d_in;
        else if( sr )
            int_q = q >> s_cnt;
        else if( sl )
            int_q = q << s_cnt;
        else int_q = q;
    end
    always @( posedge clk ) begin: register
        if (rst) q <= 0;
        else q <= int_q;
    end
endmodule
```

Figure 5.42 Shifter Verilog Code

5.3.2.1 Up-down counter. Figure 5.43 shows a 4-bit binary up-down counter. The counter has a synchronous *rst* input and a parallel load enable, *ld*. If *u_d* is **1** count-up, and if it is **0** count-down is done. When counting up, when *q* reaches **1111** adding a 1 and capturing the most significant four bits causes the count sequence to roll back to **0000** and continue the count from there.

5.3.2.2 Gray code counter. A Gray code counter, or any count sequence for which arithmetic operators cannot be used, can be implemented by a table look-up for building a conversion function. We develop our Gray code counter by use of a table look-up using an external memory file.

The next count is looked up from a memory of sixteen 4-bit entries. Each memory location contains the address of the location treated as a Gray code number, plus 1. For example, location 7 (**0111**) contains **0101**. Note that **0111** is 5 in Gray and its next count up is **0101** that is the Gray code for 6.

For the implementation of this counter, we use the style of coding in which all the combinational functions are done in one **always** block and sequential parts in another. Figure 5.44 corresponds to this style of coding. In this figure, the *combinational* block takes *ld*, *q*, and *d_in* as input. If *ld* is **1** then *im_q* becomes equal to *d_in*. If *ld* is **0**, then *q*, treated as a Gray code, is incremented and appears on *im_q*. The *register* block takes *im_q* and clocks it into the *reg* output of the counter.

Figure 5.45 shows the corresponding Verilog code. In an **initial** block, the *mem.dat* external file is read into *mem* **reg**. In the *combinational*

```
`timescale 1ns/100ps

module counter (input [3:0] d_in, input clk, rst, ld, u_d,
                output reg [3:0] q);
    always @( posedge clk ) begin
        if ( rst )
            q = 4'b0000;
        else if ( ld )
            q = d_in;
        else if ( u_d )
            q = q + 1;
        else
            q = q - 1;
    end
endmodule
```

Figure 5.43 Up-Down Counter

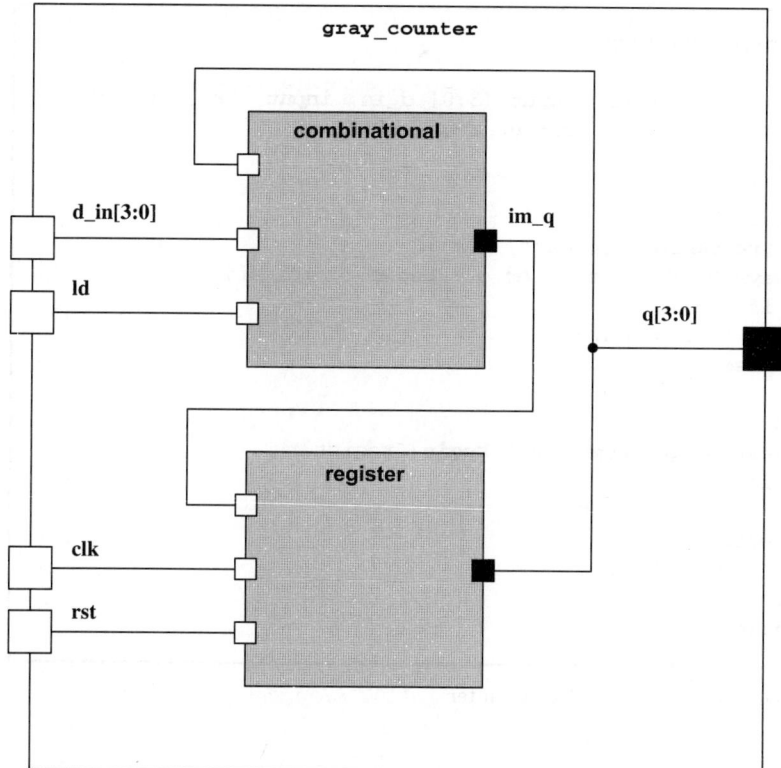

Figure 5.44 Gray Code Counter Diagram

always block, the present count value *q* is read and used as an index for *mem*. The value read from *mem* is put into *im_q*. The *register* **always** block handles clocking and synchronous resetting.

The table look-up of Fig. 5.45 makes this description adaptable to other counting sequences. Without having to recompile this code, and by just changing the contents of *mem.dat*, other count sequences can be implemented. Furthermore, because we have separated combinational and sequential parts of this design, changes in clocking or resetting mechanisms only affect the *register* block, and the *combinational* **always** block remains intact.

5.3.3 LFSR and MISR

A linear feedback shift register (LFSR) is used for pseudo random number generation. An LFSR is a shift register with feedback and XOR gates in its feedback or shift path. The initial content of the register is referred to as *seed*, and the position of XOR gates is determined by the

```
`timescale 1ns/100ps

module gray_counter (input [3:0] d_in, input clk, rst, ld,
                       output reg [3:0] q);
    reg [3:0] mem[0:15];
    reg [3:0] im_q;
    initial
        $readmemb("mem.dat", mem);
    always @( d_in or ld or q ) begin: combinational
        if( ld )
            im_q = d_in;
        else
            im_q = mem[q];
    end
    always @( posedge clk ) begin: register
        if( rst )
            q <= 4'b0000;
        else
            q <= im_q;
    end
endmodule
```

Figure 5.45 Verilog Code of a Gray Counter

polynomial (*poly*) of the LFSR. A multiple input signature register
(MISR) is like an LFSR, with parallel input and output. A MISR is used
for signature generation of multi-bit input vectors.

5.3.3.1 LFSR. Figure 5.46 shows an LFSR made of D-type flip-flops and
XOR gates in its shift path. The position of XOR gates determine the
poly of this circuit, which is *poly* = **10101**. The *seed*, which is the initial
value of the register, affects set and reset inputs of the individual flip-
flops of the shift register. The LFSR *seed* and *poly* determine bit values
that are generated on the serial output of the circuit (*sout*), as serial
input bits (*sin*) are being shifted in.

Figure 5.47 shows the LFSR Verilog code. This code describes the
structure of LFSR using XOR gates and positive edge D-type flip-flops
with asynchronous set and reset inputs. The *structural_lfsr* module
wires four flip-flops, two XOR gates in between flip-flops, and set and
reset inputs of the flip-flops are wired according to the *seed* parameter
of this module. When *init* becomes **1**, the LFSR seed is asynchronously
loaded into the register (four flip-flops).

We have used the Verilog replication construct *array of instances* and
for wiring the flip-flops of this LFSR. The *init* input is replicated four times
and ANDed with the 4-bit *seed* vector to form the set inputs of the flip-flops.

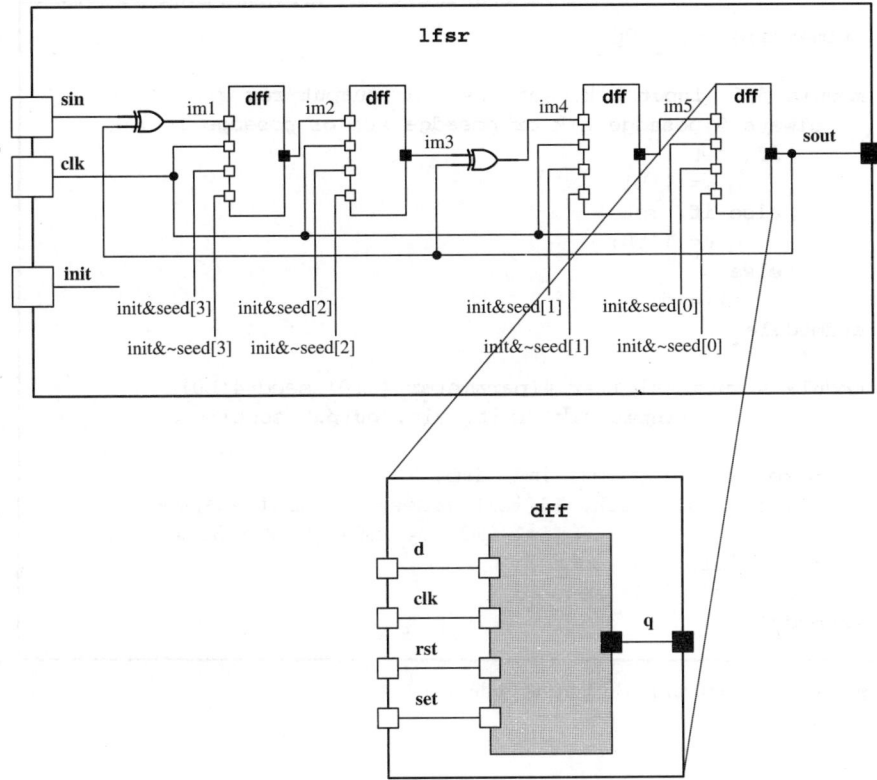

Figure 5.46   An LFSR with **10101** Polynomial

The reset inputs become active when *init* is **1** and a corresponding bit of *seed* is **0**.

Figure 5.48 shows a generic Verilog code for an LFSR. This behavioral code uses *poly* and *seed* parameters. The *poly* 4-bit parameter specifies where between flip-flops XOR gates are inserted. As in the *structural_lfsr* of Fig. 5.47, *seed* is the initial value for the register of LFSR.

An **always** block in the *behavioral_lfsr* module of Fig. 5.48 handles initialization, LFSR configuration, and shift-in and shift-out of data. The *im_data* **reg** holds the contents of the LFSR register. In the shift mode, the feedback from the right-most bit of the register (*im_data[0]*) is XORed with *sin* serial input, and is clocked into the left-most bit of the LFSR. Inputs of all remaining LFSR bits are either taken directly from flip-flops to their left (see Fig. 5.46) or from the XOR result of the feedback and the output of the flip-flop to their left. The XOR result will be taken if the corresponding *poly* bit is **1**. For example if *poly[2]* is **1**, *im_data[0]* is selected and XORed with *im_data[3]* and is used for input of *im_data[2]*.

```verilog
`timescale 1ns/100ps

module dff (input clk, set, rst, d, output reg q);
   always @(posedge clk or posedge set or posedge rst)
      if( set )
         q <= 1'b1;
      else if( rst )
         q <= 1'b0;
      else
         q <= d;
endmodule

module structural_lfsr #(parameter [3:0] seed=4'b0)
            (input clk, init, sin, output sout);

   wire im1, im2, im3, im4, im5;
   dff ff[3:0] ( clk, {4{init}}&seed, {4{init}}&~seed,
                     {im1,im2,im4,im5}, {im2,im3,im5,sout} );
   xor ( im1, sin, sout );
   xor ( im4, im3, sout );
endmodule
```

Figure 5.47   Structural LFSR Verilog Code

```verilog
`timescale 1ns/100ps

module behavioral_lfsr #(parameter [3:0] poly=0, seed=0)
            (input clk, init, sin, output reg sout );
   reg [3:0] im_data;
   always @( posedge clk or posedge init )
      begin
         if( init )
            im_data = seed;
         else
            im_data = { sin^ im_data[0],
            im_data[3:1] ^ (poly[2:0] & {3{im_data[0]}}) };
         sout = im_data[0];
      end
endmodule
```

Figure 5.48   Behavioral LFSR Code

5.3.3.2 Multiple input signature register.
A MISR is used for signature generation and data compression. Over a period of several clocks, parallel data into a MISR are compressed with the existing MISR data. The final data depends on the MISR initial data (*seed*) and its XOR and feedback structure (*poly*).

Figure 5.49 shows a MISR with a configurable polynomial (*poly*). The circuit has a reset input that initializes it to **0000**. This initial value is considered as the seed of this MISR example.

The Verilog code of Fig. 5.50 corresponds to the hardware of Fig. 5.49. An **always** block in this code handles resetting and signature generation. The generation of the signature is based on the input *poly* that configures the feedback ($d\_out[0]$) connections to the XOR gates that are between the flip-flops.

This configuration is done by the expression that appears on the right-hand side of d_out. In this expression parallel data input bits (d_in) are XORed with right shifted output data ($\{1'b0, d\_out[3:1]\}$), and are then selectively XORed with feedback coming from the right-most bit of the shift register. Selecting feedback from the right-most bit to affect input logic of a register bit is determined by the bits of *poly*. This selection is done by ANDing *poly* bits with the replication of the feedback from the right-most bit.

5.3.4 Stacks and queues

A combination of styles presented for register modeling and memory read and write operations can be used for describing stacks and queues. A queue has a memory block and read and write pointers. Read and write pointers are described as registers and counters that provide address pointers for the queue memory. Combinational logic blocks are used for providing full and empty indicators for the queue memory.

5.3.4.1 FIFO queue.
A first in first out (FIFO) queue is a queue of data such that the data that is written into it first, is read from it first. Figure 5.51 shows the block diagram of a FIFO queue.

The *pointer* block provides read and write pointers according to read and write operations into the queue. The *count* block keeps track of the number of data in the queue and issues *empty* and *full* flags. These flags are generated by combinational blocks using the present count of data as input. The *read* block uses the read pointer to read from the queue memory, and the *write* block uses the write pointer to write into this memory. The *read* and *write* blocks share the *fifo_ram* memory. Read and write operations are synchronized with the circuit clock.

The Verilog code of the FIFO queue is shown in Fig. 5.52. As shown, four **always** blocks that are sensitive to the positive edge of the clock handle writing (*write*), reading (*read*), updating read and write pointers (*pointer*),

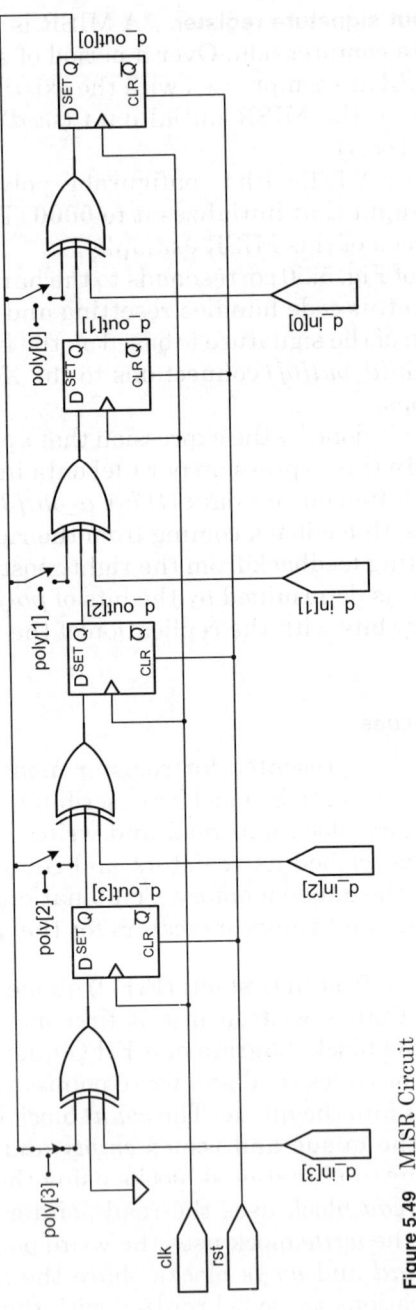

Figure 5.49 MISR Circuit

```
`timescale 1ns/100ps

module #(parameter [3:0] poly=0) misr (input clk, rst,
        input [3:0] d_in, output reg [3:0] d_out );
    always @( posedge clk )
        if( rst )
            d_out =4'b0000;
        else
            d_out = d_in ^ ({4{d_out[0]}} & poly)
                        ^ {1'b0, d_out[3:1]};
endmodule
```

Figure 5.50   MISR Verilog Code

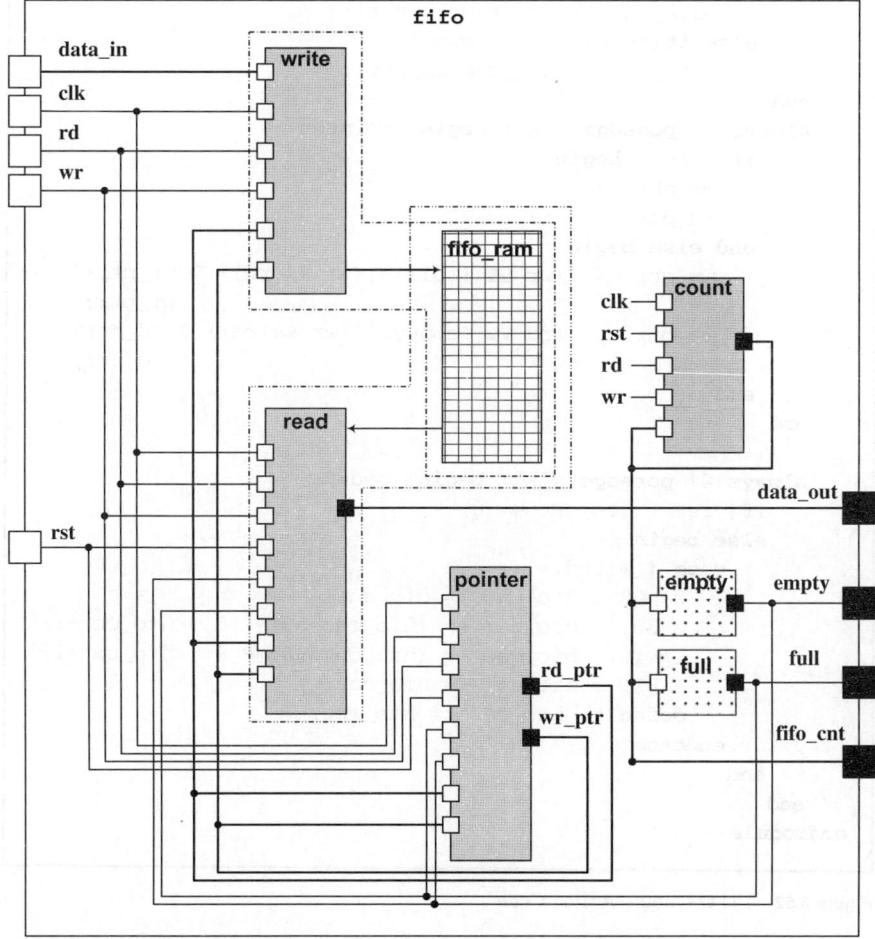

Figure 5.51   FIFO Block Diagram

```verilog
module fifo (input [7:0] data_in, input clk, rst, rd, wr,
             output empty, full, output reg [3:0]fifo_cnt,
             output reg [7:0] data_out);
    reg [7:0] fifo_ram[0:7];
    reg [2:0] rd_ptr, wr_ptr;
    assign empty = (fifo_cnt==0);
    assign full = (fifo_cnt==8);
    always @( posedge clk ) begin: write
        if(wr && !full)
            fifo_ram[wr_ptr] <= data_in;
        else if(wr && rd)
            fifo_ram[wr_ptr] <= data_in;
    end
    always @( posedge clk ) begin: read
        if(rd && !empty)
            data_out <= fifo_ram[rd_ptr];
        else if(rd && wr && empty)
            data_out <= fifo_ram[rd_ptr];
    end
    always @( posedge clk ) begin: pointer
        if( rst ) begin
            wr_ptr <= 0;
            rd_ptr <= 0;
        end else begin
            wr_ptr <= ((wr && !full)||(wr && rd)) ? wr_ptr+1 :
                                                     wr_ptr;
            rd_ptr <= ((rd && !empty)||(wr && rd)) ? rd_ptr+1 :
                                                     rd_ptr;
        end
    end

    always @( posedge clk ) begin: count
        if( rst ) fifo_cnt <= 0;
        else begin
            case ({wr,rd})
                2'b00 : fifo_cnt <= fifo_cnt;
                2'b01 : fifo_cnt <= (fifo_cnt==0) ? 0 : fifo_cnt-1;
                2'b10 : fifo_cnt <= (fifo_cnt==8) ? 8 : fifo_cnt+1;
                2'b11 : fifo_cnt <= fifo_cnt;
                default: fifo_cnt <= fifo_cnt;
            endcase
        end
    end
endmodule
```

Figure 5.52 FIFO Queue Verilog Code

and keeping the FIFO count (*count*). Concurrent with these blocks two **assign** statements issue *empty* and *full*.

The *write* block writes into *fifo_ram* if it is not full. If both *rd* and *wr* inputs are active (reading at the same time as writing), *full* is not checked and memory is written into. If none of these conditions hold, then memory is left intact. The *read* block reads from *fifo_ram* if it is not empty. If it is empty and both *rd* and *wr* inputs are active, data from the present pointer location is read into *data_out*. This output **reg** is left intact if neither read conditions are satisfied.

The *pointer* block implements two counters for read and write pointers. Separate from these pointers, the *count* **always** block performs incrementing and decrementing *fifo_cnt* depending on write and read operations being done. The *count* block uses a **case** statement with **default**, which handles ambiguous values on *wr* and *rd*. In this case, *fifo_cnt* is left intact.

5.4   State Machine Coding

Coding styles presented so far can further be generalized to cover finite state machines of any type. This section shows coding for Moore and Mealy state machines. The examples we will use are simple sequence detectors, yet they represent coding for complex control-heavy digital circuits, or the controller part of an RT level design.

5.4.1   Moore machines

A Moore machine is a state machine in which all outputs are fully synchronized with the circuit clock. In the state diagram form, each state of the machine specifies its output(s) independent of circuit inputs. In the Verilog code of a Moore state machine, only circuit state variables participate in the output expression of the circuit.

Figure 5.53 shows a **101** Moore sequence detector with its corresponding block diagram related to its Verilog coding. The machine searches for **101** on its input and when received, the output of the circuit becomes **1** and remains at this level for a complete clock period. As shown in the state diagram, when the machine reaches the *got101* state, its output becomes **1**.

The block diagram of the Verilog coding that will be used for this machine is also shown in Fig. 5.53. An **always** block that handles state transitions and clocking generates *current* state of the machine. This variable is used by an **assign** statement that generates the *z* output of the circuit.

Figure 5.54 shows the Verilog code of *moore_detector*. We have used a **localparam** declaration to assign values to the states of the machine.

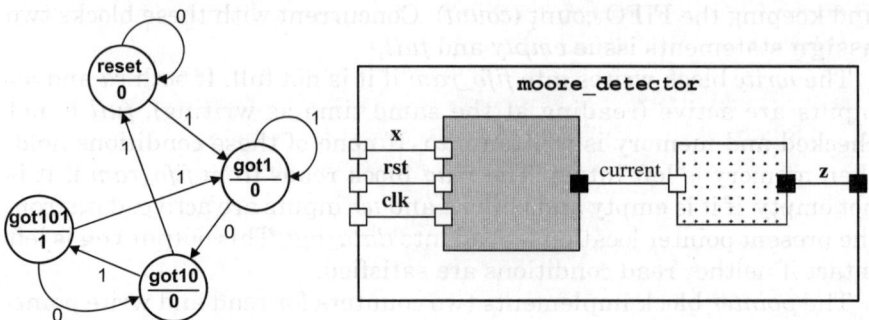

Figure 5.53   A Moore 101 Detector

Because our machine has four states, 2-bit parameters are used for the state names. Furthermore, the declaration part of the *moore_detector* module declares *current* as a 2-bit **reg**. This variable is used for holding the current state of the machine.

The **always** block of Fig. 5.54 implements a positive edge trigger sequential block with a synchronous reset (*rst*) input. If *rst* is active, *current* is set to *reset*, otherwise, a **case** statement assigns next state values to *current*. Next states of the machine are decided by the *current* state that is the **case** expression, and input values.

Each state of the machine is implemented by a **case** alternative, and its next state transitions are implemented by **if** statements conditioned by the *x* input of the circuit. Figure 5.55 shows a correspondence between the *got10* state of the machine and its Verilog coding. This state branches out to *got101* or *reset* depending on *x*. The output of the circuit is implemented by a separate **assign** statement that puts a **1** on *z* when *current* is *got101*.

Because this is a Moore machine, the condition for asserting the output of the circuit only includes the *current* variable, and circuit input(s) are not included. Figure 5.56 shows another Moore machine example. This machine searches for **110** or **101** sequences on its *x* input. The search allows overlapping sequences.

The Verilog code of the Moore machine of Fig. 5.56 is shown in Fig. 5.57. We are using `` `define `` directives for assigning values to the state names. Note that using names is only for readability purposes, and instead of using `` `define `` or **localparam**, as in the previous example, state values could be used in the **case** statement of this Verilog code.

The Verilog code shown here implements a state machine with asynchronous active high reset (*rst*) input. For this purpose, **posedge** *rst* is included in the sensitivity list of the **always** block. In this **always** block, the last **case** alternative is the **default** case that accounts for

```
`timescale 1ns/100ps

module moore_detector (input x, rst, clk, output z);
    localparam [1:0]
        reset=0, got1=1, got10=2, got101=3;
    reg [1:0] current;
    always @( posedge clk ) begin
        if( rst ) current <= reset;
        else case ( current )
            reset: begin
                if( x==1'b1 ) current <= got1;
                else current <= reset;
            end
            got1: begin
                if( x==1'b0 ) current <= got10;
                else current <= got1;
            end
            got10: begin
                if( x==1'b1 ) current <= got101;
                else current <= reset;
            end
            got101: begin
                if( x==1'b1 ) current <= got1;
                else current <= got10;
            end
            default: begin
                current <= reset;
            end
        endcase
        end

    assign z = (current==got101) ? 1 : 0;
endmodule
```

Figure 5.54   Moore Machine Verilog Code

ambiguous values of *current*, as well as those values that are not specified as valid states of this machine. Because we are using three state variables (**reg** *[2:0] current*), eight states are allowed from which only six are specified. The other two states are invalid states and are handled by the **default** case alternative.

Assignment of values to the *z* output is handled by the **assign** statement in Fig. 5.57. This output becomes **1** when *current* is `got101` or `got110`.

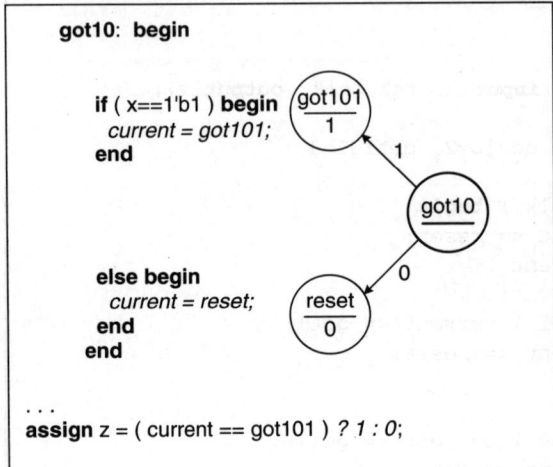

```
got10:  begin

    if ( x==1'b1 ) begin
        current = got101;
    end

    else begin
        current = reset;
    end
end

...
assign z = ( current == got101 ) ? 1 : 0;
```

Figure 5.55 Verilog Coding Correspondence with the
got10 State

5.4.2 Mealy machines

A Mealy machine is different from a Moore machine in that its output(s)
depend on its current state and inputs while in that state. State transitions, clocking, and resetting the machine are not different from those
of a Moore machine, and the same coding techniques are used for
describing them.

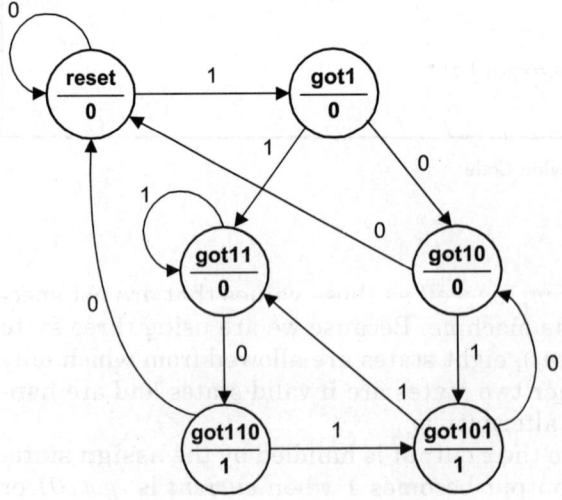

Figure 5.56 Moore Machine Detecting **110/101**

```verilog
`timescale 1ns/100ps

`define   reset    3'b000
`define   got1     3'b001
`define   got10    3'b010
`define   got11    3'b011
`define   got101   3'b100
`define   got110   3'b101

module moore_detector3 (input x, rst, clk, output z);
    reg [2:0] current;
    always @( posedge clk or posedge rst ) begin
        if( rst ) current = `reset;
        else
            case ( current )
                `reset:
                    if( x==1'b1 ) current <= `got1;
                    else current <= `reset;
                `got1:
                    if( x==1'b0 ) current <= `got10;
                    else current <= `got11;
                `got10:
                    if( x==1'b1 ) current <= `got101;
                    else current <= `reset;
                `got11:
                    if( x==1'b1 ) current <= `got11;
                    else current <= `got110;
                `got101:
                    if( x==1'b1 ) current <= `got11;
                    else current <= `got10;
                `got110:
                    if( x==1'b1 ) current <= `got101;
                    else current <= `reset;
                default:
                    current <= `got101;
            endcase
    end

    assign z = (current == `got101 || current == `got110);
endmodule
```

Figure 5.57   Verilog Code of Moore Machine Detecting 110/101

Figure 5.58 shows a **101** Mealy sequence detector and its corresponding Verilog code block diagram. This circuit has a synchronous *rst* input that resets the machine to its reset state.

The Verilog code of Fig. 5.59 corresponds to this Mealy machine. A 2-bit **localparam** construct is used for defining the states of this machine. Because the machine has three states and two state variables are used to represent them, one combination (i.e., **11**) of the state variables becomes unused. As in the previous example, the **default** in the **case** statement of this Verilog code handles this unspecified combination and ambiguous values that may appear on *current*.

The coding of the states and output of this machine are illustrated in Fig. 5.60. Each state is specified by a **case** alternative of a **case** statement for which *current* is its **case** expression. Transitions to the next states of the machine are handled by **if-else** statements. The output of the machine is set to **1** using an **assign** statement that uses a conditional expression on its right-hand side. This conditional expression uses the circuit input as well as the current state of the machine.

5.4.3   Huffman coding style

The Huffman model for a digital system characterizes it as a combinational block with feedbacks through an array of registers. According to the Huffman model, Verilog coding of digital systems uses an **always** statement for describing the register part and another concurrent statement for describing the combinational part. This coding style and the Moore machine example that we will use in this section are shown in Fig. 5.61. As shown, the *combinational* block uses *x* and *p_state* as input and generates *z* and *n_state*. The *sequential* block clocks *n_state* into *p_state*, and resets *p_state* when *rst* is active.

Figure 5.62 shows the Verilog code of the state diagram of Fig. 5.61 according to the partitioning shown. In this code a **localparam** declaration

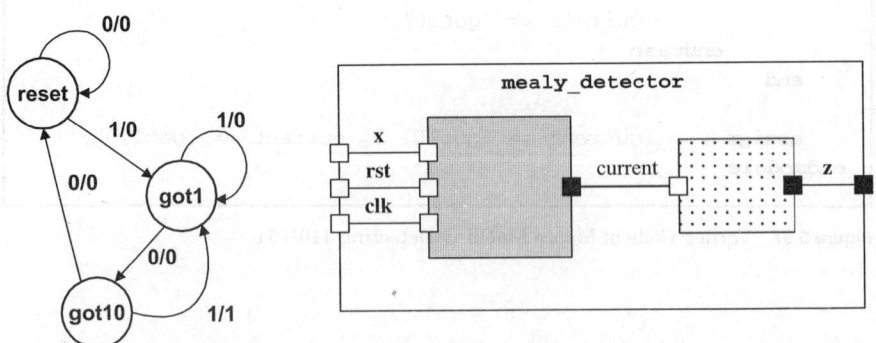

Figure 5.58   A **101** Mealy Machine

```verilog
`timescale 1ns/100ps

module mealy_detector2 (input x, rst, clk, output z);

    localparam [1:0]
        reset = 0, // 0 = 0 0
        got1 = 1, // 1 = 0 1
        got10 = 2; // 2 = 1 0

    reg [1:0] current;

    always @( posedge clk ) begin
        if (rst) current <= reset;
        else case ( current )
            reset:
                if( x==1'b1 ) current <= got1;
                else current <= reset;
            got1:
                if( x==1'b0 ) current <= got10;
                else current <= got1;
            got10:
                if( x==1'b1 ) current <= got1;
                else current <= reset;
            default:
                current <= reset;
        endcase
    end

    assign z = ( current==got10 && x==1'b1 ) ? 1'b1 : 1'b0;

endmodule
```

Figure 5.59   Verilog Code for a **101** Mealy Machine

declares the states of the machine. Following this declaration, n_state and p_state variables are declared as 2-bit **reg**s that hold values corresponding to the states of the **101** Moore detector. The *combinational* **always** block follows this **reg** declaration. Since this a purely combinational block, it is sensitive to all its inputs, namely x and p_state. Immediately following the block heading, n_state and z are set to their inactive or reset values. This is done so that these variables are always refreshed with new values and never retain their old values. As discussed before, retaining old values implies latches, which is not what we want in our combinational block.

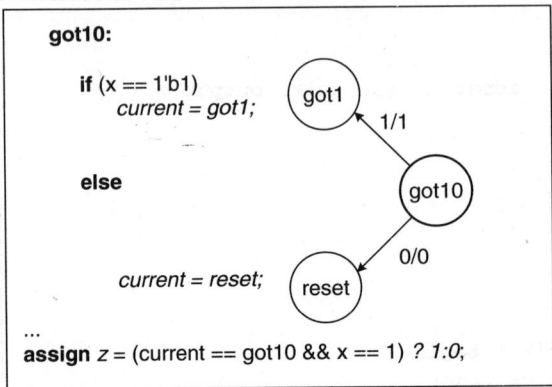

Figure 5.60   Mealy State Coding

The body of the combinational **always** block of Fig. 5.62 contains a **case** statement that uses the *p_state* input of the **always** block for its *case expression*. This expression is checked against the states of the Moore machine. As in the other styles discussed before, this **case**-statement has *case alternatives* for *reset, got1, got10,* and *got101* states.

In a block corresponding to a particular *case alternative*, based on input values, values are assigned to *n_state* and *z* output. Unlike the other styles where *current* is used both for the present and next states, here we use two different variables, *p_state* and *n_state*.

The next procedural block shown in Fig. 5.62 handles the register (sequential) part of the Huffman model of Fig. 5.61. In this part, *n_state*

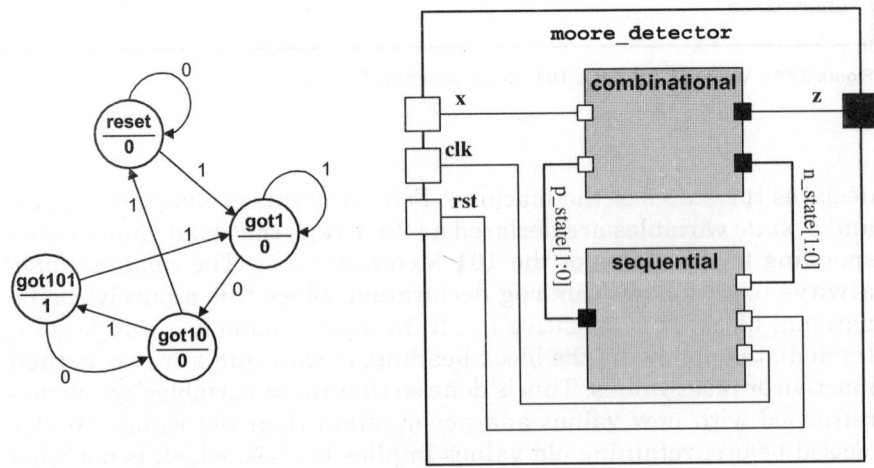

Figure 5.61   Huffman Style of Coding a State Machine

```verilog
`timescale 1ns/100ps

module moore_detector4 (input x, rst, clk, output reg z);

    localparam [1:0]
        reset=2'b00, got1=2'b01,
        got10=2'b10, got101=2'b11;

    reg [1:0] p_state, n_state;

    always @( p_state or x ) begin:combinational
        n_state = reset;
        z = 1'b0;
        case ( p_state )
            reset:
                begin
                    if( x==1'b1 ) n_state = got1;
                    else n_state = reset;
                    z = 1'b0;
                end
            got1:
                begin
                    if( x==1'b0 ) n_state = got10;
                    else n_state = got1;
                    z = 1'b0;
                end
            got10:
                begin
                    if( x==1'b1 ) n_state = got101;
                    else n_state = reset;
                    z = 1'b0;
                end
            got101:
                begin
                    if( x==1'b1 ) n_state = got1;
                    else n_state = got10;
                    z = 1'b1;
                end
            default:
                begin
                    n_state = reset;
                    z = 1'b0;
                end
        endcase
    end
```
(Continued)

Figure 5.62   Moore Detector Verilog Code According to Huffman Model

```
    always @( posedge clk ) begin:sequential
      if( rst ) p_state <= reset;
      else p_state <= n_state;
   end

endmodule
```

Figure 5.62   Moore Detector Verilog Code According to Huffman Model (*Continued*)

is treated as the register input and *p_state* as its output. On the positive edge of the clock, *p_state* is either set to the *reset* state (**00**) or is loaded with contents of *n_state*. Together, *combinational* and *sequential* blocks describe our state machine in a very modular fashion.

The advantage of this style of coding is in its modularity and defined tasks of each block. State transitions are handled by the *combinational* block and clocking is done by the *sequential* block. Changes in clocking, resetting, enabling, or presetting the machine only affect the coding of the *sequential* block. If we were to change the synchronous resetting to asynchronous, the only change we had to make was adding **posedge** *rst* to the sensitivity list of the *sequential* block.

5.4.4   A more modular style

For a design with more input and output lines and more complex output logic, the *combinational* block may further be partitioned into a block for handling transitions and another for assigning values to the outputs of the circuit. For coding both of these blocks, it is necessary to follow the rules discussed for combinational blocks in the previous chapter. Figure 5.63 shows a block diagram of this style and a Mealy sequence detector that we will use for illustrating its Verilog coding.

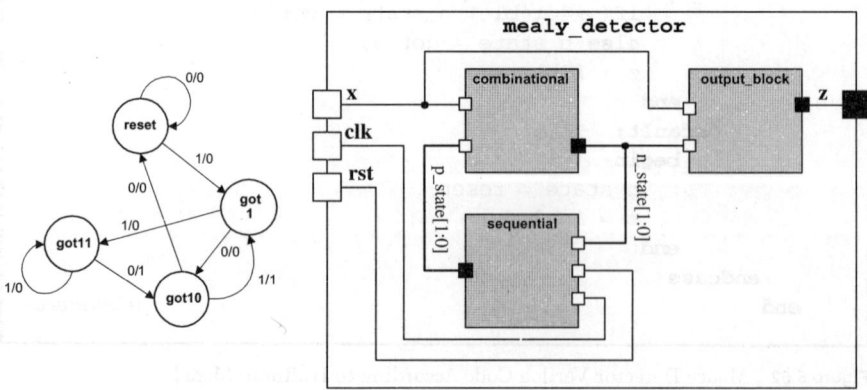

Figure 5.63   Using Three Separate Blocks for Describing a State Machine

Figure 5.64 shows the Verilog code for a Mealy machine that detects a sequence of **110** on its *x* input. This code uses two separate blocks for assigning values to *n_state* and the *z* output. In a situation like what we have in which the output logic is fairly simple, a simple **assign** statement could replace the *output_block* procedural block. In which case, *z* would have to be a **net** and not a **reg**.

The examples discussed above, in particular, the last two styles, show how combinational and sequential coding styles can be combined to describe very complex digital systems.

5.4.5   A ROM based controller

Instead of coding state transitions and output values using **case** and/or **if-else** statements, a single memory lookup can be used for representing the combinational part of a state machine. The memory lookup uses the present state of the machine and its inputs for address, and reads the next state and the controller outputs as data from the memory. The block diagram for this style of coding is the same as that of Fig. 5.61, except that the *combinational* part is implemented as a memory.

The example we use for illustrating the coding of a ROM-based controller is the Mealy machine of Fig. 5.63. The corresponding Verilog code is shown in Fig. 5.65. The memory that implements the combinational part of our circuit, as a ROM-based logic or a logic array, is loaded with the contents of *mealy.dat* at time **0**. This is done in an **initial** block using the **$readmemb** task. Combinational reading from the memory is done by an **assign** statement. The address of *mem* is formed by concatenation of the *x* input and *p_state*, and the data read from the memory is assigned to the concatenation of the *z* output and *n_state*.

Contents of the memory at any given location are circuit output(s) concatenated with the next-state of the machine. Figure 5.66 shows *mealy.dat* file for implementing the Mealy machine of Fig. 5.63.

As in the previous examples, the sequential part of our Mealy machine example is implemented by an **always** block that is sensitive to the positive edge of the clock. This block clocks *n_state* that is read from the memory into *p_state* that becomes the address input of the memory.

5.5   Sequential Synthesis

The process of synthesis involves describing a hardware in an acceptable form for the synthesis tool to recognize and then specifying a target library representing available low-power components to map to. The target library has combinational and sequential components. Exactly how a synthesizable Verilog input description translates to hardware depends on the specific target library. For example, if an input description involves a latch and the target library does not have a latch, then

```verilog
`timescale 1ns/100ps
module mealy_detector6 (input x, en, clk, rst, output reg z);

   localparam [1:0]
      reset=2'b00, got1=2'b01, got10=2'b10, got11=2'b11;

   reg [1:0] p_state, n_state;

   always @( p_state or x ) begin:combinational
      case ( p_state )
         reset:
            if( x==1'b1 ) n_state = got1;
            else n_state = reset;
         got1:
            if( x==1'b0 ) n_state = got10;
            else n_state = got11;
         got10:
            if( x==1'b1 ) n_state = got1;
            else n_state = reset;
         got11:
            if( x==1'b1 ) n_state = got11;
            else n_state = got10;
         default:
            n_state = reset;
      endcase
   end

   always @( p_state or x ) begin:output_block
      case ( p_state )
         reset:
            z=1'b0;
         got1:
            z=1'b0;
         got10:
            if( x==1'b1 ) z=1'b1;
            else z=1'b0;
         got11:
            if( x==1'b1 ) z=1'b0;
            else z=1'b1;
         default:
            z=1'b0;
      endcase
   end
```

(Continued)

Figure 5.64 A Mealy Machine Using Three Procedural Blocks

```
    always @( posedge clk ) begin:sequential
        if( rst ) p_state <= reset;
        else if( en ) p_state <= n_state;
    end

endmodule
```

Figure 5.64 A Mealy Machine Using Three Procedural Blocks (*Continued*)

a latch will be build using gates or logic functions that are available in the target library.

Verilog models described in this chapter, except those with **initial** statements loading a memory, are synthesizable. In this section, we will go back and look at several typical styles of coding and discuss the kind of hardware that they synthesize to.

5.5.1   Latch models

Except for the delay values, the latch Verilog description of Fig. 5.14 is synthesizable. If a target hardware library contains a D-latch, it will be used for mapping this description, otherwise it will be built by wiring existing target hardware parts.

```
`timescale 1ns/100ps

module mealy_detector7 (input x, clk, rst, output z);
    localparam [1:0]
        reset=2'b00, got1=2'b01, got10=2'b10, got11=2'b11;

    reg [1:0] p_state;
    wire [1:0] n_state;
    reg [2:0] mem[0:7];
    initial
        $readmemb( "mealy.dat", mem );
    assign { z, n_state } = mem[{ x, p_state }];
    always @( posedge clk ) begin:sequential
        if( rst ) p_state <= reset;
        else p_state <= n_state;
    end

endmodule
```

Figure 5.65 ROM-Based State Machine Coding

```
000
010
000
110
001
011
101
011
```

Figure 5.66   *mealy.dat* Memory File

Figure 5.67 shows an Altera field programmable gate array (FPGA) logic element that has several logic gates, a look-up table, and a flip-flip. The implementation of a D-latch is highlighted in this diagram. The gray areas are those parts of the logic element (LE) that are actually used for our latch. As shown, the latch is implemented by programming the look-up table (the rectangular box with *A*, *B*, *C*, *D* inputs). The output of this table is the latch output and also feeds back to the table to cause the latching action. Note in this figure that the flip-flop (the rectangular box on the right) of the FPGA logic element is not used.

5.5.2   Flip-flop models

The Verilog code of Fig. 5.17 shows a synthesizable D-type flip-flop (ignoring delays). The FPGA logic element used for realization of this flip-flop is shown in Fig. 5.68.

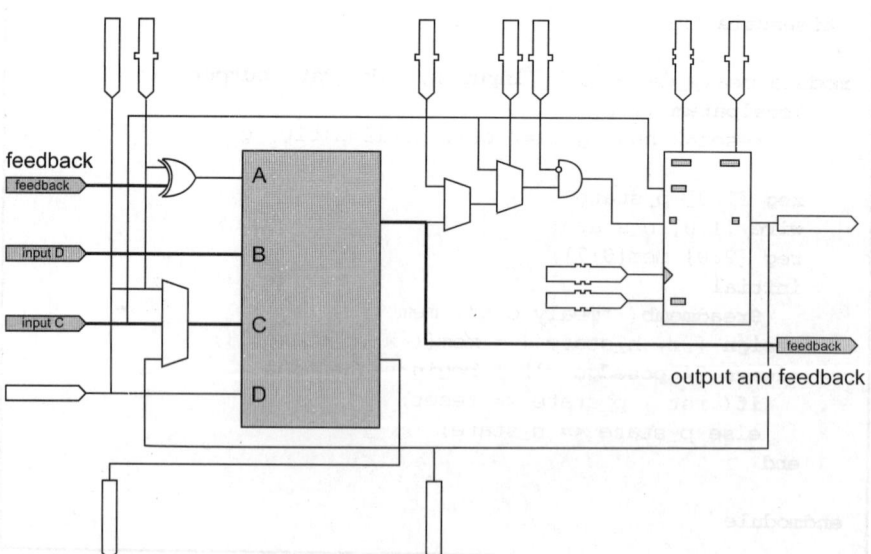

Figure 5.67   FPGA Latch Implementation

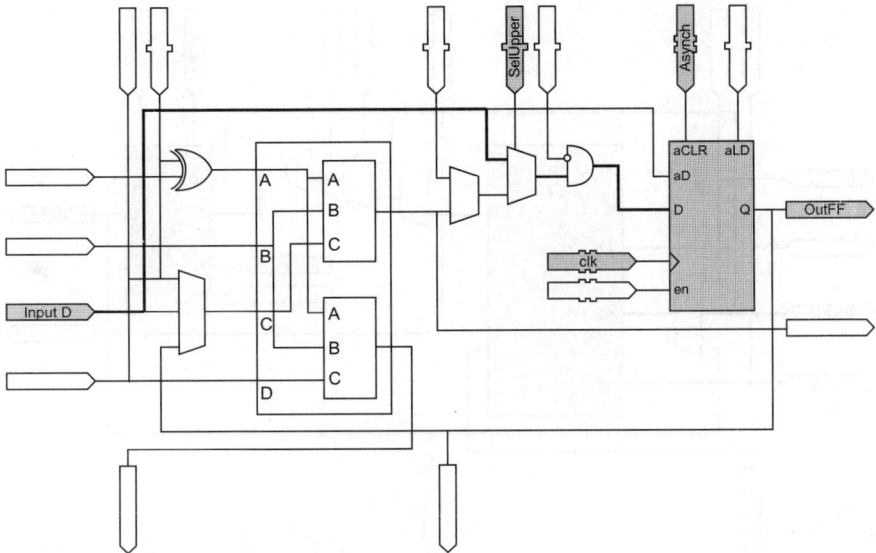

Figure 5.68   FPGA Flip-Flop Implementation

As shown, the FPGA cell contains a flip-flop that corresponds to the required Verilog behavior, and is therefore used for implementing our Verilog flip-flop description. If the FPGA flip-flop does not exactly match the required behavior, e.g., requiring a JK flip-flop implementation, the logic around the flip-flop would be used for realization of the correct behavior.

The flip-flop of the logic element of Fig. 5.68 has asynchronous preset and clear inputs (pins on the top part of the flip-flop), and would be utilized if the input Verilog description required them. However, if a synchronous control is required, the logic in the FPGA logic element, i.e., the look-up table, will be used. Figure 5.69 shows the realization of a flip-flop with synchronous reset input.

As compared with the logic element of Fig. 5.68, the logic element of Fig. 5.69 uses its look-up table for bringing in a synchronous **0** into the flip-flop for resetting it.

5.5.3   Memory initialization

As discussed previously in this section, **initial** statements and memory initialization tasks are not synthesizable. Therefore the Gray counter of Fig. 5.45 or the state machine of Fig. 5.65 is not accepted by most present commercial synthesis tools. On the other hand, most synthesis tools provide a mechanism for specifying ROM based logic. For example, in

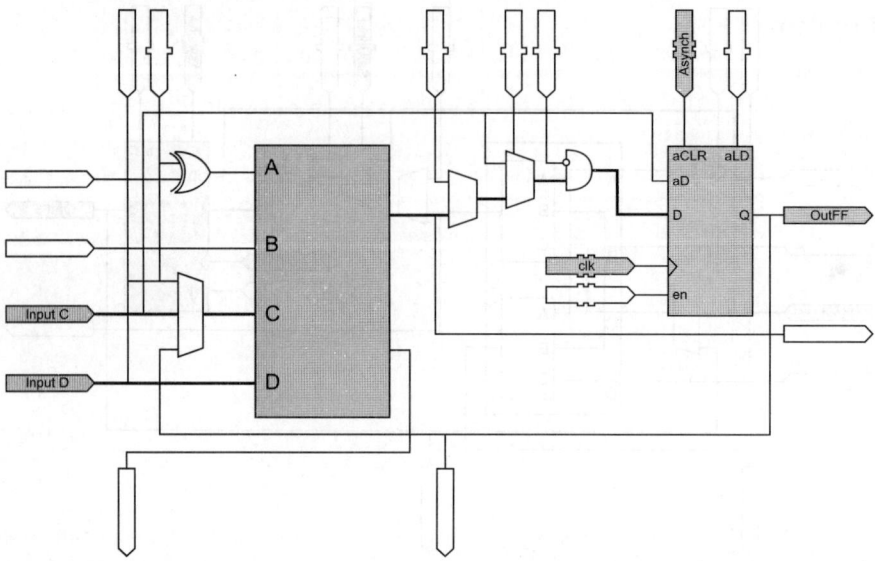

Figure 5.69   Logic Element Used for a Flip-Flop with Synchronous Reset

Altera's Quartus environment a memory block and its initialization file can be specified outside of a Verilog module. This memory is directly mapped to the FPGA memory.

5.5.4   General sequential circuit synthesis

Sequential circuits consist of a combinational and a register part. For the synthesis of the combinational parts, rules discussed in Chap. 4 must be followed. For the synthesis of the register parts, clocking rules and rules regarding synchronous and asynchronous controls must be observed. Discussions in Sec. 5.5.2 about flip-flop synthesis, apply to individual bits of the register part of a sequential circuit.

5.6   Summary

This chapter presented Verilog description of sequential circuits. We presented memory elements at the gate, boolean and behavioral levels. Most of this chapter, however, concentrated on the behavioral description of sequential circuits. We discussed sequential circuits as simple as latches and flip-flops and as complex as queues and sequence detectors. We showed that we follow the same basic rules for description of these circuits. Most styles of coding presented in this chapter were synthesizable. The last section of this chapter showed synthesis correspondence of several typical styles.

Problems

5.1 Show Verilog code of an SR-latch with active low set and reset inputs.

5.2 Write a negative edge trigger D-type flip-flop with an asynchronous active-low reset, a synchronous active-high set, and an active-high clock enable input.

5.3 Write a 32 word register file in Verilog which reads its data from a file named *data.mem*. It has a 5-bit address input named *index* and a read input signal. When the read input is 1, the output of the memory is equal to the data written in the *index* address of the register file, otherwise the output equals 8'bZZ.

5.4 Show synthesizable Verilog code for a register unit that performs operations shown below. The unit has a 3-bit mode *(md)* input, an asynchronous reset *(rs)* input, a 1-bit output control *(oc)* input, and an 8-bit bidirectional *io* bus. The internal register drives the *io* bus when *oc* is 1 and *md* is not **111**.

> md=000: does nothing
>
> md=001: right shift the register
>
> md=010: left shift the register
>
> md=011: up count, Gray (000, 001, 011, 010, 110, 111, 101, 100)
>
> md=100: down count, Gray (opposite of the above)
>
> md=101: complement register contents
>
> md=110: swap right and left 4 bits
>
> md=111: parallel load

5.5 Design an 8-bit shifter with parallel input and output, *sin_l*, *sin_r* input lines, and *shift* and *parallel* control signals. When the *shift* signal is issued, data on *sin_l* and *sin_r* lines will be entered to the most and least significant bits of the register respectively, and third and fourth bits of the register will be put on a 2-bit output line named *out*. No shifting will be done if the unit is in parallel mode.

5.6 Show the Verilog code of the following up-down counter. The counter has a *u* input that controls its count direction. If *u* is 1, it counts **010, 011, 101, 011, 111, 001** ... . If *u* is **0**, it counts this same sequence in the opposite direction.

5.7 Show Verilog code for a 4-bit counter that counts the following sequence: **0100, 0001, 1011, 1010, 0111, 1111, 0111, 0000, 1000**. Write this according to the Huffman model in which the logic part is separate from the register part. Code the logic part by a memory, *mem*, that is initialized in a procedural block using the **$readmemh** system task. This task should read the contents of *mem.dat* file and load it into *mem*. Show the complete Verilog code and contents of the memory file.

5.8 Write Verilog description for an 8-word FILO stack with 8-bit input and output data lines and *rd* and *wr* input signals. The stack has a *full* output flag, that when it becomes 1, the stack cannot accept any more data.

5.9   Show the Verilog code for a Moore **100110** detector using the Huffman style of coding.

5.10   Show the design of a Mealy machine sequence detector that detects the **100** and **001** sequences on its serial input. Provide an asynchronous reset input that starts the detector in its first state.

5.11   A sequential circuit has a synchronizing input, s, and two data inputs a and b that are treated as a 2-bit binary number ($\{a,b\}$). After a start sequence of **110** on s, the machine starts adding data on the data inputs and produces the modulo-4 add result of all the data received on the data inputs. This continues until an end sequence, **011**, is received on s. At this time the machine goes into the halt state and ignores all the data on its a and b inputs. While in this state, the machine searches for the start sequence on s. The start and end sequences do not overlap. Write a Verilog description for this circuit.

5.12   A simple sequential multiplier is developed by adding A to itself B number of times. Both A and B inputs are 8-bit unsigned numbers. The circuit has a *start* input that becomes **1** for one clock when the inputs are valid. The inputs will not be valid when this signal is **0**. After the start, the circuit sets its *done* output to **0** and starts the multiplication process. When done, it puts the result on R and sets *done* to **1**. Given the following interface, write the complete code of this multiplier.

```
module mult (a, b, start, clk, r, done);
input [7:0] a, b;
input start, clk;
output [15:0] r;
output done;
reg [7:0] abuf, bbuf; // use these if you like
reg [15:0] r;
 . . .
endmodule
```

5.13   A data collector module has seven data registers of length 15 driving its seven 15-bit outputs. This module has *ser_data* serial input, *data_ready* input indicating when the stream of data is available on *ser_data*, and the *busy* output. A 112 ($112 = 7*15 + 4 + 3$) bit packet starts on *ser_data* when *data_ready* becomes **1**. A packet starts with its header that includes 4 bits for m, followed by 3 bits for n, and it is then followed by $n*m$ data bits. The data bits of a packet constitute n m-bit data words. Bits of data words are arranged from least to most. There are always fewer than 8 data words ($n <= 7$) and each data-word has less than 16 bits ($m <= 15$). After the header, the first m bits are those of data-word 1, immediately followed by m bits of data-word 2, and at the end, m bits of data-word n. Your module should separate the data words and put them in its 15 registers. The registers should be left-filled with **0**s for $m<15$. For n of less than 7, only the first n registers will be filled and the rest will be filled with zeros. Write a synthesizable *DataCollector* module.

5.14 Write an always statement to count the number of **1**s in a given **reg** *[255:0] InVec*. Declare all necessary registers and wires. The system clock (*clk*) is available for you to use.

5.15 An important issue in CPU modeling and testing is modeling of its memory. Because of machine memory limitations, a CPU memory must be modeled to use external files. In this problem you are to write a memory model that initially reads hexadecimal data from an external file, *mem_file.dat*, and holds it in its internal buffer. Data being read from the memory will be read from the buffer and data being written into the memory will be written into the buffer and into the external memory file, *mem_file.dat*, at the same time. Assume the data in *mem_file.dat* is for the first 256 words of the memory. You are to implement this memory for addresses 0 to 255 only, and reading from addresses outside of this range will return **X**s. The memory model will ignore writing outside of this range. Use the module declaration shown below.

```
module memory (mem_wr, mem_rd, databus, adbus);
input mem_wr, mem_rd;
inout [7:0] databus ;
input [15:0] adbus;
. . .
endmodule
```

Suggested Reading

Brown, S., and Z. Vranesic, *Fundamentals of Digital Logic with Verilog Design*, McGraw-Hill, New York, 2002, ISBN: 0-07-283878-7.

IEEE Std 1364-2001, *IEEE Standard Verilog Language Reference Manual*, SH94921-TBR (print) SS94921-TBR (electronic), ISBN 0-7381-2827-9 (print and electronic), 2001.

Navabi, Z., *Verilog Computer-Based Training Course*, CBT CD with hardcopy User's manual, McGraw-Hill, 2002, ISBN 0-07-137473-6.

Nelson, V. P., H. T. Nagle, B. D. Carroll, and et al., *Digital Logic Circuit Analysis & Design*, Prentice-Hall, Inc., New Jersey, 1996.

6

Component Test and Verification

The previous chapters discussed Verilog for describing combinational and sequential circuits. Except in a few cases, where we dealt with timing of modules, language constructs we discussed were synthesizable. This chapter discusses the use of the Verilog language for testing design modules. We will see that timing and display procedures become more important when dealing with testbench modules.

This chapter shows how Verilog language constructs can be used for application of data to a module under test (MUT), and how module responses can be displayed and checked. In the first part of this chapter data application and response monitoring are discussed. In the second part, we discuss the use of assertion verification for giving a better observability to our design modules. Advanced utilization of external files for testing will be put off until Chap. 8, where system design and test is described.

6.1 Testbench

Verilog simulation environments provide tools for graphical or textual display of simulation results. Some simulation environments go further, and provide graphical tools for editing input test data to a design module that is being tested. Such tools are referred to as waveform editors, and are usually good for small designs. They become too complex to use for a design with many busses and control signals. Another problem with waveform editors is that each simulation environment uses a different procedure for waveform editing, and moving from one simulator to another requires relearning a whole new set of procedures.

This problem can be alleviated by use of Verilog testbenches. A Verilog testbench is a Verilog module that instantiates an MUT, applies data to it, and monitors its output. Because a testbench is in Verilog, it can go from one simulation environment to another. A module and its corresponding testbench form a simulation model in which MUT is tested for the same input data regardless of what simulation environment is used.

To facilitate development of testbenches, some simulation environments provide testbench tools that automatically generate a template testbench. Such tools also provide ways of inserting templates for generation of test data for applying them to MUT. Using templates is helpful, but a designer must understand testbenches and language constructs that are used for testing a design module. In the next two subsections basics of testbenches are discussed.

6.1.1 Combinational circuit testing

Developing a testbench for a combinational circuit is straight forward, however selection of data and how much testing should be done depends on the MUT and its functionality.

Chapter 4 presented a simple arithmetic logical unit (ALU) (Fig. 4.63) that we use here to test. Module header and declarations of its ports are repeated in Fig. 6.1 for reference. The *alu_4bit* module is a four function ALU. Data inputs are *a* and *b*, and its function input is *f*. In addition to its *y* data output, the ALU generates parity (*p*), overflow (*ov*), and compare outputs.

A testbench for *alu_4bit* is shown in Fig. 6.2. Variables corresponding to inputs and outputs of the module under test are declared in the testbench. Variables connecting to the inputs are declared as **reg** and outputs as **wire**. Instantiation of *alu_4bit* shown in the testbench associates local **reg**s and **wire**s with the ports of this module.

Variables that are associated with the inputs of *alu_4bit* have been given initial values when declared. Application of data to the *b* data input and *oe* output-enable of ALU are done in an **initial** statement. For

```
module alu_4bit (a, b, f, oe, y, p, ov, a_gt_b, a_eq_b,
                 a_lt_b);
   input [3:0] a, b;
   input [1:0] f;
   input oe;
   output [3:0] y;
   output p, ov, a_gt_b, a_eq_b, a_lt_b;

   // . . . .

endmodule
```

Figure 6.1   *alu_4bit* Module Declaration

```
module test_alu_4bit;
    reg [3:0] a=4'b1011, b=4'b0110;
    reg [1:0] f=2'b00;
    reg oe=1;
    wire [3:0] y;
    wire p, ov, a_gt_b, a_eq_b, a_lt_b;

    alu_4bit cut( a, b, f, oe, y, p, ov, a_gt_b, a_eq_b,
                  a_lt_b );

    initial begin
        #20 b=4'b1011;
        #20 b=4'b1110;
        #20 b=4'b1110;
        #80 oe=1'b0;
        #20 $finish;
    end
    always #23 f = f + 1;

endmodule
```

Figure 6.2   Testbench for *alu_4bit*

the first 60 ns every 20 ns, a new value is assigned to *b*. The **initial** block then waits for 80 ns, disables the ALU output by setting *oe* to **0**, and after 20 ns it finishes the simulation. This last 20 ns wait, allows effects of the last input change to be shown in simulation results.

Application of data to the *f* input of *alu_4bit* is done in an **always** statement. Starting with the initial value of **0**, *f* is increment by 1 every 23 ns.

The **$finish** statement in the **initial** block of the testbench is reached at 160 ns. At this time all active procedural blocks stop and simulation terminates. Figure 6.3 shows simulation results of the *alu_4bit* module.

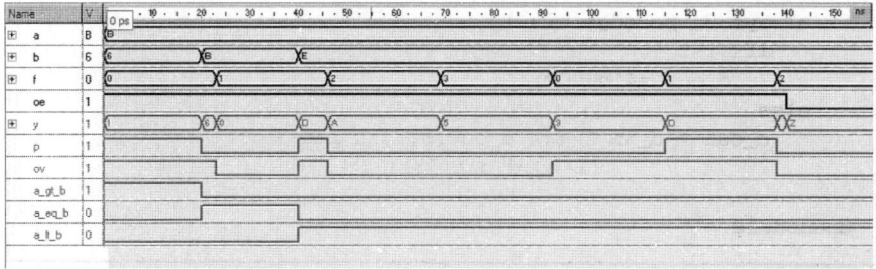

Figure 6.3   ALU Simulation Results

Throughout the simulation a remains constant, and b changes from 6 to B and then to E at 40 ns. The f function input changes every 23 ns causing various ALU functions to be examined. At 140 ns, oe changes to **0**, causing the y output become **Z**.

6.1.2   Sequential circuit testing

Testing sequential circuits involves synchronization of circuit clock with other data inputs. We use the *misr* module of Chap. 5 for an example here. This circuit, repeated in Fig. 6.4 for reference, has a clock input, a reset, data inputs, and outputs.

The circuit has a *poly* parameter that determines its signature and data compression. With each clock a new signature will be calculated with the new data and existing *misr* register data.

Figure 6.5 show a testbench for *misr*. As before, variables corresponding to the ports of MUT are declared in the testbench. When *misr* is instantiated, these variables are connected to its actual ports. Our *misr* instance also includes specification of its *poly* parameter.

The **initial** block of this testbench generates a positive pulse on *rst* that begins at 13 ns and ends at 63 ns. The timing is so chosen to cover at least one positive clock edge, so that the synchronous *rst* input can initialize the *misr* register. The *d_in* data input begins with x, and is initialized to *4'b1000* while *rst* is **1**.

In addition to the **initial** block, the *test_misr* module includes two **always** blocks that generate data on *d_in* and *clk*. Clock is given a periodic signal that toggles every 11 ns. The *misr d_in* input is assigned a new value every 37 ns. In order to reduce chance of several inputs changing at the same time, we usually use prime numbers for timing of sequential circuit inputs.

```
module #(parameter [3:0] poly=0) misr (input clk, rst,
        input [3:0] d_in, output reg [3:0] d_out );

    always @( posedge clk )
        if( rst )
            d_out =4'b0000;
    else
        d_out = d_in ^ ({4{d_out[0]}} & poly) ^
                    {1'b0,d_out[3:1]};

endmodule
```

Figure 6.4   *misr* Sequential Circuit

```
module test_misr;
    reg clk=0, rst=0;
    reg [3:0] d_in;
    wire [3:0] d_out;

    misr #(4'b1100) MUT ( clk, rst, d_in, d_out );

    initial begin
        #13 rst=1'b1;
        #19 d_in=4'b1000;
        #31 rst=0'b0;
        #330 $finish;
    end

    always #37 d_in = d_in + 3;

    always #11 clk = ~clk;

endmodule
```

Figure 6.5 A Testbench for *misr*

As shown in Fig. 6.6, starting at 40 ns with this and every positive edge of *clk*, a new signature is generated in *misr*. Since prior to time 80 ns, *misr* is reset to **0**, the first signature that happens at 80 ns is the same as *d_in*.

6.2 Testbench Techniques

Various Verilog coding techniques for generation of test data and observing circuit responses are discussed in this section. We use state machines of Chap. 5 for our test modules. The first example is a **101** Moore detector circuit depicted in Fig. 6.7.

We have used a coding style that is somewhat different than that used in Chap. 5. The *z* output becomes **1** in state *d* when a sequence of **101** is detected on *x*. The circuit has a synchronous reset input.

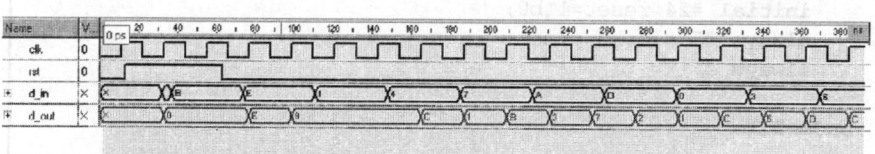

Figure 6.6 Testing *misr*

```
module moore_detector (input x, rst, clk, output z );

    parameter [1:0] a=0, b=1, c=2, d=3;
    reg [1:0] current;

    always @( posedge clk )
        if ( rst )  current = a;
        else case ( current )
            a : current = x ? b : a ;
            b : current = x ? b : c ;
            c : current = x ? d : a ;
            d : current = x ? b : c ;
            default : current = a ;
        endcase

    assign z = (current==d) ? 1'b1 : 1'b0;

endmodule
```

Figure 6.7   101 Moore Detector for Test

6.2.1   Test data

A testbench for *moore_detector* of Fig. 6.7 is shown in Fig. 6.8. As before, our testbench is a module with no ports. Within this module, four procedural blocks provide data for testing the state machine. Variables connected to inputs of MUT and used on the left-hand sides in the procedural blocks are declared as **reg**.

```
module test_moore_detector;
    reg x, reset, clock;
    wire z;

    moore_detector MUT ( x, reset, clock, z );

    initial begin
        clock=1'b0; x=1'b0; reset=1'b1;
    end
    initial #24 reset=1'b0;
    always #5 clock=~clock;
    always #7 x=~x;

endmodule
```

Figure 6.8   Basic Data Generation

Instead of initializing **reg** variables when they are declared, we have used an **initial** block for this purpose. It is important to initialize variables, like *clock*, for which their old values are used for determining their new values. If not done so, *clock* would start with value **X** and complementing it would never change its value. The **always** block shown generates a periodic signal with a period of 10 ns on *clock*.

Following the **always** block producing *clock*, another **always** block generates a periodic signal on *x* with a period of 14 ns. The waveform generated on *x* may or may not be able to test our machine for a correct **101** sequence. However, periods of *clock* and *x* can be changed to make this happen. With the timing used here, the *moore_detector* output becomes **1** at 55 ns, and every 70 ns from then on.

6.2.2 Simulation control

Another testbench for the circuit of Fig. 6.7 is shown in Fig. 6.9. Although, Verilog constructs are used differently, data and clock applied to MUT by this testbench are the same as those of Fig. 6.8. However, if the simulation of the previous testbench is not interrupted, or stopped, it runs forever. The testbench of Fig. 6.9 corrects this problem by adding another **initial** block that stops the simulation at 189 ns.

Simulation control tasks are **$stop** and **$finish**. The first time the flow of a procedural block reaches such a task, simulation stops or finishes. A stopped simulation can be resumed, but a finished one cannot.

Another testbench for the state machine of Fig. 6.7 is shown in Fig. 6.10. This testbench combines the **initial** blocks of deactivating *reset* and simulation control into one **initial** block. The timing is adjusted to terminate simulation at 189 ns, the same as that of Fig. 6.9.

```
module test_moore_detector;
    reg x=0, reset=1, clock=0;
    wire z;

    moore_detector MUT ( x, reset, clock, z );

    initial #24 reset=1'b0;
    always #5 clock=~clock;
    always #7 x=~x;
    initial #189 $stop;

endmodule
```

Figure 6.9   Testbench with **$stop** Simulation Control

```
module test_moore_detector;
    reg x=0, reset=1, clock=0;
    wire z;

    moore_detector MUT ( x, reset, clock, z );

    initial begin
        #24 reset=1'b0;
        #165 $finish;
    end
    always #5 clock=~clock;
    always #7 x=~x;

endmodule
```

Figure 6.10   Testbench with **$finish** Simulation Control

6.2.3   Limiting data sets

Instead of setting simulation time limit, a testbench can put a limit on the number of data put on inputs of a MUT. This will also be able to stop simulation from running indefinitely.

Figure 6.11 shows a testbench for our famous *moore_detector* MUT. This testbench uses **$random** to generate random data on the *x* input of the circuit. **repeat** statements in the **initial** blocks cause *clock* to toggle 13 times every 5 ns, and *x* to receive random data 13 times every 7 ns. Instead of a deterministic set of data to guarantee a deterministic test state, random data is used here. This strategy makes it easier

```
module test_moore_detector;
    reg x=0, reset=1, clock=0;
    wire z;

    moore_detector MUT ( x, reset, clock, z );

    initial #24 reset=1'b0;
    initial repeat(13) #5 clock=~clock;
    initial repeat(10) #7 x=$random;

endmodule
```

Figure 6.11   Testbench Using **repeat** to Limit Data Sets

to generate data, but makes analysis of circuit output more difficult, due to unpredictable inputs. In large circuits, random data is more useful for data inputs than for control signals. The testbench of Fig. 6.11 stops at 70 ns.

6.2.4 Applying synchronized data

The previous examples of testbenches for MUT used independent timings for the clock and data. Where several sets of data are to be applied, synchronization of data with the system clock becomes difficult. Furthermore, changing the clock frequency would require changing the timing of all data inputs of the module being tested.

The testbench of this section (Fig. 6.12), that is written for the *moore_detector* of Fig. 6.7, uses an event control statement to synchronize data applied to x with the clock that is generated in the testbench. The *clock* signal is generated in an **initial** statement using the **repeat** construct. Another **initial** statement is used for generation of random data on x. As shown in this **initial** statement, a **forever** loop that continuously repeats its statement is used here. This loop waits for the positive edge of the *clock*, and 3 ns after the clock edge, a new random data is generated for x. The stable data after the positive edge of the clock will be used by *moore_detector* on the next leading edge of the clock. This technique of data application guarantees that changing of data and clock do not coincide.

The 3 ns delay used here makes it possible to use this same testbench for simulating post-synthesis designs as well as behavioral descriptions like that of Fig. 6.7. In a post-synthesis simulation, in which component models with actual delay values are used, testbench delays

```
module test_moore_detector;
    reg x=0, reset=1, clock=0;
    wire z;

    moore_detector MUT ( x, reset, clock, z );

    initial #24 reset=0;
    initial repeat(13) #5 clock=~clock;
    initial forever @(posedge clock) #3 x=$random;

endmodule
```

Figure 6.12 Synchronizing Data with Clock

allow propagation of test signals to complete before application of other test signals.

6.2.5 Synchronized display of results

The technique used in the previous section can be used for synchronized observation of MUT outputs or internal signals.

Figure 6.13 shows another testbench for our *moore_detector*. In this testbench, 1 ns after the positive edge of the clock, that is when the circuit output is supposed to have its new stable value, the *z* output is displayed using the **$displayb** task.

As in the testbench of Fig. 6.12, the delays used in this testbench make it usable for *moore_detector* after it has been synthesized.

Using hierarchical naming, this testbench can be used for displaying internal variables and signals of MUT. A testbench that is developed for observing states of *moore_detector* is shown in Fig. 6.14. This testbench uses **$monitor** to display the *current* **reg** of *moore_detector* of Fig. 6.7, and an **always** block to display its output. The *current* state and *z* output are displayed when they receive new values.

Except the last two procedural statements, (an **initial** and an **always**) the rest of this testbench is the same as that of Fig. 6.13. The **initial** statement containing **$monitor** is responsible for displaying *MUT.current*, which is *current* of *moore_detector* addressed by its hierarchical name. The **initial** statement starts the **$monitor** task in the background. Display occurs when this task is started and when an event occurs on one of the variables of the task's arguments. The **%b** and **%t** format specifications cause the value of *MUT.current* to be displayed in binary and that of **$time** to be displayed with its time unit.

```
module test_moore_detector;
    reg x=0, reset=1, clock=0;
    wire z;

    moore_detector MUT ( x, reset, clock, z );

    initial #24 reset=0;
    initial repeat(13) #5 clock=~clock;
    initial forever @(posedge clock) #3 x=$random;
    initial forever @(posedge clock) #1 $displayb(z);

endmodule
```

Figure 6.13 Testbench Displaying Output

```
module test_moore_detector;
    reg x=0, reset=1, clock=0;
    wire z;

    moore_detector MUT ( x, reset, clock, z );

    initial #24 reset=0;
    initial repeat(19) #5 clock=~clock;
    initial forever @(posedge clock) #3 x=$random;
    initial $monitor("New state is %d and occurs at %t",
                        MUT.current, $time);
    always @(z) $display("Output changes at %t to %b",
                        $time, z);

endmodule
```

Figure 6.14 Testbench Displays Design Variables when they change

The last procedural statement of Fig. 6.14 is an **always** statement that is sensitive to z. This statement encloses a **$display** task that displays values of z and the times that this output changes. Figure 6.15 shows the output generated by running the testbench of Fig. 6.14. The result shown was obtained by repeating the clock toggling 19 times instead of 13. This allowed more data to be applied to our MUT.

6.2.6 An interactive testbench

For the next series of testbenches we use a different state machine. This is a **1101** Moore detector with start (*start*) and reset (*rst*) control inputs. If *start* becomes **0** while searching for **1101**, the machine resets to its initial state. As shown in Fig. 6.16 this circuit has five states, and its output becomes **1** when it reaches state *e*.

```
New state is x and occurs at                0
Output changes at              50 to 0
New state is 0 and occurs at               50
New state is 1 and occurs at              250
New state is 2 and occurs at              850
Output changes at             950 to 1
New state is 3 and occurs at              950
```

Figure 6.15 Test Results of Testbench of Fig. 6.14

```
module moore_detector (input x, start, rst, clk,
                       output z );
    parameter a=0, b=1, c=2, d=3, e=4;

    reg [2:0] current;

    always @( posedge clk )
        if ( rst )   current <= a;
        else if ( ~start ) current <= a;
            else case ( current )
                a : current <= x ? b : a ;
                b : current <= x ? c : a ;
                c : current <= x ? c : d ;
                d : current <= x ? e : a ;
                e : current <= x ? c : a ;
                default: current <= a;
            endcase

    assign z = (current==e);

endmodule
```

Figure 6.16   Moore Sequence Detector Detecting **1101**

The testbench for this state machine is an interactive one. In the **initial** block shown in Fig. 6.17, the testbench communicates with MUT. The x input and *clock* are generated by two **always** blocks. A continuous periodic signal is generated on *clock*, and periodic random data is assigned to x.

Initially, **0** and **1** are placed on *reset* and *start* to get the machine started. Following this, a **wait** statement waits for z to become **1** as a result of the MUT reacting to values of x and *clock*. After this happens, *start* is set to **0** and back to **1** after 13 ns to restart the machine. Following this first round of activity, a **repeat** statement repeats the process of starting the machine and waiting for z to become **1** three more times. At the end, after 50 ns the testbench stops the simulation using a **$stop** task.

A portion of the waveform resulted by the testbench of Fig. 6.17 is shown in Fig. 6.18. In addition to the ports of *moore_detector* of Fig. 6.16, its *current* state is also displayed in this figure.

Another interactive testbench for *moore_detector* of Fig. 6.16 is shown in Fig. 6.19. As in Fig. 6.17, this testbench applies random data to x and periodic data to *clock*. The testbench uses hierarchical naming to access

```
module test_moore_detector;
    reg x=0, start, reset=1, clock=0;
    wire z;

    moore_detector MUT ( x, start, reset, clock, z );

    initial begin
        #24 reset=1'b0; start=1'b1;
        wait(z==1'b1);
        #11 start=1'b0;
        #13 start=1'b1;
        repeat(3) begin
            #11 start=1'b0;
            #13 start=1'b1;
            wait(z==1'b1);
        end
        #50 $stop;
    end
    always #5 clock=~clock;
    always #7 x=$random;

endmodule
```

Figure 6.17   An Interactive Testbench

the *e* parameter and *current* variable within MUT. The **$display** and **$strobe** tasks, shown in Fig. 6.19, are used to observe the output of *moore_detector* when the machine enters state *e*.

In the *Output_Display* **always** block when *current* becomes *e* (both *current* and *e* are inside *MUT*), output *z* is displayed by **$display** and **$strobe** tasks. While the **$strobe** task waits for all simulation events to complete before displaying its parameters, the **$display** task displays its parameters as soon as the program flow reaches it. Since we are not delaying the flow of this **always** block after detection of state *e*, the **$display** task

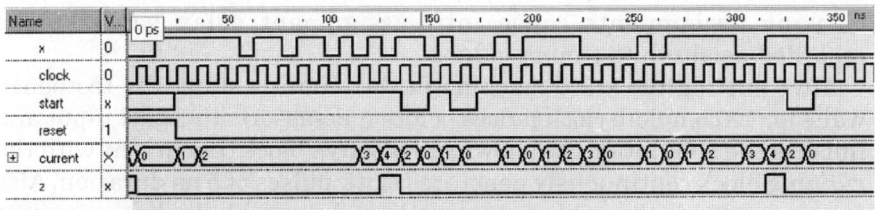

Figure 6.18   Waveform Resulted by the Interactive Testbench

```
module test_moore_detector;
    reg x=0, start, reset=1, clock=0;
    wire z;

    moore_detector MUT ( x, start, reset, clock, z );

    initial begin
        #24 reset=1'b0; start=1'b1;
    end
    always begin : Output_Display
        wait (MUT.current == MUT.e);
        $display ("$display task shows: The output is %b ",
                    z);
        $strobe ("$strobe task shows: The output is %b ", z);
        #2 $stop;
    end
    always #5 clock=~clock;
    always #7 x=$random;

endmodule
```

Figure 6.19   Interactive Testbench Using Display Tasks

displays the old value of *z*. On the other hand, after *e* is detected, a simulation cycle later, *z* becomes **1** and **$strobe** displays this output correctly.

6.2.7 Random time intervals

We have shown how **$random** can be used for generation of random data. The testbench we are discussing in this section uses random wait times for assigning values to *x*.

Figure 6.20 shows a testbench for the **1101** sequence detector that uses **$random** for its delay control. As shown, the *running* **initial** statement applies appropriate values to *reset* and *start* for the system to start its search for the **1101** sequence. In this procedural block non-blocking assignments cause intra-assignment delay values to be regarded as absolute timing values.

After putting the state machine in the running state, the testbench waits for 13 complete clock pulses before it de-asserts the *start* input and finishes the simulation. As shown, an **always** block concurrent with the *running* block continuously generates clock pulses of 5 ns duration. Also concurrent with these blocks is another **always** block that generates random data on *t*, and uses *t* to delay assignment of random values to *x*.

```
module test_moore_detector;
   reg x, start, reset, clock;
   wire z;

   reg [3:0] t;

   moore_detector MUT ( x, start, reset, clock, z );

   initial begin:running
      clock <= 1'b0; x <= 1'b0;
      reset <= 1'b1; reset <= #7 1'b0;
      start <= 1'b0; start <= #17 1'b1;
      repeat (13) begin
         @( posedge clock );
         @( negedge clock );
      end
      start=1'b0;
      #5;
      $finish;
   end
   always #5 clock=~clock;
   always begin
      t = $random;
      #(t) x=$random;
   end
endmodule
```

Figure 6.20 Testbench using Random Time Intervals

This block generates data on the *x* input for as long as the **$finish** statement in the *running* block is not reached.

6.2.8 Buffered data application

None of the testbenches discussed so for applied a given set of test data to the circuit input(s). The testbench we are discussing here uses a buffer to hold data to be applied to the MUT data input. We take a predefined series of bits and assign them to the *x* input of *moore_detector*.

As shown in Fig. 6.21, the 19-bit *buffer* is initialized with test data. In an **always** statement each bit of this buffer is shifted out onto the *x* input of *moore_detector* 1 ns after the positive edge of the *clk* clock. As data is shifted, *buffer* is rotated in order for the applied buffered data to be able to repeat. Start and stop control of the state machine are done in another **initial** block in this testbench.

```
module test_moore_detector;
    reg x=0, rst, start, clk=0;
    wire z;
    reg [18:0] buffer;

    moore_detector MUT ( x, start, rst, clk, z );

    initial buffer = 19'b0001101101111001001;
    initial begin
        rst=1'b1; start=1'b0;
        #29 rst=1'b0;
        #29 start=1'b1;
        #500 $stop;
    end
    always @(posedge clk) #1 {x, buffer} = {buffer, x};
    always #5 clk = ~clk;

endmodule
```

Figure 6.21   Testbench Applying Buffered Data

With this testbench we are sure a correct sequence is applied to our MUT. This way, we can more easily check for our expected results. Generally, the more effort we put into generating our test data, the easier it will be to analyze the output results. Random or pseudo-random data generation is easy, but requires a significant time analyzing the produced output.

6.3   Design Verification

The previous section discussed test techniques for testing a Verilog design. We presented several methods of test data generation and test application, and suggested ways of observing and inspecting test results. Stimuli generation and response analysis require significant efforts on the part of a hardware designer. Learning correct test techniques is good, but automation of either of these procedures will be very useful for a design engineer.

Formal verification is a way of automating design verification by eliminating testbenches and problems associated with their data generation and response observation. In formal verification, a designer writes properties to check his or her design. Formal verification tools do not perform simulation, but come up with a Yes/No answer for every property the design is being checked for. Although this method of design verification helps discover many design errors, most designs still need testbench development and simulation for validating that their Verilog code indeed functions as expected. In other words, an all "Yes" answers to design properties checked by formal verification tools is still not enough.

Instead of eliminating data generation and response observation (like the formal verification tools), a step in the direction of automating design validation is to reduce or eliminate efforts needed for analyzing output responses. For this purpose *assertion verification* is used. Assertion verification adds monitors to a design to improve its observability. While the design is being simulated with its testbench data, assertion monitors that represent certain design properties continuously check for correct design behavior by validating these properties. If the simulation data leads into conditions that indicate to an assertion monitor that the design is misbehaving, the monitor is said to fire to alert the designer of the problem.

As mentioned, we still need to develop a testbench and careful planning of test inputs for the design being tested is needed in assertion verification. But, in many cases, assertions automatically check to make sure events that occur in the design are as expected. This significantly reduces the need for processing long output lists or waveforms.

6.4   Assertion Verification

Unlike simulation that a testbench or a human has to interpret the results, in assertion verification, in-code monitors take the responsibility of issuing a message if something happens that is not expected. In Verilog, these monitors are modules, and they are instantiated in a design to check for certain design properties. Instantiating an assertion module is not to be regarded as instantiation of a hardware module. Instead, this kind of instantiation is more like an always-active procedure that continuously checks for events in the design module.

The present set of assertion monitors are available in a library that is referred to as open verification library (OVL). Designers can develop their own set of assertions, and use them in their designs. The existing monitors check for values of signals, relation of several signals with each other, sequence of events, and expected patterns on vectors or groups of signals. For using assertions, a designer compiles OVL and his or her own assertion library into a simulation library and makes this library available to designs being verified. When a design is developed, assertions are placed at key points in the design to check for key functionalities. When the design is being simulated as a stand-alone component, or in a hierarchy of a larger design, the monitors check signals for their expected values. If a signal does not have a value expected by a monitor, the assertion monitor displays a message and the time that the discrepancy (violation of the property) has occurred. Usually, such messages appear in the simulation report area, *transcript*, or *console*.

6.4.1 Assertion verification benefits

Ways in which placement of assertion monitors in a design are helpful are discussed here.

Designer discipline. When a designer places an assertion in a design, he or she is disciplining him/herself to look into the design more carefully and extract properties.

Observability. Assertions add monitoring points to a design that make it more observable.

Formal verification ready. Assertions correspond to properties that are used in formal verification tools. Having inserted assertion monitors to a design, readies it for verification by a formal verification tool.

Executable comments. Assertion monitors can be regarded as comments that explain some features or behavior of a design. These comments produce messages when the behavior they are explaining is violated.

Self-contained designs. A design with assertion monitors has the design description and its test procedure all in one Verilog module.

6.4.2 Open Verification Library

OVL is available from Accellera, and other EDA organizations. The Language Reference Manual (LRM), user's manual, and Verilog and VHDL code of the library are also available from these organizations. The list of the presently available assertions is shown in Fig. 6.22, and Appendix E has their complete description and their parameters.

assert_always	assert_always_on_edge
assert_change	assert_cycle_sequence
assert_decrement	assert_delta
assert_even_parity	assert_fifo_index
assert_frame	assert_handshake
assert_implication	assert_increment
assert_never	assert_never_at_x_or_z
assert_next	assert_no_overflow
assert_no_transition	assert_no_underflow
assert_odd_parity	assert_one_cold
assert_one_hot	assert_proposition
assert_quiescent_state	assert_range
assert_time	assert_transition
assert_unchange	assert_width
assert_win_change	assert_win_unchange
assert_window	assert_zero_one_hot

Figure 6.22 Assertions

An assertion is placed in code like a module instantiation. As shown in Fig. 6.23, assertion module name comes first. This is followed by *static_parameters* like vector size and options. Following this, Verilog module instantiation requires an *instance_name* for which any unique name is allowed. The last part of an assertion monitor includes reference and monitor signals, and other dynamic arguments. Dynamic arguments are module ports and are also referred to as assertion ports.

Typical static parameters are severity of failure, vector size, number of clocks, time frame specification (in terms of clock cycles), and the displayed failure message. Reference clock, starting signal, reset signal, and the test expression are some of typical dynamic arguments for assertion monitors. The details of parameters and arguments of OVL assertion monitors are discussed in Appendix E. Examples of some OVL assertion monitors and their application are presented in the next section.

6.4.3 Using assertion monitors

This section shows several examples of using assertion monitors. Like the section on testbenches, we show Verilog design examples and their testing procedures. With assertion monitors, testing procedures include insertion of assertion monitors.

6.4.3.1 assert_always. The general format for **assert_always** assertion monitor is:

```
assert_always
        #( severity_level, property_type,
           msg, coverage_level )
        instance_name ( clk, reset_n, test_expr )
```

This assertion continuously checks its *test_expr* to make sure it is always true on the edge of the specified clock (*clk*). If the test expression fails, the assertion fires and its corresponding message (*msg*) is displayed.

As an example consider the binary coded decimal (BCD) counter of Fig. 6.24. This counter counts between 0 and 9. The assertion monitor shown here uses *severity_level* 1 to issue an error and continue simulation if assertion is fired. The reader is encouraged to refer to Appendix E for a detailed description of this assertion.

```
assert_name
        #(static_parameters)
            instance_name
                  (dynamic_arguments);
```

Figure 6.23 Assertion Module Instantiation

```
module BCD_Counter (input rst, clk, output reg [3:0] cnt);

    always @(posedge clk) begin
       if (rst || cnt >= 10) cnt = 0;
       else cnt = cnt + 1;
    end

    assert_always #(1, 0, "Err: Non BCD Count", 0)
    AA1 (clk, 1'b1, (cnt >= 0) && (cnt <= 9));

endmodule
```

Figure 6.24   BCD with assert_always

As shown in the dynamic arguments of the invocation of **assert_always** in Fig 6.24, the test expression is *(cnt >= 0) && (cnt <= 9)*, and it is being monitored on the rising edge of the clock *(clk)*. The monitor checks that on every rising edge of *clk*, *cnt* must be between 0 and 9. The second dynamic argument *(1'b1)* indicates that the assertion is to be monitored at all times.

A testbench for the BCD counter is shown in Fig. 6.25. Note that even though checking simulation results is done in a semi-automatic fashion, test data generation is still done manually by the designer. Proper verification of the design depends on development and insertion of good monitors and quality of test data.

```
module BCD_Counter_Tester;
    reg r, c;
    wire [3:0] count;

    BCD_Counter UUT (r, c, count);

    initial begin
        r = 0; c = 0;
    end
    initial repeat (200) #17 c= ~c;
    initial repeat (03) #807 r= ~r;

endmodule
```

Figure 6.25   BCD Counter Testbench

6.4.3.2 assert_change. The **assert_change** monitor verifies that within a given number of clocks after the start event, the test expression changes. This assertion uses the format shown below.

```
assert_change
          #( severity_level, width, num_cks,
             action_on_new_start, property_type,
             msg, coverage_level )
          instance_name ( clk, reset_n, start_event, test_expr )
```

As an example, see the *Walking_One* module of Fig. 6.26. This is a shift register that walks a **1** with every clock. A **1** is loaded into the left-most bit of the register with the *rst* reset signal. The **assert_change** monitor is discussed below, and the other assertion shown in this figure will be described later.

Parameters of the **ssert_change** monitor used in this figure specify 1 for the length of the test expression and 7 for the number of clocks that change is to occur. The arguments of this monitor specify the falling edge of the clock (*~clk*), *rst* value of **0** for the activity period, and *rst* becoming **0** for the start of the count (*start_event*). As shown, the test expression (*test_expr*) is *wo*. The parameters and arguments specified as such, check that from the time that *rst* becomes **0** (i.e., *rst == 0*) and while it remains **0** (i.e., *~rst*), it takes at most 7 negative clock edges for *wo[0]* to change.

Figure 6.27 shows the testbench for the *Walking_One* module. Note here again, that it is the responsibility of the testbench developer to make sure enough data is applied to cause design errors to trigger assertion monitors.

Figure 6.28 shows test results of the *Walking_One* module. The last eight waveforms shown are signals driven by to *wo[7]* to *wo[0]*, respectively. When *rst* becomes **1**, the falling edge of the clock puts a **1** into

```
module Walking_One (input rst, clk, output reg [7:0] wo);
   always @(negedge clk) begin
      if (rst) wo <= 8'b10000000;
      else wo <= {wo[0], wo[7:1]};
   end

   assert_change #(1, 1, 7, 0, 0, "Err: Bit 0 is not changing", 0)
                 AC1 (~clk, ~rst, (rst==0), wo[0]);
   assert_one_hot #(1, 8, 0, "Err: Multiple active bits", 0)
                 AOH (~clk, ~rst, wo);

endmodule
```

Figure 6.26 Walking One Circuit with Assertions

```
module Walking_One_Tester ();
    reg rst=0, clk=0;
    wire [7:0] walking;

    Walking_One MUT (rst, clk, walking);

    initial repeat (223) #7 clk= ~clk;
    initial repeat (15) #109 rst= ~rst;

endmodule
```

Figure 6.27 *Walking_One* Testbench

wo[7]. When *rst* becomes **0**, this **1** starts walking, and it takes 7 clock edges for this **1** to walk to bit **0** of *wo*. The **assert_change** monitor verifies this.

6.4.3.3 assert_one_hot. The **assert_one_hot** assertion monitor checks that while the monitor is active, only one bit of its n-bit test expression is **1**. Syntax, parameters, and arguments of this monitor are as shown below.

```
assert_one_hot
          #( severity_level, width, property_type,
             msg, coverage_level )
          instance_name ( clk, reset_n, test_expr )
```

The *Walking_One* module of Fig. 6.26 invokes an **assert_one_hot** monitor. The test expression is *wo* and its width is 8, as specified in the

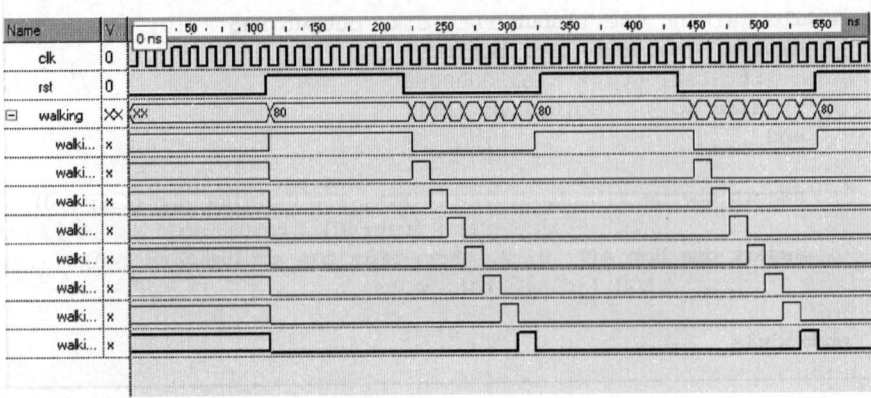

Figure 6.28   Walking-One Test Results

assertion's parameter list. The *~rst* argument of this assertion makes the checking active only when *rst* is **0**.

As another example of an **assert_one_hot** invocation, consider the Gray code counter of Fig. 6.29. The *mem.dat* file that contains consecutive Gray code numbers is read into *mem*, and with each clock, the next Gray count is looked up from *mem*.

In order to check for the correct Gray code sequencing, we have used some auxiliary logic to prepare the test expression for the **assert_one_hot** assertion. The auxiliary logic, shown at the end of the Verilog module of Fig. 6.29, holds the old count of the counter in the declared *old* **reg**. The test expression of the **assert_one_hot** assertion monitor becomes the exclusive-or of the old (*old*) and the present count (*q*). Since consecutive Gray code numbers are only different in one bit, their XOR must be one-hot.

6.4.3.4 assert_cycle_sequence. The **assert_cycle_sequence** shown below is a very useful assertion for verifying state machines.

```
module gray_counter (input [3:0] d_in, input clk, rst, ld,
                        output reg [3:0] q);
    reg [3:0] mem[0:15];
    reg [3:0] im_q;
    initial $readmemb("mem.dat", mem);

    always @( d_in or ld or q ) begin: combinational
        if( ld )
            im_q = d_in;
        else
            im_q = mem[q];
    end

    always @( posedge clk ) begin: register
        if( rst )
            q <= 4'b0000;
        else
            q <= im_q;
    end

    reg [3:0] old; always @(posedge clk) old <= q;
    assert_one_hot #(1, 4, 0, "Err: Not Gray", 0)
                AOH (~clk, ~rst, (old ^ q));
endmodule
```

Figure 6.29 Gray Code Counter

```
assert_cycle_sequence
        #( severity_level, num_cks, necessary_condition,
           property_type,
           msg, coverage_level )
        instance_name ( clk, reset_n, event_sequence )
```

This assertion checks for a sequence of events in a given number of clocks. As with other assertion monitors, this checker has an enabling input that is usually driven by the inactive level of a circuit's reset input.

We use the state machine of Fig. 6.30 to demonstrate the use of this assertion monitor. As shown, after the reset state, i.e., a, the machine searches for **110**. When received, the next two clocks take the machine back to state a, while transiting through state e. In state e, the z output of the circuit becomes **1**.

The Verilog code corresponding to the state machine of Fig. 6.30 is shown in Fig. 6.31. This code also includes assertion monitors for its testing. The **assert_cycle_sequence** monitor shown in this code is setup to check if the machine reaches states d and then state e, then the next clock will move the machine into state a.

This transition path is highlighted in the state diagram of Fig. 6.30 and is verified by the timing diagram of Fig. 6.32, which shows that if the machine enters state 3, then its next state is 4 and then state 0 is entered. As the Verilog code of this circuit shows states 3, 4, and 0 correspond to d, e, and a respectively.

As shown in the Verilog code of the *Sequencing_Machine*, the second parameter of **assert_cycle_sequence** is 3, which corresponds to the number of states in sequence. The third parameter is 0 which configures

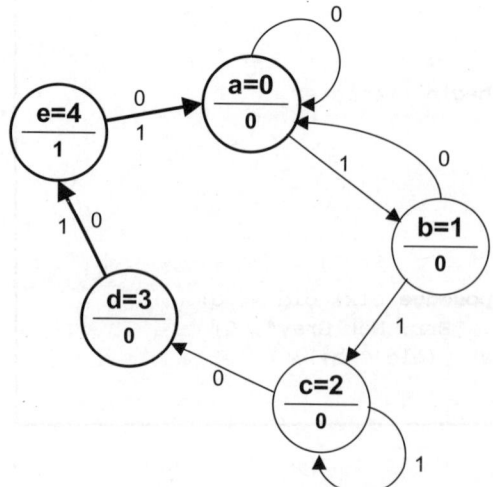

Figure 6.30   A Sequencing State Machine

```
module Sequencing_Machine (input x, start, rst, clk,
                                output z );
   parameter a=0, b=1, c=2, d=3, e=4;

   reg [2:0] current;

   always @( posedge clk )
      if ( rst )  current <= a;
      else if ( ~start ) current <= a;
         else case ( current )
            a : current <= x ? b : a ;
            b : current <= x ? c : a ;
            c : current <= x ? c : d ;
            d : current <= e ;
            e : current <= a ;
            default: current <= a;
         endcase
      assign z = (current==e);

      assert_cycle_sequence
         #(1, 3, 0, 0, "Err: State sequence not followed", 0)
         ACS (clk, ~rst, {(current==d), (current==e),
            (current==a)});

      assert_next #(1, 2, 1, 0, 0, "Err: Output state not
                     reached", 0)
                  AN1 (clk, ~rst, (current==c && x==0), (z==1));

endmodule
```

Figure 6.31   Verilog Code of *Sequencing_Machine*

this assertion monitor for checking that the last state of the sequence is reached if the previous states are reached in the specified sequence. If a value of 1 were used for this parameter, this assertion monitor would be configured for checking the sequencing of all remaining states if the first state were reached.

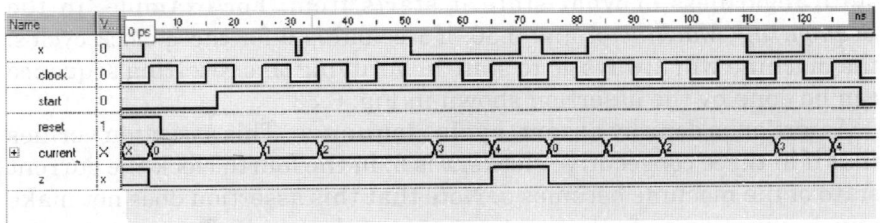

Figure 6.32   *Sequencing_Machine* State Transitions

The arguments of **assert_cycle_sequence** indicate monitoring on the rising edge of the clock while *rst* is 0. The sequence of states to verify are concatenated to form the third argument of this assertion monitor. This argument is referred to as the *event sequence* of the assertion.

6.4.3.5   assert_next.   The **assert_next** assertion uses the syntax shown below, and verifies that starting and an ending events occur with a specified number of clocks in between.

```
assert_next
        #( severity_level, num_cks, check_overlapping,
           check_missing_start, property_type,
           msg, coverage_level )
        instance_name ( clk, reset_n, start_event, test_expr )
```

We can use this assertion monitor to verify traveling time of a walking **1** from one bit position of the *Walking_One* module to another bit position. Another example use of this assertion monitor is shown in the Verilog code of *Sequencing_Machine* of Fig. 6.31. In this design we are verifying that there are two clock cycles between the time that *current* becomes *c* while *x* is **0**, and the time that *z* becomes **1**. The start expression is *(current == c && x == 0)* and the test expression is *(z == 1)*. The second parameter of this assertion (*num_cks*) specifies the number of clock cycles between the events.

6.4.4   Assertion templates

In the previous section we discussed the use of assertion monitors and showed several assertions and related examples. The examples concentrated on common usage of these assertions. In this section we focus on hardware features designers may need to verify, and show how assertions can be used for their verification.

6.4.4.1   Reset sequence.   Often controllers have a resetting sequence that with certain sequence of inputs, the machine ends up in a certain state regardless of what state it starts from. For example, in the *Sequencing_Machine* of Fig. 6.30, if *x* remains **0** for three clock cycles, the machine will always go to state *a*. Verifying this resetting sequence can be done by the assertion shown in Fig. 6.33.

Assertion is done by **assert_cycle_sequence**. This assertion verifies that if in three consecutive clocks, *x* is **0**, in the fourth clock the current state of the machine becomes *a*. Note that this assertion does not make any assumption as to the starting state of the machine.

```
module Sequencing_Machine (input x, start, rst, clk, output z );
    parameter a=0, b=1, c=2, d=3, e=4;
    // . . .

    assert_cycle_sequence
    #(1, 4, 0, 0, "Err: Resetting does not occur", 0)
    ACS2 (clk, ~rst, {(x==0), (x==0), (x==0), (current==a)});
    // . . .

endmodule
```

Figure 6.33   Assertion Reset Sequence

6.4.4.2   Initial resetting. For verification of many sequential circuits, it becomes necessary to check for resetting the circuit using a synchronous or asynchronous reset input. For an example for this situation, consider the Mealy machine of Fig. 6.34. The **assert_next** assertion monitor is always active because we are passing a *1'b1* value to its enabling argument. The assertion checks if *rst* is 1, then the next *current* state becomes *reset*. For the description shown in Fig. 6.34, the assertion passes and never fires.

6.4.4.3   Implication. A useful assertion for checking expected events, or events implied by other events, is the **assert_implication** assertion. As shown below, this assertion has an expression for *antecedent* and one for *consequence*.

```
assert_implication
        #( severity_level, properety_type,
           msg, coverage_level )
    instance_name ( clk, reset_n,
                        antecedent_expr, consequence_expr )
```

The **assert_implication** assertion monitor checks on the specified clock edge for the antecedent expression (*antecedent_expr*) to be true. If it is, then it checks for the consequence expression (*consequence_expr*) to be true. If so, it will stay quiet, otherwise it will fire.

An example for **assert_implication** is shown in Fig. 6.35. This assertion is written for the Mealy machine of Fig. 6.34. The assertion checks the output value in the *got10* state while *x* is 1. It reads as: it is implied that *z* is 1 when *current* is *got10* and *x* is 1.

The assertion shown in this example is always active because of the *1'b1* argument. On the rising edge of the clock, if *current* is found to be *got10* and *x* is 1, it expects the *z* output to be 1 on the same clock edge; otherwise the message of its *msg* parameter will be displayed.

```
module mealy_detector (input x, rst, clk, output z);
    localparam [1:0]
        reset   = 0,    // 0 = 0 0
        got1    = 1,    // 1 = 0 1
        got10   = 2;    // 2 = 1 0
    reg [1:0] current;
    always @( posedge clk ) begin
        if (rst) current <= reset;
        else case ( current )
            reset: if( x==1'b1 ) current <= got1;
                   else current <= reset;
            got1:  if( x==1'b0 ) current <= got10;
                   else current <= got1;
            got10: if( x==1'b1 ) current <= got1;
                   else current <= reset;
            default: current <= reset;
        endcase
    end
    assign z = ( current==got10 && x==1'b1 ) ? 1'b1 : 1'b0;

    assert_next
        #(1, 1, 1, 0, 0, "Err: Machine does not reset
           properly", 0)
        AN1 (clk, 1'b1, rst, (current==reset));

endmodule
```

Figure 6.34   Hard-Reset Assertion

6.4.4.4 Valid states.

In the sequential circuit testing it often becomes necessary to check for the machine's valid states and issue a warning if the machine enters an invalid state. If the states of the machine being tested are consecutive binary numbers, the **assert_no_overflow** assertion monitor shown below can be used for this purpose.

```
assert_no_overflow
        #( severity_level, width, min, max, property_type,
           msg, coverage_level )
        instance_name ( clk, reset_n, test_expr )
```

Consider for example the Mealy machine of Fig. 6.34. This machine has two state variables, allowing four states. Of the four possible states, it is using only three, i.e., *reset, got1, got10*. The assertion shown in Fig. 6.36 fires if the Mealy machine ever enters the machine's invalid state.

The **assert_no_overflow** assertion in Fig. 6.36 uses 1, 2, 0, and 2 for its first four parameters. The second parameter (2) is the width of the

```
module mealy_detector2 (input x, rst, clk, output z);
    // . . .

    assert_implication
            #(1, 0, "Err: Output not asserted", 0)
        AI1 (clk, 1'b1, (current==got10 && x), (z==1));

    // . . .
endmodule
```

Figure 6.35 Asserting Implication

vector being tested. The third and fourth parameters are the range of values (*min* to *max*) the test expression can take. This assertion makes sure *current* never exceeds 2.

We have shown several applications of assertion monitors. We have kept this discussion simple and have only shown direct use of several assertions. Combining several assertion monitors, and using auxiliary logic to adapt assertions to a specific design, provide a very complete design validation environment.

6.5 Text Based Testbenches

Verilog has an extensive set of tasks for reading and writing external files. These tasks include tasks for opening and closing files, positioning a pointer in a file, writing or appending a file, and reading files. Appendix B has a complete list of these tasks and a brief description of each.

Instead of presenting our text-based testbenches in this chapter, we postpone this topic until Chap. 8, where more complete systems are discussed. In there, several large examples will be presented and their corresponding testbenches are discussed. In all systems designed in

```
module mealy_detector2 (input x, rst, clk, output z);
    // . . .

    assert_no_overflow #(1, 2, 0, 2, 0, "Err: Invalid state", 0)
            ANV1 (clk, 1'b1, current);

    // . . .
endmodule
```

Figure 6.36 Checking for Invalid States

Chap. 8, we take advantage of external data files for reading test data, and our design results are also written into external files.

6.6 Summary

This chapter discussed the use of Verilog constructs for developing testbenches. In the first part of the chapter we focused on data generation and response analysis by use of Verilog. In the second part, we showed how assertion monitors could be used for reducing efforts needed for response analysis of a unit-under-test. Developing good testbenches for complex designs requires design observability given to designers by use of assertions. Appendix E discusses assertion monitors that are distributed by Accellera at the time of writing of this book. At this time OVL version 1.0 is released and updates will be available from Accellera.

Problems

6.1 Write an interactive testbench for the multiplier of Prob. 5.12. Your testbench must be complete with clock and proper timing and multiplier handshaking. Use the following interface for the multiplier,

```
module mult (a, b, start, clk, r, done);
    input [7:0] a, b;
    input start, clk;
    output [15:0] r;
    output done;
    . . . . .
endmodule
```

6.2 Add a part to the testbench of Prob. 6.1 to calculate the expected result and compare it with that of the multiplier result. Every time *start* becomes 1 pick up the data that is being applied to the multiplier, perform the multiplication by the * operator, wait for the multiplier to complete its multiplication, and then compare the expected result with the multiplier result. If they are not the same an error signal should be issued.

6.3 To the testbench of Prob. 6.2 add a part that will issue a display message each time the error signal becomes 1.

6.4 Using **assert_always** assertion monitor in the testbench of Prob. 6.1, continuously check the result every time *done* becomes 1. Display an error message if the result is wrong.

6.5 Rewrite the assertion of the BCD counter of Fig. 6.24. Instead of **assert_always**, use **assert_no_overflow** assertion monitor to check the BCD counting.

Suggested Reading

Accellera, Open Verification Library: *Assertion Monitor Reference Manual*, www. accellera.org, v1.0, 2005.

Bening, L., and H.D. Foster, *Principles of Verifiable RTL Design Second Edition–A Functional Coding Style Supporting Verification Processes in Verilog*, 2d ed. Springer, Boston, 2001, ISBN: 0792373685.

Foster, H.D., A.C. Krolnik, and D.J. Lacey, *Assertion-Based Design*, 1st ed. Springer, Boston, 2003, ISBN: 1402074980.

IEEE Std 1364-2001, *IEEE Standard Verilog Language Reference Manual*, SH94921-TBR (print) SS94921-TBR (electronic), ISBN 0-7381-2827-9 (print and electronic), 2001.

Lam, W. K., *Hardware Design Verification: Simulation and Formal Method-Based Approaches*, Prentice Hall PTR, New Jersey, 2005, ISBN: 0131433474.

Navabi, Z., *Digital Design and Implementation with Field Programmable Devices*, Kluwer Academic Publishers, Boston, 2005, ISBN: 1-4020-8011-5.

Navabi, Z., *Verilog Computer-Based Training Course*, CBT CD with hardcopy User's manual, McGraw-Hill, New York, 2002, ISBN 0-07-137473-6.

7

Detailed Modeling

The previous chapters presented Verilog from a design point of view. Except in a few cases, most of the constructs and styles of coding that we discussed were synthesizable. Although describing hardware for synthesis covers the majority of cases that a hardware description language (HDL) like Verilog is used in industry, there are also cases that an existing hardware needs to be modeled. Take for example, developing components of an application-specific integrated circuits (ASIC) library for post-synthesis simulation, or modeling very large-scale integration (VLSI) components to be used in a hierarchical multi-level design description. In such cases, there may be a need to go beyond gate level details of a structure and develop models to exhibit timing signal strengths, and signal values that are less abstract than just **0** and **1**.

There are two facilities in Verilog for detailed modeling beyond what we have already discussed: transistor (or switch) level modeling and signal strengths. These facilities are independent, but can be combined to facilitate modeling hardware for a very detailed simulation. This chapter presents switch level modeling and signal strengths in Verilog. We show simulation results that can be obtained by hardware models that take advantage of such language facilities.

7.1 Switch Level Modeling

Usually, transistor level modeling is referred to modeling hardware structures using transistor models with analog input and output signal values. On the other hand, gate level modeling refers to modeling hardware structures using gate models with digital input and output signal values. Between these two modeling schemes is what is referred to as switch level modeling. At this level, a hardware component is described at the transistor level, but transistors only exhibit digital behavior and their input, and output signal values are only limited to digital values. At the switch level, transistors behave as on-off switches. Verilog uses a 4-value logic value system, so Verilog switch input and output signals can take any of the four **0**, **1**, **Z**, and **X** logic values. Switch constructs, their simulation behavior and simulation of hardware constructs based on such switches will be discussed here.

7.1.1   Switch level primitives

Figure 7.1 shows Verilog switch and pull primitives. Switches are unidirectional or bidirectional and resistive or nonresistive. For each group we have those primitives that switch on with a positive gate (like an NMOS transistor) and those that switch on with a negative gate (like a PMOS transistor). Switching on means that logic values flow from input of a transistor to its output. Switching off means that the output of a transistor is at **Z** level regardless of its input value.

A unidirectional transistor passes its input value to its output when it is switched on. A bidirectional transistor conducts both ways. A resistive structure reduces the strength of its input logic when passing it to its output. Strengths will be discussed in the next section. In addition to switch level primitives, Fig. 7.1 lists pull-primitives that are used as pull-up and pull-down resistors for tri-state outputs.

Figure 7.2 shows standard switches, pull primitives, and tri-state gates that behave like **nmos** and **pmos**. Instantiations of these primitives and their corresponding symbols are also shown. **cmos** is a unidirectional transmission gate with a true and complemented control lines. **nmos** and **pmos** are unidirectional pass gates representing NMOS and PMOS transistors respectively. Not shown in Fig. 7.2 are **rcmos**, **rnmos**, and **rpmos** that are the resistive versions of **cmos**, **nmos**, and **pmos**. When such a

SWITCHES	Unidirectional	Bidirectional	PULL-GATES
Standard	cmos nmos pmos	tran tranif1 tranif0	pullup
Resistive	rcmos rnmos rpmos	rtran rtranif1 rtranif0	pulldown

Figure 7.1   Switch Level Primitives

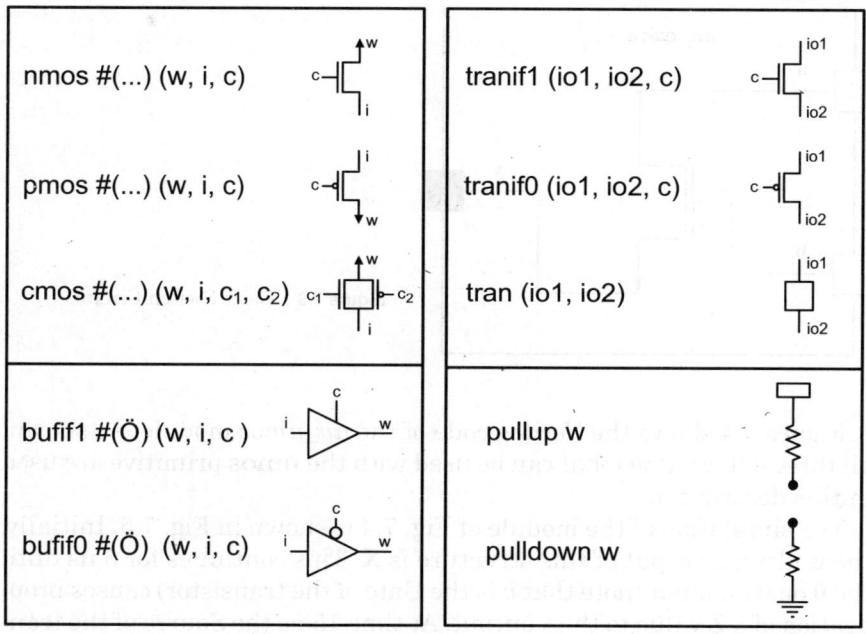

Figure 7.2   Switch Level Primitive Instantiations

resistive switch conducts, the strength of its output signal is one or two levels below that of its input signal.

Delay values for transition to **1**, transition to **0**, and transition to **Z** can be specified in the # (*to-1, to-0, to-Z*) format for unidirectional switches. Bidirectional **tran** switches shown in Fig. 7.1 are functionally equivalent to unidirectional switches shown in the adjacent column of this figure. The difference is that they have a control input and two inout ports. When conducting, the two inout ports are connected and logic values flow in both directions. Furthermore, bidirectional switches cannot have delay values.

Figure 7.2 also shows instantiation of **pullup** and **pulldown** primitives. As mentioned before, these primitives pull a **Z** value on a tri-state gate output to a **1** and **0**, respectively. Pull-gates cannot be given delay values.

Using several examples, the following sections show how the primitives discussed above can be used for modeling standard logic structures such as logic gates and memory elements.

7.1.2   The basic switch

The behavior of the basic switch can be illustrated by simulating the **nmos** circuit shown in Fig. 7.3.

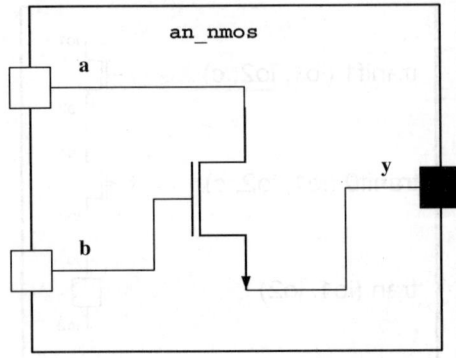

Figure 7.3   Basic Switch Circuit

Figure 7.4 shows the Verilog code of the *an_nmos* module. As shown, all three delay values that can be used with the **nmos** primitive are used in this description.

The simulation of the module of Fig. 7.4 is shown in Fig. 7.5. Initially, the undriven output of this structure is **X**. This continues for 5 ns until the **0** on its *b* input (note that *b* is the Gate of the transistor) causes propagation of a **Z** value to the *y* output. At time 15 ns the Source of the transistor (input *a*) changes to **1**. This does not change the output of the transistor since its Gate (input *b*) is still **0**. However at time 30 ns when *b* changes to **1**, the **1** on the input *a* transfers to *y* after the specified 3 ns delay. Other changes on *y* happen with a 4 ns and a 5 ns delay for its changing to **0** and to **Z** respectively.

If fewer than three delay values are used when a unidirectional switch is instantiated, the minimum of all specified delay values will be used for the missing values.

7.1.3   CMOS gates

A 2-input NAND gate using NMOS and PMOS transistors is shown in Fig. 7.6. The inputs of the **pmos** primitives are tied to *Vdd* to supply logic **1** to the output, and the input of the lower **nmos** primitive is tied to *Gnd* to supply logic **0** to the output. For unidirectional switches, the switch input is the transistor Source, its output is the Drain, and the

```
`timescale 1ns/10ps
module an_nmos ( input a, b, output y );
    nmos #(3, 4, 5) (y, a, b);
endmodule
```

Figure 7.4   Verilog Code of a Basic Switch

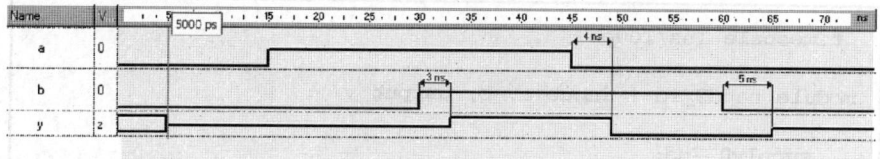

Figure 7.5   Simulation of the Basic Switch

switch control input is the transistor Gate input. The input-output arrangement of switches (shown in Fig. 7.6) are such that the input sides of the **nmos** switches are on the *Gnd* side and the input sides of the **pmos** switches are on the *Vdd* side. This arrangement is justified by an actual transistor level circuit because the Source of an NMOS transistor feeds logic **0** to the output (discharging the output throught *Gnd*), and the Source of a PMOS transistor feeds logic **1** to the output (charging it through *Vdd*) of a CMOS gate.

Figure 7.7 shows the Verilog code of the *nand2_1d* circuit. **Net** declarations **supply0** and **supply1** declare *Gnd* and *Vdd* **nets**. The *im1*

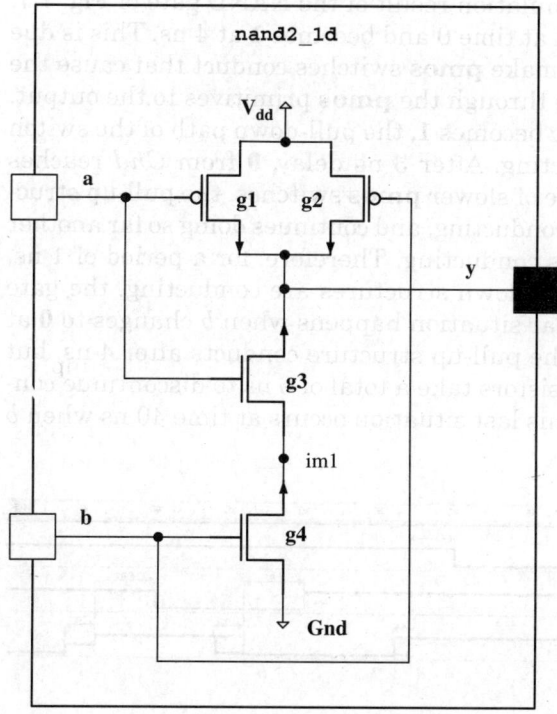

Figure 7.6   Switch Level 2-input NAND Gate

```verilog
`timescale 1ns/100ps

module nand2_1d ( input a, b, output y );

    supply0 Gnd;
    supply1 Vdd;

    wire im1;

    pmos #(4) g1 ( y, Vdd, a );
    pmos #(4) g2 ( y, Vdd, b );
    nmos #(3) g3 ( y, im1, a );
    nmos #(3) g4 ( im1, Gnd, b );

endmodule
```

Figure 7.7   CMOS 2-input NAND

wire connects output of *g4* **nmos** to the input of *g3*. The output of the circuit is driven simultaneously by outputs of *g1*, *g2*, and *g3*.

Figure 7.8 shows the simulation result of the NAND gate of Fig. 7.7. The output *y* starts with **X** at time 0 and becomes **1** at 4 ns. This is due to the **0** input values that make **pmos** switches conduct that cause the flow of *Vdd* supply voltage through the **pmos** primitives to the output. At time 20 ns when input *a* becomes **1**, the pull-down path of the switch level circuit starts conducting. After 3 ns delay, **0** from *Gnd* reaches output *y*. However, because of slower **pmos** switches, the pull-up structure has not discontinued conducting, and continues doing so far another 1 ns after the pull-down is conducting. Therefore, for a period of 1 ns, when both pull-up and pull-down structures are conducting, the gate output becomes **X**. A similar situation happens when *b* changes to **0** at time 30 ns. In this case, the pull-up structure conducts after 4 ns, but two series pull-down transistors take a total of 6 ns to discontinue conducting. The opposite of this last situation occurs at time 40 ns when *b*

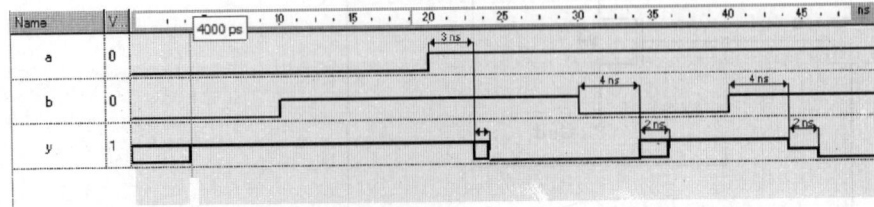

Figure 7.8   NAND Gate Detailed Simulation

becomes **1**. Because of delay differences in the pull-up and pull-down paths, the *y* output becomes float (**Z**) for a period of 2 ns before it reaches its stable value of **0** at time 46 ns.

The simulation report obtained from the switch level model of the CMOS NAND gate includes far more details than simulating a 2-input **nand** primitive. The price we are paying for this detailed simulation is the simulation performance.

As another example of a CMOS gate structure, consider the 4-input AND-OR-INVERT (AOI) gate of Fig. 7.9. A pull-down structure using

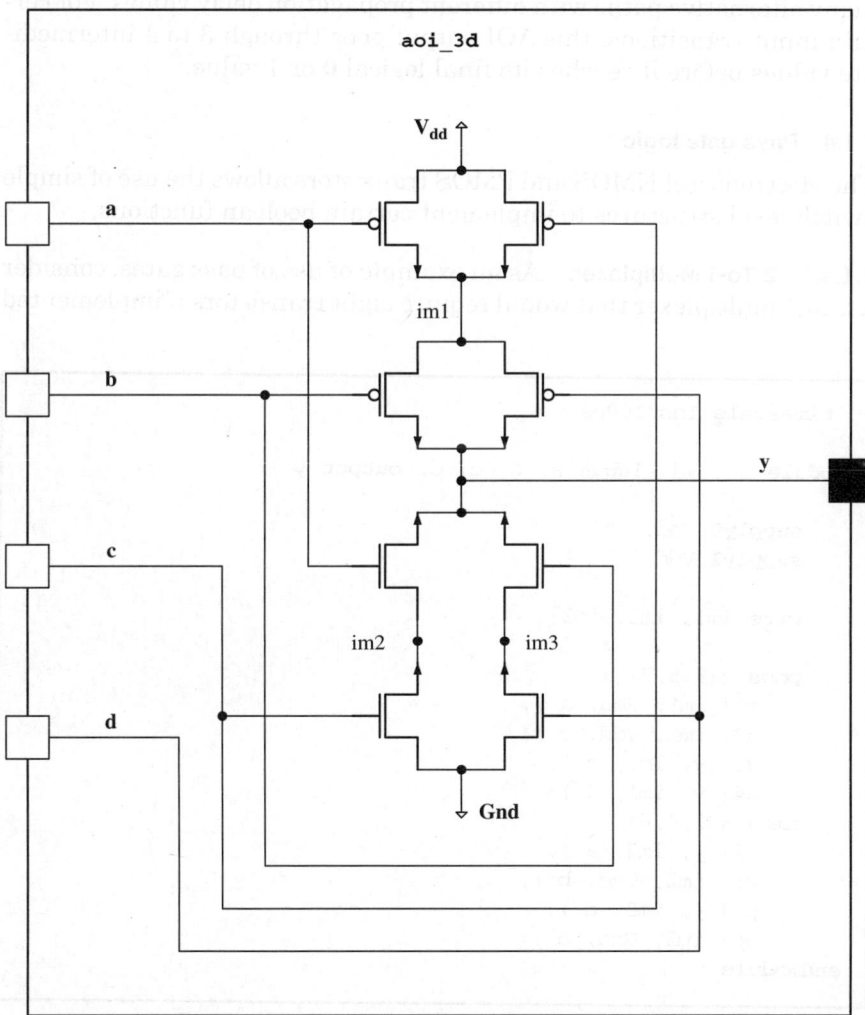

Figure 7.9 AOI Gate

NMOS transistor and its complementary pull-up structure with PMOS transistors is shown here.

The Verilog code of Fig. 7.10 is the switch level model for this AOI gate. As shown, three delay values are used for **nmos** and **pmos** primitives of this model. These values are for to-**1**, to-**0** and to-**Z** transitions, and transitions of the switch outputs to the **X** value use the minimum of the three delay values specified here.

Analysis of the timing of the output of this AOI gate is more complex than that of the NAND gate of Fig. 7.7. This is partly due to the fact that we are using three delay values for AOI switches. Another factor contributing to the complexity of timing of this circuit is the existence of many alternative paths with different propagation delay values. For certain input transitions, this AOI output goes through 3 to 4 intermediate values before it reaches its final logical **0** or **1** value.

7.1.4 Pass gate logic

The electronics of NMOS and PMOS transistors allows the use of simple switch level structures to implement certain boolean functions.

7.1.4.1 2-To-1 Multiplexer. As an example of use of pass gates, consider a 2-to-1 multiplexer that would require eight transistors if implemented

```
`timescale 1ns/100ps

module aoi_3d (input a, b, c, d, output y );

    supply0 Gnd;
    supply1 Vdd;

    wire im1, im2, im3;

    pmos #(3,5,7)
        g1( im1, Vdd, a ),
        g2( im1, Vdd, b ),
        g3( y, im1, c ),
        g4( y, im1, d );
    nmos #(2,4,6)
        g5( y, im2, a ),
        g6( im2, Gnd, b ),
        g7( y, im3, c ),
        g8( im3, Gnd, d );
endmodule
```

Figure 7.10   AOI Verilog Code

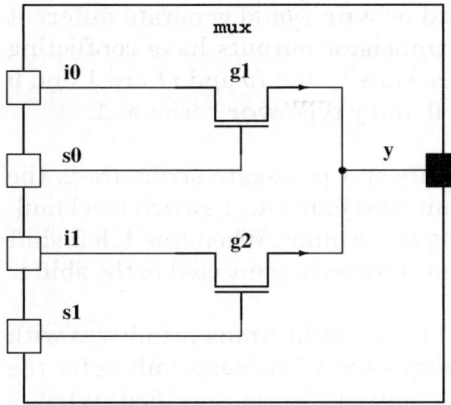

Figure 7.11   2-to-1 Multiplexer
Using Pass Gates

as a CMOS structure with a pull-up and a pull-down (similar to the AOI of Fig. 7.9). This circuit can be realized with only two transistors using pass gates. Figure 7.11 shows such a multiplexer.

In this circuit, *g1* or *g2* conducts when *s0* or *s1* is **1**, which allows *i0* or *i1* to propagate to *y*. Because of the transistor threshold values, and their resistance, the *y* output is at a weaker logic level than if this structure was build as a CMOS complex gate. Verilog models this strength reduction using its signal strength modeling that will be discussed in the next section.

The Verilog code of the *mux* circuit of Fig. 7.11 is shown in Fig. 7.12. Note the direction of unidirectional transistors, and note that output *y* is driven by the outputs of both switches. Output *y* is declared as a **net** of **wire** type. Therefore, conflicting values on transistor outputs will be handled by the **wire** resolution function. If both *g1* and *g2* conduct and *i0* and *i1* are **1** and **0** respectively, the *y* output becomes **X**.

```
`timescale 1ns/100ps

module mux (input i0, i1, s0, s1, output y );
    wire y;
    nmos #(4)
        g1( y, i0, s0 ),
        g2( y, i1, s1 );

endmodule
```

Figure 7.12   Switch Level 2-to-1 Multiplexer

Declaring *y* as a **net** of type **wand** or **wor** would generate different results than the **wire** type when transistor outputs have conflicting values. For example, if *s0* and *s1* are both **1**, and *i0* and *i1* are **1** and **0** respectively, *y* type **wand** becomes **0** and *y* type **wor** becomes **1**.

7.1.4.2 4-Bit shifter.   Another example of a pass-gate structure is the shifter shown in Fig. 7.13. This circuit uses four 2-to-1 switch level multiplexers and an inverter for decoding the *ls* input. When *ls* is **1**, left shift takes place, otherwise, the 4-bit input is directly connected to the shifter output.

The *shifter* Verilog code of Fig. 7.14 uses eight **nmos** primitives with a 3 ns delay value, and **pmos** and **nmos** with 5 ns delay values for the inverter. Since the **net** type for the outputs is not specified, **wire** is assumed. This makes each bit of this shifter behave like the multiplexer of Fig. 7.12.

The timing diagram of Fig. 7.15 shows the simulation run of the shifter. Notice that when a data input (*i3, i2, i1, i0*) changes (e.g., *i1* become **0** at 40 ns), its effect appears on the corresponding output after 3 ns. On the other hand, when *ls* changes, the outputs change after going through a transitional value of **X** or **Z** for 5 ns. This is because of the inverter delay of 5 ns. When *ls* become 1 (e.g., time 20 ns), both switches driving a changing output conduct, and cause a 5 ns period of **X** on that output. The opposite of this happens when *ls* becomes **0** (see time 60 ns). In this case, all transistors that drive *shifter* outputs go into the off mode for a period of 5 ns, causing *y3, y2, y1,* and *y0* to become **Z** for 5 nanoseconds.

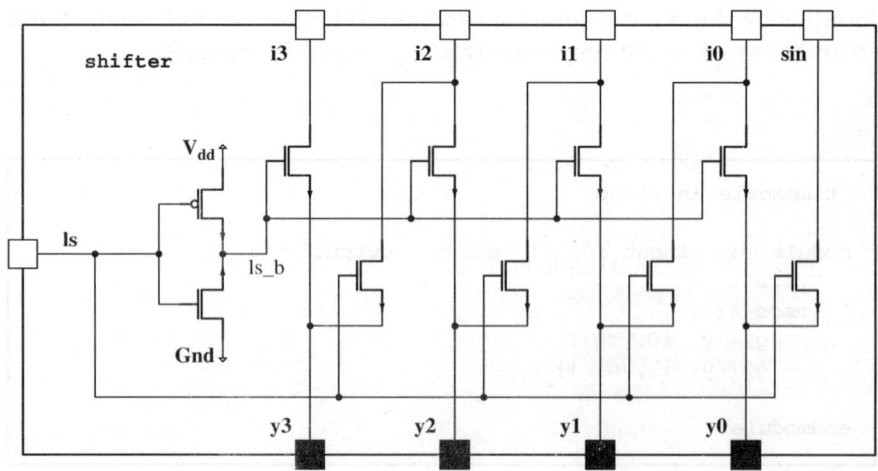

Figure 7.13 Switch Level Shifter

```
`timescale 1ns/100ps

module shifter (input i3, i2, i1, i0, sin, ls,
                output y3, y2, y1, y0 );
    supply1 Vdd;
    supply0 Gnd;

    nmos #(3)
        ( y0, sin, ls ),
        ( y0, i0 , ls_b ),
        ( y1, i0, ls ),
        ( y1, i1 , ls_b ),
        ( y2, i1, ls ),
        ( y2, i2 , ls_b ),
        ( y3, i2, ls ),
        ( y3, i3 , ls_b );
    nmos #(5) ( ls_b, Gnd, ls );
    pmos #(5) ( ls_b, Vdd, ls );
endmodule
```

Figure 7.14   Shifter Verilog Code

7.1.4.3   Barrel shifter.

Figure 7.16 shows the design of a 4-bit barrel shifter. The i ports are the data inputs and y ports are the outputs. Data inputs appear on outputs 0, 1, 2, or 3 positions shifted to the right. Shift positions are determined by $l0$, $l1$, $l2$, or $l3$. If $l0$ is **1**, no shifting is done. If $l1$ is **1**, y3, y2, y1, and y0 receive $i0$, $i3$, $i2$, and $i1$ respectively.

Figure 7.17 shows the Verilog code of the barrel shifter. This code uses nested **generate loop**s and **if** statements for wiring the two-dimensional switch array of Fig. 7.16. This Verilog code uses *SIZE* for the size of the barrel shifter.

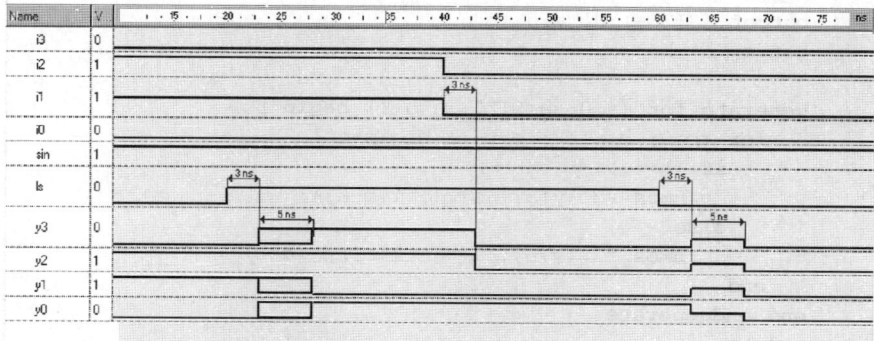

Figure 7.15   Shifter Simulation Run

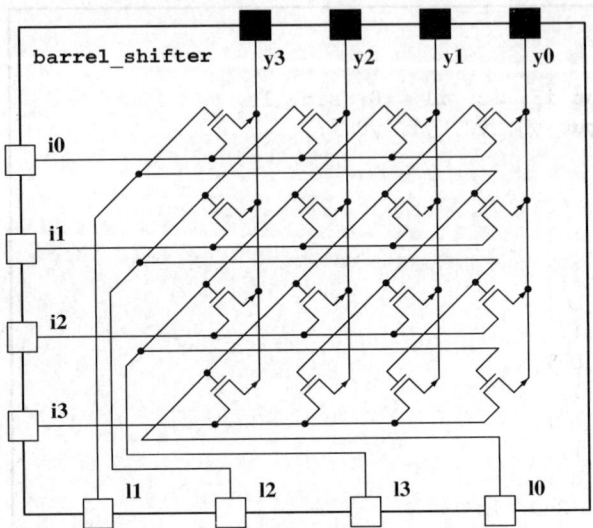

Figure 7.16 Switch Level Barrel Shifter

7.1.5 Switch level memory elements

Static memory circuits can be built using standard gate level structures as discussed in Chap. 5. Switch level versions of these structures can be built using CMOS gates described in Sec. 7.1.3. Describing dynamic and pseudo-static memory elements require utilization of gate capacitances

```
`timescale 1ns/100ps

module barrel_shifter (i, l, y);
   parameter SIZE = 4;
   input [SIZE-1:0] i, l;
   output [SIZE-1:0] y;
   genvar a, b;

   generate for (a=0; a<SIZE; a=a+1) begin:row
      for (b=0; b<SIZE; b=b+1) begin:col
         if (b>=a)
            nmos #2 (y[a], i[b], l[b-a]);
         else
            nmos #2 (y[a], i[b], l[SIZE-(a-b)]);
      end
   end endgenerate
endmodule
```

Figure 7.17 Barrel Shifter Verilog Code

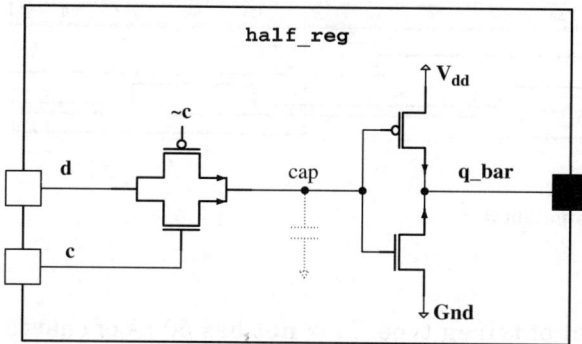

Figure 7.18 Switch Level Half-Register

and use of bidirectional or unidirectional switches. This section shows several switch level memory elements that use gate capacitors for storage.

7.1.5.1 Half register.

A simple half-register can be built using a pass transistor or a CMOS transmission gate to charge gate capacitance of NMOS and PMOS transistors. Figure 7.18 shows *half_reg* that uses a CMOS transmission gate and an inverter. When *c* becomes active, the current from *d* charges *cap*. This charge is complemented by the NMOS-PMOS inverter structure and appears on *q_bar*. The *q_bar* output holds its value for as long as *cap* holds its charge. When *cap* loses its charge, *q_bar* will go into an unknown state. Therefore it is required that this structure is clocked frequently to refresh its charge holding capacitor. The Verilog code of this structure that uses two **nmos** and two **pmos** primitives are shown in Fig. 7.19. Modeling the gate capacitances is done

```
`timescale 1ns/100ps

module half_reg (input d, c, output q_bar);

    supply0 Gnd;
    supply1 Vdd;

    trireg #( 0, 0, 50 ) cap;

    cmos #(0,0,5) ( cap, d, c, ~c );
    nmos #(3) ( q_bar, Gnd, cap );
    pmos #(4) ( q_bar, Vdd, cap );

endmodule
```

Figure 7.19   Half Register Verilog Code

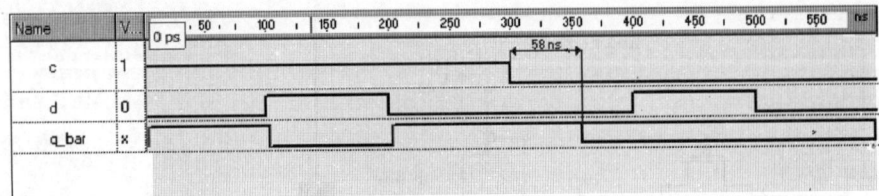

Figure 7.20   Half-Register Simulation Run

by declaring *cap* as a **net** of **trireg** type. This **net** has 50 ns of charge holding capacity. As shown, *cap* is used for the output of **cmos** and input of the output inverter. Without the **trireg** declaration, this structure would not have any storage capability and would only function as a combinational logic.

The timing diagram of Fig. 7.20 demonstrates the storage capability of the *half_reg* module. While *c* is **1**, the complement of *d* appears on *q_bar* with delay values that are associated with **cmos**, **nmos**, and **pmos** primitives. When *c* becomes **0**, *q_bar* holds its last value for the specified **trireg** value plus switch delays as specified in this module. As shown, *q_bar* becomes **X**, 58 ns after *c* becomes **0**. It is expected that a new clock pulse arrives before *q_bar* loses its retained data.

7.1.5.2   Pseudo-static d-latch.   Switch level structure of a pseudo-static D-latch is shown in Fig. 7.21. When *c* is **1**, data on *d* passes through the input CMOS transmission gate, and after passing through two inverters it reaches the *q* output.

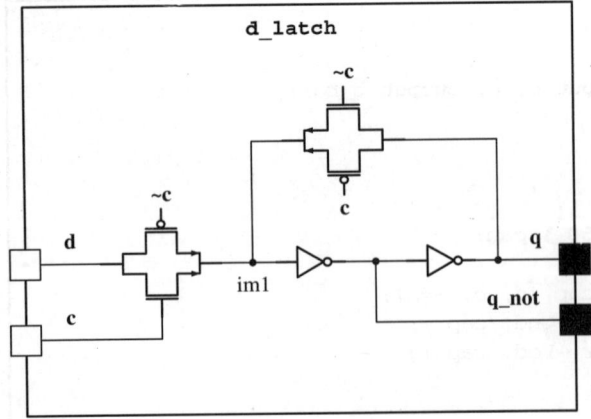

Figure 7.21   Pseudo-Static Transparent Latch

```
`timescale 1ns/100ps

module d_latch (input d, c, output q);

    cmos #(0,0,3)
        ( im1, d, c, ~c ),
        ( im1, q, ~c, c );
    not #(5)
        ( q_not, im1 ),
        ( q, q_not );

endmodule
```

Figure 7.22 Pseudo-Static Latch Verilog Code

When c becomes **0**, the d input is disconnected from q, and instead the CMOS gate on the feedback path feeds the current q back to itself through the inverter pair.

The Verilog code of Fig. 7.22 corresponds to the transparent latch of Fig. 7.21. As shown here, we do not need a **trireg** node, because clocking and refreshing are complementary, and $im1$ is either driven by d or q. Note that $im1$ is the node that this structure uses for holding its state.

7.1.5.3 Cross-couple SRAM memory.   A six-transistor static memory element is shown in Fig. 7.23. When wr becomes **1**, data from d is stored in the cross-couple inverters. When rd becomes **1**, the complement of this data appears on q_bar. When writing into the memory cell, the strength of the output of the **nmos** primitive that drives d into $im\hat{1}$ must be higher than the strength of the **not** primitive driving this node. This requires

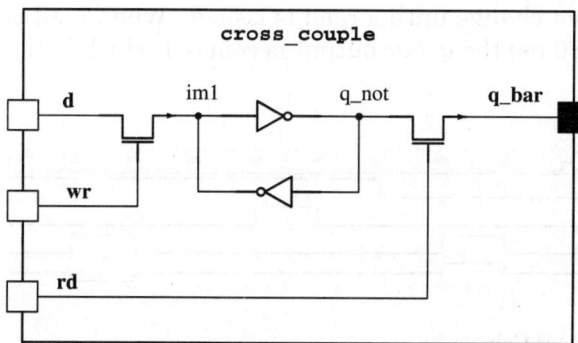

Figure 7.23 Static Memory with Cross-Couple Inverters

```
`timescale 1ns/100ps

module cross_couple (input d, wr, rd, output q_bar);

    wire #(5) q_not;
    wire #(3) im1, q_bar;

    nmos
        ( im1, d, wr ),
        ( q_bar, q_not, rd );
    not ( q_not, im1 );
    not (pull0, pull1) ( im1, q_not );

endmodule
```

Figure 7.24   Verilog Code for SRAM Cell

a special electronics for "over driving" the inverter output. Figure 7.24 shows the Verilog code of the SRAM cell. Cell delays are described by associating delay values with **wire** declaration of internal nodes of the cell. The memory cell uses two **nmos** primitives for the read and write hardware, and two **not** primitives for the cross-couple structure.

In Verilog, the default strength for the outputs of primitives used here is **strong**. If *d* is driven by a **strong** logic, the write **nmos** must be able to over-drive the existing cell value. Therefore, we have used **pull** strength (**pull1**, **pull0**) for the inverter driving the same node (*im1*) as this **nmos**.

Figure 7.25 shows the simulation run of the *cross_couple* module. As the timing diagram shows, *wr* becomes **1** at 20 ns. This causes the **0** on *d* to be written into the latch internal node, *im1*, and its complement to go into *q_not*. While this is happening, the *q_bar* output of this structure remains at **Z**, and does not change until a read is issued. When read is issued (*rd* becomes **1** at 80 ns) the *q_bar* output becomes **1** which is the

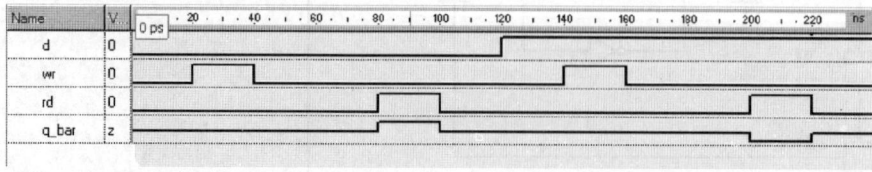

Figure 7.25   Simulating the SRAM Cell

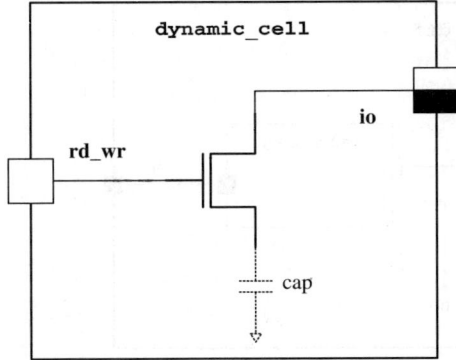

Figure 7.26 One-Transistor Dynamic Cell

complement of the input data put into this latch at time 20 ns. The timing diagram of Fig. 7.25 shows another write and read cycle; writing a **1** and reading its complement are demonstrated at 140 ns and 200 ns respectively.

7.1.5.4 Dynamic cell.   Figure 7.26 shows a one-transistor dynamic RAM cell. The *cap* capacitance is where this cell stores its data. When *rd_wr* is **0**, the cell data remains in *cap*. Reading is done by setting *rd_wr* to **1** and sensing *io*. Writing is done by setting *rd_wr* to **1** and driving *io* with input data.

Figure 7.27 shows the Verilog code for this dynamic memory cell. The *io* signal is declared as **inout** and issued for input and output of this cell. The charge holding capacitance is modeled by declaring *cap* as a **trireg net**. The bidirectional transistor is modeled by the **tranif1** primitive. The first two ports of this bidirectional primitive are its **inout** ports, and the third is its control.

```
`timescale 1ns/100ps

module dynamic_cell (inout io, input rd_wr);

    trireg #(0, 0, 50) cap;

    tranif1 #(5) ( cap, io, rd_wr );

endmodule
```

Figure 7.27 Dynamic Memory Cell

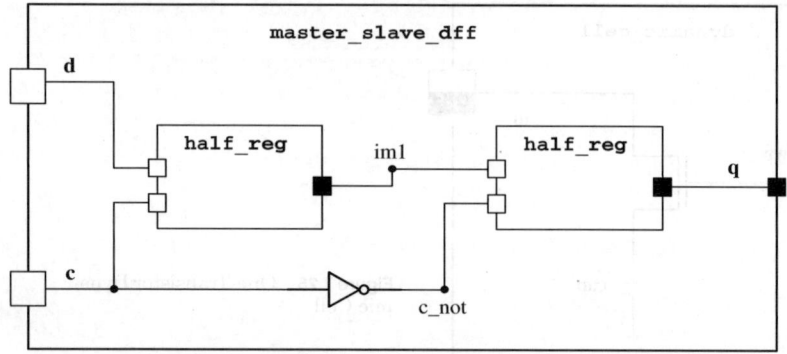

Figure 7.28 Master Slave Flip-Flop

When writing into this cell, *rd_wr* must be driven with a **1**, and input data **0** or **1** must be placed on *io*. When reading this cell, *rd_wr* must be **1**, and *io* must externally be set to **Z**. In this case, the cell data (**0** or **1**) will override the tri-state *io* value.

7.1.5.5 Master-slave flip-flop.

Using two *half_reg* modules (Fig. 7.18) and an inverter, builds a master-slave D-type flip-flop (Fig. 7.28). When *c* is **1**, complement of *d* goes into *im1*, and when *c* becomes **0** the complement of *im1* appears on *q*. With a slight timing difference, the functionality of this circuit is like that of a falling edge trigger flip-flop.

The simulation of *master_slave_dff* module of Fig. 7.29 is shown in Fig. 7.30. As shown, the *q* output takes the value of *d* with a slight delay after the negative edge of the clock. This diagram also shows a temporary transitional *q* value of **Z** at 119 ns. The changing of *q* goes through **Z** or **X** transitional value, because of the difference in delay values of the **nmos** and **pmos** transistors at the output of the slave half-register (see Fig. 7.19, **nmos** uses 3 ns, **pmos** uses 4 ns).

```
`timescale 1ns/100ps

module master_slave_dff (input d, c, output q);

    half_reg master ( d, c, im1 );
    half_reg slave ( im1, c_not, q );
    not ( c_not, c );

endmodule
```

Figure 7.29 Master-Slave Verilog Code

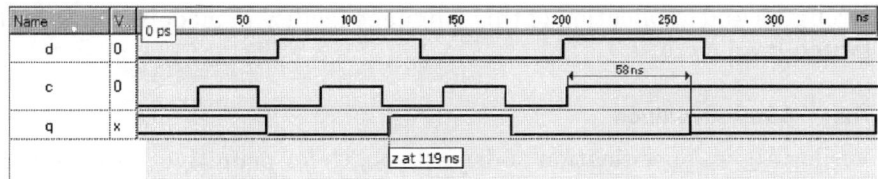

Figure 7.30   Simulation of Dynamic Master-Slave D Flip-flop

Another behavior of this module, which is also contributed to the way half-registers behave, is the fact that if the flip-flop is not clocked for a relatively long time, it loses its data. Note in Fig. 7.30 that clock c stops ticking at 200 ns. After 58 ns, the flip-flop output goes into unknown **X** state. The 58 ns discharge time is also shown in the simulation of the half-register in Fig. 7.20.

Figure 7.31 shows a 4-bit register that uses four instances of the master-slave flip-flop of Fig. 7.29. As in the case of the individual flip-flops, this register loses its stored data if it is not clocked for over 50 ns.

The discussions and examples of switch level modeling presented in this section demonstrated that detailed simulation data can be obtained by using such models. Although we only covered basic logic primitives, the concepts presented can be used in modeling dynamic logic and various clocked gates. The next section presents signal strengths which is still another feature for low level modeling.

7.2   Strength Modeling

The four-value logic in Verilog provides an adequate precision for most logic level simulations. The previous section showed that more precise simulation data can be obtained by using switch level models for the basic logic level constructs. Another feature of Verilog for a more precise simulation data is signal strength. This section discusses logic

```
`timescale 1ns/100ps

module register_4 (input [3:0] d, input c, output [3:0] q);
    genvar i;
    generate for (i=0; i<4; i=i+1) begin:bits
        master_slave_dff ff ( d[i], c, q[i] );
    end endgenerate

endmodule
```

Figure 7.31   A 4-bit Register

strengths and application of this language facility in modeling gate and switch level circuits.

7.2.1   Strength values

Verilog allows specification of drive strength for primitive gate outputs and **net**s. Gate output or **net** signal strength values are specified in a set of parenthesis that include a strength value for logic **0** and one for logic **1**. Allowable drive strengths for logic **0** (i.e., *strength0*) are **supply0**, **strong0**, **pull0**, **weak0**, and **highz0**. Similarly, allowable strengths for logic **1** (*strength1*) are **supply1**, **strong1**, **pull1**, **weak1**, and **highz1**. Strength values can appear in any order in the set of parenthesis that follows a primitive name, a **net** declaration, or the **assign** keyword. The default strengths for a gate output or a **net** are **strong0** and **strong1** for logic **0** and logic **1**, respectively. Charge strengths, representing the strength of a capacitive **net**, are also supported in Verilog. Charge strength values are **large**, **medium**, and **small**.

7.2.1.1   Primitive strengths.   Gate output drive strengths are specified after the primitive name when the primitive is instantiated. The example below shows a **nand** primitive with **pull0** and **pull1** output strength values.

```
nand (pull0, pull1) # (3, 5) n1 (w, a, b, c);
```

Figure 7.32 shows strength values for outputs of various built-in gate types. Strength values in this figure are listed in the descending order, i.e., **supply0** is the highest and **highz0** is the lowest strength for logic **0**.

7.2.1.2   Net strength.   Two types of **net** strengths are *drive_strength* and *charge_strength*. For **wire** and **tri** type **net**s, strength values represent

	LOGIC GATES	PULL GATES		SWITCHES
		Pullup	Pulldown	
Strength0	supply0 strong0 pull0 weak0 highz0		supply0 strong0 pull0 weak0	No strength
Strength1	supply1 strong1 pull1 weak1 highz1	supply1 strong1 pull1 weak1		No strength

Figure 7.32   Primitive Gate Output Strength Values

their drive strengths, while for storage **net**s, i.e., **trireg**, charge strength values can be specified. Possible values for drive strength are **supply0**, **strong0**, **pull0**, **weak0**, **highz0**, **supply1**, **strong1**, **pull1**, **weak1**, and **highz1**. Charge strength values are **large**, **medium**, and **small**.

For the situations that a logic block is specified by a continuous assignment, **net** drive strengths can be specified in a set of parenthesis after the **assign** keyword, or with the **net** declaration that declares the output of the logic block. For example, the *w* output of the following logic block has **(pull0, pull1)** strength values. The strength values specified for *w* are one level lower than the default **strong0** and **strong1** values.

```
assign (pull0, pull1) # (3, 5) w = s ? A: B;
```

Drive strengths of a **net** specify its logic **1** and **0** drive powers. For as long as a **net** is not used as a switch input or not involved in a multi-driver resolution, its weak or strong logic values do not affect the output value of the logic using the **net**. However, strengths of a **net** used as a switch input affect the strength of the output of the switch. Furthermore, strengths of several signals driving a **net**, affect the way **net** logic values are resolved. Details of how drive strength values affect logic values will be discussed in the sections that follow. An example drive strength for a **wand** type **net** is shown below. The declared **net** is *sim* with **pull0** and **supply1** strength values. **pull0** is one level below the **strong0** default strength for logic **0**, and **supply1** is one level higher than the default strength for logic **1**.

```
wand (pull0, supply1) sim;
```

The strength of storage **net**s specify how weak or strong the charge capability of the **net** is. Three strength values, **large**, **medium**, and **small** are used for these **net** types, and the default is **medium**. In a capacitive network, charge strengths propagate from a larger **trireg** net to a smaller **trireg** net. Causing a connection (e.g., with a bidirectional switch) between a **small** undriven **trireg** and a **large** undriven **trireg** of a different logic value, makes the value of the **small trireg** logic value to be overwritten by that of the **large trireg**. An example **trireg** declaration with **small** charge strength is shown below. This level of charge strength is one level below the default level.

```
trireg (small) #(3, 2, 50) cap1;
```

Figure 7.33 shows drive-strength values for **wire** and **tri net**s and charge-strength values for the **trireg net** type.

wire (tri), wand (triand), wor (trior), tri0, tri1		*trireg*	
Strength values	*Level*	*Strength values*	
supply0	7		↑ Strongest
strong0	6		
pull0	5		
	4	**large (0)**	
weak0	3		
	2	**medium (0)**	
	1	**small (0)**	
highz0	0		
highz1	0		Weak
	1	**small (1)**	
	2	**medium (1)**	
weak1	3		
	4	**large (1)**	
pull1	5		
strong1	6		
supply1	7		▼ Strongest

Figure 7.33   **net** Types and their Strengths

7.2.2   Strength used in resolution

Strength values are instrumental in deciding logic values of signals with multiple drivers. If a **net** has multiple drivers of the same strength values, its **net** type, i.e., **wire**, **wand**, or **wor** decides on the final value of the **net**. However, if drive strengths of multiple drivers of a **net** are different, conflicts are resolved by taking the logic value that has a higher strength. In this case resolution type of the **net** (i.e., **wire**, **wand**, or **wor**) does not play any role in determining the **net** value.

Examples that follow show how strength values are altered and how they contribute to logic values of **net**s with multiple drivers. Figure 7.34 shows a network of buffers producing two drivers for every output of the circuit. In two Verilog codes for this circuit we show that resolution functions are only used when drive strengths cannot determine output values.

Figure 7.35 shows a Verilog module that corresponds to the diagram of Fig. 7.34. The $z1$, $z2$, $z3$ outputs are **net**s of types **wire**, **wand**, and **wor** respectively. These outputs are driven by **buf** primitives that have **pull1** and **weak0** output strengths. As shown in the simulation run of Fig. 7.36, in spite of different **net** types of $z1$, $z2$, and $z3$, waveforms generated on these outputs are the same. The reason is, since our Verilog model uses **pull1** for logic **1** and **weak0** for logic **0**, and **pull** is stronger

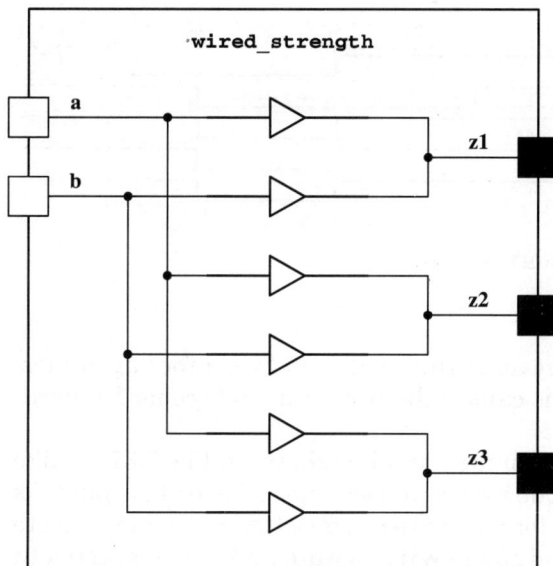

Figure 7.34   Buffers Producing Multiple Drivers

```
`timescale 1ns/100ps

module wired_strength (input a, b, output z1, z2, z3);

    wire z1;
    wand z2;
    wor z3;

    // Wired logic
    buf (pull1, weak0) (z1, a);
    buf (pull1, weak0) (z1, b);h
    // Wired-and logic
    buf (pull1, weak0) (z2, a);
    buf (pull1, weak0) (z2, b);
    // Wired-or logic
    buf (pull1, weak0) a3 (z3, a);
    buf (pull1, weak0) b3 (z3, b);

endmodule
```

Figure 7.35   Verilog Code of Buffer Circuit, Ignoring Resolutions

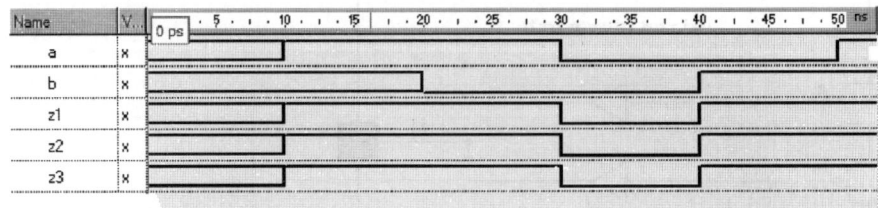

Figure 7.36 Output Waveforms: **pull1, weak0**

than **weak**, in the case that an output is driven by conflicting values, the value **1** overrides **0**. This causes the **net** value to become **1** regardless of its type.

Another Verilog code for the buffer circuit is shown in Fig. 7.37. Unlike the code of Fig. 7.35, this code uses equal strengths for **buf** outputs. As will be seen, this situation forces the use of resolution functions. Note that resolutions of $z1$, $z2$, and $z3$ are **wire**, **wand**, and **wor** respectively.

The simulation run of this circuit (Fig. 7.38) reveals that the three outputs ($z1$, $z2$, and $z3$) are different. This is because **buf** outputs have equal **pull1** and **pull0** strengths for logic **1** and **0**. Therefore conflicting values driving these outputs will be resolved by the resolutions specified

```
`timescale 1ns/100ps

module wired_strength (input a, b, output z1, z2, z3);

    wire z1;
    wand z2;
    wor z3;

    // Wired logic
    buf (pull1, pull0) (z1, a);
    buf (pull1, pull0) (z1, b);
    // Wired-and logic
    buf (pull1, pull0) (z2, a);
    buf (pull1, pull0) (z2, b);
    // Wired-or logic
    buf (pull1, pull0) (z3, a);
    buf (pull1, pull0) (z3, b);

endmodule
```

Figure 7.37 Verilog Code of Buffer Circuit Using Resolutions

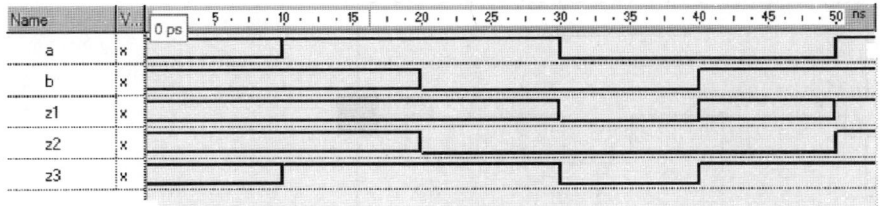

Figure 7.38   Output Waveforms: **pull1, pull0**

for each output. Since *z1*, *z2*, and *z3* use **wire**, **wand**, and **wor** resolutions, these functions are used for determining output values.

7.2.3   Strength reduction

Switch outputs cannot have strength specification. Instead they pass the strength of their inputs to their output with certain changes that depend on the switch type.

The **nmos, pmos,** and **cmos** switches pass the strength of their inputs to their outputs, except that a **supply** strength is reduced to **strong**. The same applies to **tran, tranif1, tranif0,** except that passing or reducing strengths occur across their bidirectional ports. Consider for example, the **nmos** instantiation shown below. This instantiation uses the *sim* **net** example presented in Sec. 7.2.1.2 with **pull0** and **supply1** strength values. Since this signal is used at the input of an **nmos** switch with a Gate input value of **1**, the strength of the *sim_out* output of the **nmos** switch becomes (**pull0, strong1**).

```
nmos T1 ( sim_out, sim, 1 );
```

Resistive switches, i.e., **rnmos, rpmos, rcmos, rtran, rtranif1,** and **rtranif0,** reduce **supply** strength by two to **pull**, and reduce other strength levels by one. A **weak** strength, for which there is no lower drive strength, is reduced to a **medium** capacitor. Charge strengths are also reduced by one across these structures. The **small** charge strength and **highz** drive strength are not affected by resistive switches. As an example for resistive reduction, consider the **rnmos** instantiation shown below. This instantiation uses the *sim* **net** example used above with **pull0** and **supply1** strength values. Since this signal is used at the input of an **rnmos** switch with a Gate input value of **1**, the strength of the *sim_out* output of the **rnmos** switch becomes (**weak0, pull1**).

```
rnmos T2 ( sim_out, sim, 1 );
```

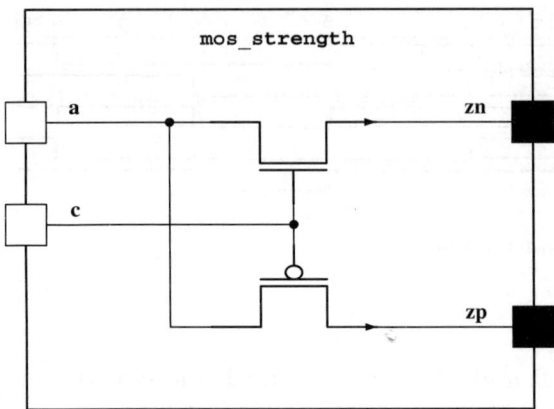

Figure 7.39   MOS Pass Circuit

An MOS circuit for demonstrating strength reduction in Verilog is shown in Fig. 7.39. Figure 7.40 shows a Verilog code corresponding to this diagram. This code uses **nmos** and **pmos** primitives that pass default **strong0** and **strong1** strength values to their outputs.

Simulating this circuit with the testbench of Fig. 7.41, results in the display report of Fig. 7.42. The **%v** format, in the **$monitor** command of Fig. 7.41, displays signal strength values. The *a* input of the circuit of Fig. 7.40 has **strong0** and **strong1** default strength values. As expected (shown in Fig. 7.42), these values propagate to **nmos** and **pmos** outputs. Note in this figure that *St1* and *St0* are used for **strong1** and **strong0**. As shown, at 200 ns and 300 ns, *zp* is *St1* and *St0*, respectively.

Strength reduction is demonstrated in the example of Fig. 7.43. The Verilog code of this example uses **rnmos** and **rpmos** resistive switches for implementing the circuit of Fig. 7.39.

```
`timescale 1ns/100ps

module mos_strength (a, c, zn, zp);
    input a, c;
    output zn, zp;

    nmos (zn, a, c);
    pmos (zp, a, c);

endmodule
```

Figure 7.40   Verilog Code Using Nonresistive Switches

```
`timescale 1ns/100ps

module test_mos_strength;
    reg a, c;
    wire zn, zp;

    mos_strength cut (a, c, zn, zp);

    initial begin
        #10 a = 1;
        #10 c = 0;
        #10 a = 0;
        #10 c = 1;
        #10 a = 1;
        #10 c = 0;
        #10 $stop;
    end
    initial
        $monitor ("At time %t   zn: %v, zp: %v", $time, zn, zp);
endmodule
```

Figure 7.41   Testing *mos_strength*

```
            At time                  0 zn: StX, zp: StX
            At time                100 zn: StH, zp: StH
            At time                200 zn: HiZ, zp: St1
            At time                300 zn: HiZ, zp: St0
            At time                400 zn: St0, zp: HiZ
            At time                500 zn: St1, zp: HiZ
            At time                600 zn: HiZ, zp: St1
```

Figure 7.42   Nonresistive Display Report

```
`timescale 1ns/100ps

module mos_strength (a, c, zn, zp);
    input a, c;
    output zn, zp;

    rnmos (zn, a, c);
    rpmos (zp, a, c);

endmodule
```

Figure 7.43   Verilog Code Using Resistive Switches

```
At time                    0   zn: PuX,  zp: PuX
At time                  100   zn: PuH,  zp: PuH
At time                  200   zn: HiZ,  zp: Pu1
At time                  300   zn: HiZ,  zp: Pu0
At time                  400   zn: Pu0,  zp: HiZ
At time                  500   zn: Pu1,  zp: HiZ
At time                  600   zn: HiZ,  zp: Pu1
```

Figure 7.44   Resistive Display Report

Simulating this circuit using the testbench of Fig. 7.41 produces the report shown in Fig. 7.44. As shown, strength values for logic **1** and **0** at the output of the resistive circuit are **pull**, which is one strength level lower than **strong**. At 200 ns and 300 ns when a propagates to zp its strengths are reduced from **strong1** and **strong0** to **pull1** and **pull0**. Similarly, at 400 ns and 500 ns, zn output strengths become **pull0** and **pull1**.

Sections 7.2.2 and 7.2.3 briefly discussed several simple cases of strength modeling in Verilog. The concept of ambiguous strength, which is defined as a strength value consisting of more than one level, also exists in Verilog. This concept and more details of charge and drive strengths can be found in the standard IEEE Language Reference Manual for Verilog.

7.3   Summary

This chapter discussed some of the Verilog features to generate detailed models for components of a cell library or an existing set of parts. These features are not synthesizable and their only purpose is to generate accurate models. The first part of this chapter discussed switch level Verilog models. We discussed unidirectional and bidirectional switches. We also showed the use of capacitive nodes as a means of storing memory data. More complex switch level structures can be modeled using concepts presented here. The second part of this chapter gave a brief presentation of strength modeling. Combining switches and signal drives and charge strengths enable Verilog modelers to write fairly accurate models for VLSI cells.

Problems

7.1   Generate a switch level description for an AOI gate using NMOS and PMOS transistors [w = (a.b + c.d)`].

7.2   Use the AOI gate of Problem 1 to generate a quad 2-to-1 multiplexer.

7.3 Generate a switch level description for a MAJ (Majority) gate using NMOS and PMOS transistors (w = a.b + a.c + b.c).

7.4 After minimization of the following function, show its transistor level CMOS implementation. Use a single complex gate for realizing this function. Available to you are all inputs and their complements. Your implementation should use a minimum number of transistors.

f (a, b, c, d) = a' . c' . d' + b . d'

Suggested Reading

Brown, S., and Z. Vranesic, *Fundamentals of Digital Logic with Verilog Design*, McGraw-Hill, New York, 2002, ISBN: 0-07-283878-7.

IEEE Std 1364-2001, *IEEE Standard Verilog Language Reference Manual*, SH94921-TBR (print) SS94921-TBR (electronic), ISBN 0-7381-2827-9 (print and electronic), 2001.

Uyemura, J.P. *Introduction to VLSI Circuits and Systems*, John Wiley & Sons, Hoboken, New Jersey, USA, 2002, ISBN: 0471127043.

Weste, N.H.E, and D. Harris, *CMOS VLSI Design: A Circuits and Systems Perspective (3rd Edition)*, 3rd ed, Addison Wesley, Boston, MA, 2004, ISBN: 0321149017.

8

RT Level Design and Test

This chapter discusses design and test of complete systems. Topics discussed in Chaps. 4, 5, and 6 are used here to describe systems for synthesis and develop testbenches for these complete systems. We will show how a design is partitioned into its datapath and controller and how these components are described in Verilog. In a complex design we will show further partitioning of the datapath of the design into its individual registers, busses, and logic units. For testing our designs, we show how interactive testing and use of files are utilized for testing actual systems. The chapter begins with specification and design of a sequential multiplier. We will then discuss a simple processor to familiarize readers with design methodology that we are promoting for larger systems. The last section of this chapter uses our design methodology to design, code and test a CPU with a typical architecture.

8.1 Sequential Multiplier

Our first example of RT level system design is an add-and-shift sequential multiplier, with 8-bit A and B inputs and a 16-bit result. The block diagram of this circuit is shown in Fig. 8.1. This multiplier has an 8-bit bidirectional I/O for inputting its A and B operands, and outputting its 16-bit output one byte at a time.

Multiplication begins with the *start* pulse. On the clock edge that *start* is 1, operand A is on the *databus* and in the next clock, this bus will contain operand B. The two operands appear on the bus in two consecutive clock pulses. After accepting these data inputs, the multiplier begins its multiplication process and when it is completed, it starts sending the result out on the *databus*. When the least-significant byte is placed on *databus*, the *lsb_out* output is issued, and for the most-significant byte,

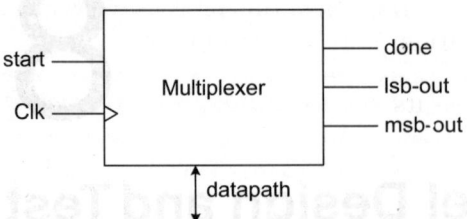

Figure 8.1 Multiplier Block Diagram

msb_out is issued. When both bytes are outputted, *done* becomes **1**, and the multiplier is ready for another set of data.

The multiplexed bidirectorial *databus* is used to reduce the total number of input and output pins of the multiplier.

8.1.1 Shift-and-add multiplication process

When designing multipliers there is always a compromise to be made between how fast the multiplication process is done and how much hardware we are using for its implementation. A simple multiplication method that is slow, but efficient in use of hardware is the shift-and-add method. In this method, depending on bit *i* of operand *A*, either operand *B* is added to the collected partial result and then shifted to the right (when bit *i* is **1**), or (when bit *i* is **0**) the collected partial result is shifted one place to the right without being added to *B*. This method can better be understood by considering how binary multiplication is done manually. Figure 8.2 shows manual multiplication of two 8-bit binary numbers.

We start considering bits of *A* from right to left. If a bit value is **0** we select **00000000** to be added with the next partial product, and if it is a **1**, the value of *B* is selected. This process repeats, but each time **00000000** or *B* is selected, it is written one place to the left with respect to the previous value. When all bits of *A* are considered, we add all calculated values to come up with the multiplication result.

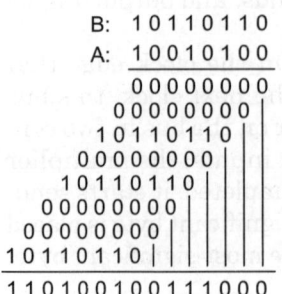

Figure 8.2 Manual Binary Multiplication

Understanding hardware implementation of this procedure becomes easier if we make certain modifications to this procedure. First, instead of moving our observation point from one bit of A to another, we put A in a shift-register, always observe its right-most bit, and after every calculation, we move it one place to the right, making its next bit accessible. Second, for the partial products, instead of writing one and the next one to its left, we move the partial product to the right as we are writing it. Finally, instead of calculating all partial products and adding them up at the end, we add a newly calculated partial product to the previous one and write the calculated value as the new partial result.

Therefore, if the observed bit of A is **0**, **00000000** is to be added to the previously calculated partial result, and the new value should be shifted one place to the right. In this case, since the value being added to the partial result is **00000000**, adding is not necessary, and only shifting the partial result is sufficient. This process is called *shift*. However, if the observed bit of A is **1**, B is to be added to the previously calculated partial result, and the calculated new sum must be shifted one place to the right. This is called *add-and-shift*.

Repeating the above procedure, when all bits of A are shifted out, the partial result becomes the final multiplication result. We use a 4-bit example to clarify the above procedure. As shown in Fig. 8.3, $A = $ **1001** and $B = $ **1101** are to be multiplied. Initially at time 0, A is in a shift-register with a register for partial results (P) on its left.

At the time 0, because $A[0]$ is **1**, the partial sum of $B + P$ is calculated. This value is **01101** (shown in the upper part of time 1) and has 5 bits

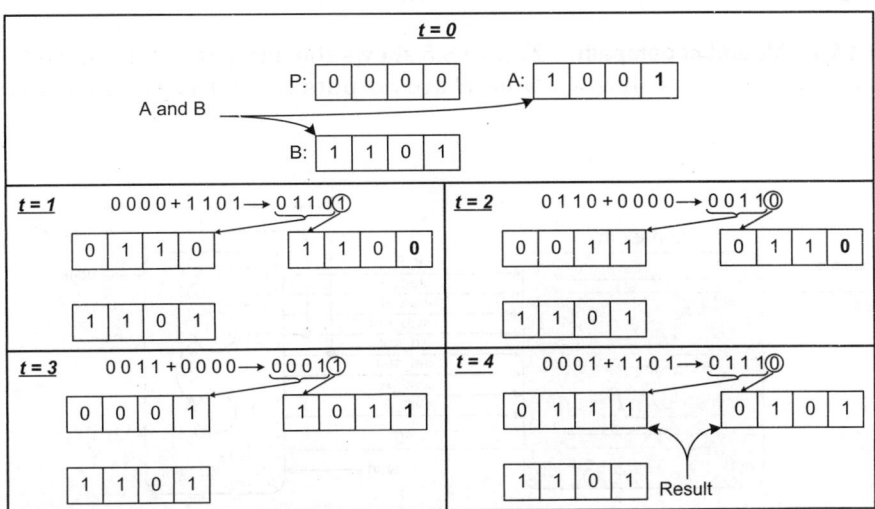

Figure 8.3 Hardware Oriented Multiplication Process

to consider the carry bit. The right most bit of this partial sum is shifted into the A register, and the other bits replace the old value of P. When A is shifted, **0** moves into the $A[0]$ position. This value is observed at time 1. At this time, because $A[0]$ is **0**, **0000** + P is calculated (instead of $B + P$). This value is **00110**, the right most bit of which is shifted into A, and the rest replace P. This process repeats 4 times. At the end of the 4th cycle, the least significant 4 bits of the multiplication result become available in A and the most-significant bits in P. The example used here performed 9*13 and 117 is obtained as the result of this operation.

8.1.2 Sequential multiplier design

The multiplication process discussed in the previous section justifies the hardware implementation that is being discussed here.

8.1.2.1 Control data partitioning. The multiplier has a datapath and a controller. The data part consists of registers, logic units, and their interconnecting busses. The controller is a state machine that issues control signals for control of what gets clocked into the data registers.

As shown in Fig. 8.4, the data path registers and the controller are triggered with the same clock signal. On the rising edge of the system clock, the controller goes into a new state. In this state, several control signals are issued, and as a result the components of the datapath start reacting to these signals. The time given for all activities of the datapath to stabilize is from one edge of the clock to another. Values that are propagated to the inputs of the datapath registers are clocked into these registers with every positive edge of the clock.

8.1.2.2 Multiplier datapath. Figure 8.5 shows the datapath of the sequential multiplier. As shown, P and B are outputs of 8-bit registers and A

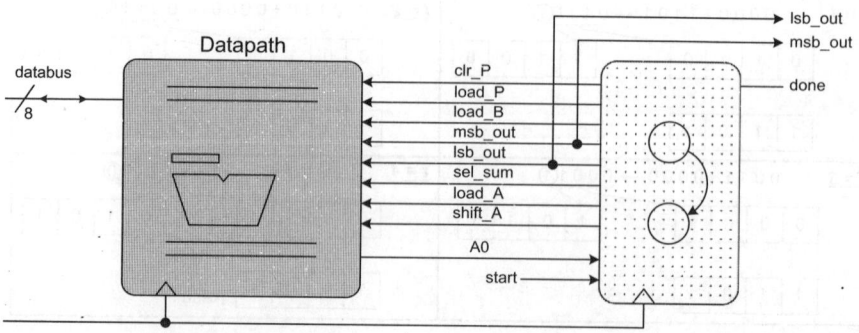

Figure 8.4 Datapath and Controller

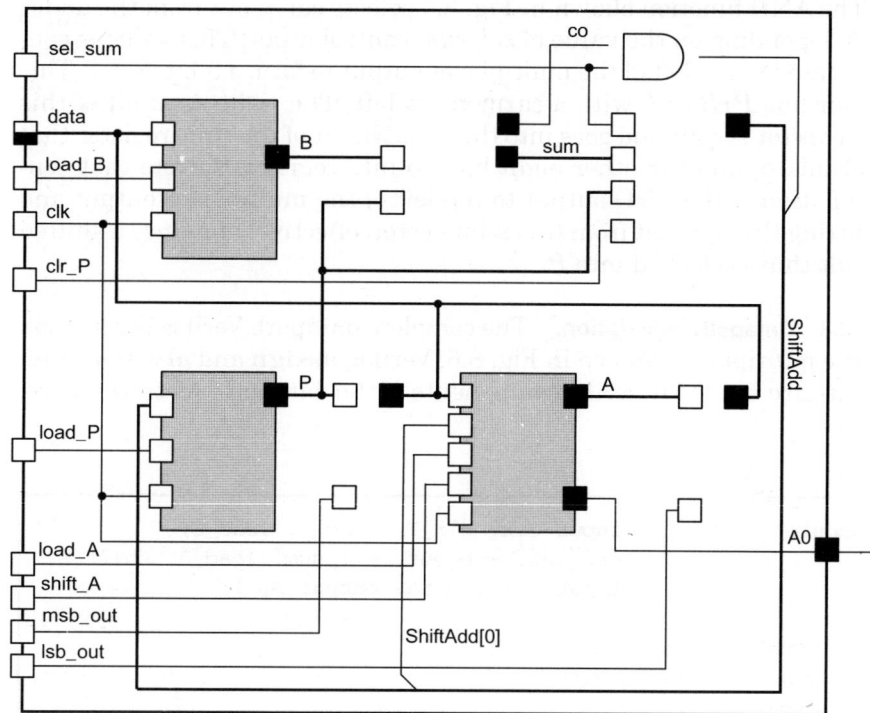

Figure 8.5 Multiplier Block Diagram (Verilog code correspondence)

is the output of an 8-bit shift-register. These components are implemented with **always** statements in the Verilog code of the multiplier. An adder, a multiplexer and two tri-state buffers constitute the other components of this datapath. These components are implemented with **assign** statements.

Control signals that are outputs of the controller and inputs of the datapath (Fig. 8.4), are named according to their functionalities like loading registers, shifting, etc. These signals are shown in the corresponding blocks of Fig. 8.5 next to the data component that they control.

The input *databus* connects to the inputs of *A* and *B* to load multiplier and multiplicand into these registers. This bi-directional bus is driven by the outputs of *P* and *A* through tri-state buffers. These tri-states become active when multiplication result is ready.

The output from *B* and *P* are added to form *co* and *sum* to be put in *P* if adding is to take place. Otherwise, *P* is put on *ShiftAdd* to be shifted, while being put back into *P*. *ShiftAdd* is the multiplexer output that selects *sum* or *P*. The *sel_sum* control input determines if *sum* or *P* is to go on the multiplexer output.

The AND function shown in Fig. 8.5 selects carry-out from the adder or **0** depending on the value of *sel_sum* control input. This value is concatenated to the left of the multiplexer output to form a 9-bit vector. This vector has *P+B* or *P* with a carry to its left. The right-most bit of this 9-bit vector is split and goes into the serial input of the shift-register that contains *A*, and the other eight bits go into register *P*. Note that concatenation of the AND output to the left of the multiplexer output and splitting the right bit from this 9-bit vector, effectively produce a shifted result that is clocked into *P*.

8.1.2.3 Datapath description.   The complete datapath Verilog description of the multiplier is shown in Fig. 8.6. Verilog **assign** and **always** statements are used to describe components of the datapath. As shown here,

```
module datapath ( input clk, clr_P, load_P, load_B,
                  msb_out, lsb_out, sel_sum, load_A, shift_A,
                  inout [7:0] data, output A0 );

    wire [7:0] sum, ShiftAdd;
    reg [7:0] A, B, P;
    wire co;
    always @( posedge clk ) if (load_B) B <= data;

    always @( posedge clk )
        if (load_P) P <= {co&sel_sum, ShiftAdd[7:1]};

    assign { co, sum } = P + B;

    always @( posedge clk )
        case ( { load_A, shift_A } )
            2'b01 : A <= { ShiftAdd[0], A[7:1] };
            2'b10 : A <= data;
            default : A <= A;
        endcase
    assign A0 = A[0];

    assign ShiftAdd = clr_P ? 8'h0 : ( ~sel_sum ? P : sum );

    assign data = lsb_out ? A : 8'hzz;
    assign data = msb_out ? P : 8'hzz;

endmodule
```

Figure 8.6 Datapath Verilog Code

the first two **always** statements represent registers B and P for operand B and the partial result, P. The **assign** statement that comes next in this figure represents the 8-bit adder. This adder adds P and B.

Another component of our multiplier datapath is an 8-bit shift-register for operand A of the multiplier. This shift-register either loads A with *data* (controlled by *load_A*) or shifts its contents (controlled by *shift_A*). An **always** statement to implement this shift-register is shown in Fig. 8.6. Following this statement, an **assign** statement representing the multiplexer for selection of *sum* or P is shown in the Verilog code of the datapath. This statement puts *8'h0* on *ShiftAdd* if *clr_P* is active. We will use this enabling feature of the multiplexer for resetting P at the start of the multiplication process.

The last two **assign** statements of Fig. 8.6 represent two sets of tristate buffers driving the bidirectional *data* bus of the datapath. As shown, if *lsb_out* is **1**, A (the least-significant byte of result) drives *data* and if *msb_out* is **1**, P (the most-significant byte) drives *data*.

8.1.2.4 Multiplier controller.   The multiplier controller is a finite state machine that has two starting states, eight multiplication states, and two ending states. States and their binary assignments are shown in Fig. 8.7. In the `idle` state the multiplier waits for *start* while loading A. In `init`, it loads the second operand B. In `m1` to `m8`, the multiplier performs add-and-shift of $P+B$, or $P+0$, depending on $A0$. In the last two states (`rslt1` and `rslt2`), the two halves of the result are put on *databus*.

The Verilog code of the controller is shown in Fig. 8.8. This code declares signals that connect to *datapath* ports, and uses a single **always** block to issue control signals and make state transitions. At the beginning of this **always** block all control signal outputs are set to their inactive

```
`define idle    4'b0000
`define init    4'b0001
`define m1      4'b0010
`define m2      4'b0011
`define m3      4'b0100
`define m4      4'b0101
`define m5      4'b0110
`define m6      4'b0111
`define m7      4'b1000
`define m8      4'b1001
`define rslt1   4'b1010
`define rslt2   4'b1011
```

Figure 8.7   Multiplier Control States

```verilog
module controller ( input clk, start, A0,
           output reg clr_P, load_P, load_B, msb_out,
                      lsb_out, sel_sum,
           output reg load_A, Shift_A, done );
   reg [3:0] current;

   always @ ( posedge clk ) begin
      clr_P = 0; load_P = 0; load_B = 0; msb_out = 0;
      lsb_out = 0;
      sel_sum = 0; load_A = 0; Shift_A = 0; done = 0;

      case ( current )
         `idle :
            if (~start) begin
               current <= `idle;
               done = 1;
            end else begin
               current <= `init;
               load_A = 1; clr_P = 1; load_P = 1;
            end
         `init : begin
               current <= `m1;
               load_B = 1;
            end
         `m1, `m2, `m3, `m4, `m5, `m6, `m6, `m7, `m8 : begin
               current <= current + 1;
               Shift_A = 1; load_P = 1; if (A0) sel_sum = 1;
            end
         `rslt1 : begin
               current <= `rslt2;
               lsb_out = 1;
            end
         `rslt2 : begin
               current <= `idle;
               msb_out = 1;
            end
         default : current <= `idle;
      endcase
   end

endmodule
```

Figure 8.8   Verilog Code of Controller

values. This eliminates unwanted latches that may be generated by a synthesis tool for these outputs.

The 4-bit *current* variable represents the currently active state of the machine. When *current* is `idle` and *start* is **0**, the *done* output remains high. In this state if *start* becomes **1**, control signals *load_A*, *clr_P* and *load_P* become active to load *A* with *databus* and clear the *P* register. Clearing *P* requires *clr_P* to put **0**'s on the *ShiftAdd* of the datapath and loading the **0**'s into *P* by asserting *load_P*.

In `m1 to `m8 states, *A* is shifted, *P* is loaded, and if A0 is **1**, *sel_sum* is asserted. As discussed in relation to *datapath, sel_sum* controls shifted *P+B* (or shifted *P+0*) to go into *P*. In the result states, *lsb_out* and *msb_out* are asserted in two consecutive clocks in order to put *A* and *P* on the data bus respectively.

8.1.2.5 Top-level code of the multiplier.   Figure 8.9 shows the top-level *Multiplier* module. The *datapath* and *controller* modules are instantiated here. The input and output ports of this unit are according to the block diagram of Fig. 8.1. This description is synthesizable, and can be used in any FPGA device programming environment for synthesis and device programming.

8.1.3 Multiplier testing

This section shows an auto-check interactive testbench for our sequential multiplier. Several forms of data applications and result monitoring are

```
module Multiplier ( input clk, start,
                    inout [7:0] databus,
                    output lsb_out, msb_out, done );

   wire clr_P, load_P, load_B, msb_out,
        lsb_out, sel_sum, load_A, Shift_A;

   datapath dpu( clk, clr_P, load_P, load_B,
                 msb_out, lsb_out, sel_sum, load_A, Shift_A,
                 databus, A0 );

   controller cu( clk, start, A0, clr_P, load_P, load_B,
                  msb_out, lsb_out, sel_sum, load_A, Shift_A,
                  done );

endmodule
```

Figure 8.9 Top-Level Multiplier Code

```
timescale 1ns/100ps

module test_multiplier;
    reg clk, start, error;
    wire [7:0] databus;
    wire lsb_out, msb_out, done;
    reg [7:0] mem1[0:2], mem2[0:2];
    reg [7:0] im_data, opnd1, opnd2;
    reg [15:0] expected_result, multiplier_result;
    integer indx;

    Multiplier uut ( clk, start, databus, lsb_out, msb_out, done );

    initial begin: Apply_Data   ... end          // Figure 8.11
    initial begin: Apply_Start ... end           // Figure 8.12
    initial begin: Expected_Result ... end       // Figure 8.13
    always @(posedge clk)
        begin: Actual_Result ... end             // Figure 8.14
    always @(posedge clk)
        begin: Compare_Results ... end           // Figure 8.15
    always #50 clk = ~clk;
    assign databus=im_data;
endmodule
```

Figure 8.10   Multiplier Testbench Outline

demonstrated by this example. The outline of the *test_multiplier* module
is shown in Fig. 8.10.

In the declarative part of this testbench inputs and outputs of the multiplier are declared as **reg** and **wire**, respectively. Since *databus* of the
multiplier is a bidirectional bus, it is declared as **wire** for reading it, and
a corresponding *im_data* **reg** is declared for writing into it. An **assign**
statement drives *databus* with *im_data*. When writing into this bus from
the testbench, the writing must be done into *im_data*, and after the completion of writing the bus must be released by writing *8'hZZ* into it.

Other variables declared in the testbench of Fig. 8.10 are *expected_result*
and *multiplier_result*. The latter is for the result read from the multiplier,
and the former is what is calculated in the testbench. It is expected that
these values are the same.

The testbench shown in Fig. 8.10, applies three rounds of test to the
Multiplier module. In each round, data is applied to the module under
test and results are read and compared with the expected results. These
are tasks performed by this testbench:

- Read data files *data1.dat* and *data2.dat* and apply data to *databus*
- Apply *start* to start multiplication
- Calculate the expected result
- Wait for multiplication to complete, and collect the calculated result
- Compare expected and calculated results and issue error if they do not match

These tasks are timed independently, and at the same time, an **always** block generates a periodic signal on *clk* that clocks the multiplier.

8.1.3.1 Reading data files. Figure 8.11 shows the *Apply_Data* **initial** block that is responsible for reading data and applying them to *im_data*, which in turn goes on *databus*. Hexadecimal data from *data1.dat* and *data2.dat* external files are read into *mem1* and *mem2*. In each round of test, data from *mem1* and *mem2* are put on *im_data*. Data from *mem2* is distanced from that of *mem1* by 100 ns. This way, the latter is interpreted as data for the *A* operand and the former for the *B* multiplication operand. After placing this data, *8'hzz* is put on *im_data*. This releases the *databus* so that it can be driven by the multiplier when its result is ready.

8.1.3.2 Applying start. Figure 8.12 shows an **initial** block in which variable initializations take place and *start* signal is issued. Using a **repeat** statement, three 100 ns pulses distanced by 1400 ns are placed on *start*.

8.1.3.3 Calculating expected result. Figure 8.13 shows an **initial** block that reads data that is placed on *databus* by the *Apply_Data* block (Fig. 8.11),

```
initial begin: Apply_Data
    indx=0;
    $readmemh ( "data1.dat", mem1 );
    $readmemh ( "data2.dat", mem2 );
    repeat (3) begin
        #300 im_data = mem1 [indx];
        #100 im_data = mem2 [indx];
        #100 im_data = 8'hzz;
        indx = indx+1;
        #1000;
    end
    #200 $stop;
end
```

Figure 8.11 Reading Data Files

```
initial begin: Apply_Start
   clk=1'b0; start=1'b0; im_data=8'hzz;
   #200 ;
   repeat(3) begin
      #50  start = 1'b1;
      #100 start = 1'b0;
      #1350;
   end
end
```

Figure 8.12   Initializations and Start

and calculates the expected multiplication result. After *start*, when *databus* is updated, the first operand is read into *opnd1*. The next time *databus* changes, *opnd2* is read. The expected result is calculated using these operands.

8.1.3.4 Reading multiplier output.   When the multiplier completes its task, it issues *msb_out* and *lsb_out* to signal that it has readied the two bytes of the result. The **always** block of Fig. 8.14 is triggered by the rising edge of the circuit clock. After a clock edge, if *msb_*out or *lsb_*out is **1**, it reads the *databus* and puts in its corresponding position in *multiplier_result*.

8.1.3.5 Comparing results.   Figure 8.15 shows the **always** block that is responsible for comparing actual and expected multiplication results. After the active edge of the clock, if *done* is **1**, then comparing *multiplier_result* and *expected_result* takes place. If values of these variables do not match *error* is issued.

   The self-running testbench presented here verifies RT level operation of our multiplier. This design is synthesizable and because of the timing

```
initial begin: Expected_Result
   error=1'b0;
   repeat(3) begin
      wait ( start==1'b1 );
      @( databus );
      opnd1=databus;
      @( databus );
      opnd2=databus;
      expected_result = opnd1 * opnd2;
   end
end
```

Figure 8.13   Calculating Expected Result

```
always @(posedge clk) begin: Actual_Result
    if (msb_out) multiplier_result[15:8] = databus;
    if (lsb_out) multiplier_result[7:0] = databus;
end
```

Figure 8.14   Reading Multiplier Results

used in our testbench, it can also be used for the post-synthesis description of our multiplier.

This section showed a complete design of a system with a well-defined datapath and a controller. The design demonstrates top-down design, data/control partitioning, and a complete flow from design to test of a system. This flow will be used in the sections that follow to illustrate how Verilog can be used for design of systems that access memory for instructions and data.

8.2   von Neumann Computer Model

The previous section was our first step in showing how a complete design could be put together and tested in Verilog. In this section we take our presentation of complete system design one step further, and present design, implementation, and testbench development for a hardware based on the *von Neumann* computer model.

8.2.1   Processor and memory model

The von Neumann computer model is based on a processor using instructions and data from a single memory. Using a sequencer (see Fig. 8.16), the processor fetches instructions from its memory. An instruction has an opcode indicating the function it is supposed to perform. Using this opcode, the processor performs its proper operation. Such an operation may involve reading or writing data from or to the memory, for which the same memory as that of the instructions will be used.

Our design example in this section is a simply von Neumann model with memory accessing mechanism for instructions and data. This design is at the RT level and has a separate datapath and control unit.

```
always @(posedge clk) begin: Compare_Results
    if (done)
    if (multiplier_result != expected_result) error = 1;
    else error = 0;
end
```

Figure 8.15   Comparing Results

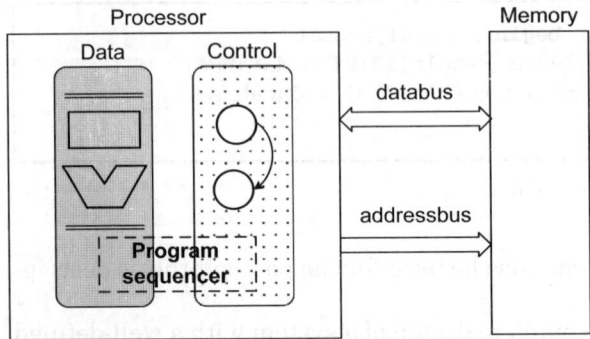

Figure 8.16 von Neumann Process Model

In designing the datapath we partition it into its subcomponents and describe each subcomponent separately. This includes the instruction sequencer part of the datapath that is the program counter component.

In testing this simple model we will show how a memory model can be written in Verilog and how external files can be used for representing memories for read and write. Verilog file handling tasks will be used for this purpose. To demonstrate some of Verilog testbench facilities, our memory model will be developed to read processor instructions in a mnemonic format and convert them to binary for the processor hardware to access.

8.2.2 Processor model specification

The example that we use is a simple adding machine, which we refer to as *AddingCPU*. It must be mentioned, that we are using this example to demonstrate our design and test methodologies. This specific example has very little practical value, if it were to be designed for a real application, much simpler coding than what we are presenting here could be used. The techniques presented here will be used for the design and test of the processor of the next section.

Our Adding CPU reads **Load**, **Add**, **Store**, and **Jump** instructions from its memory, and depending on the instruction it reads, loads data, performs addition, stores data into memory, or jumps to another memory location. Figure 8.17 shows the overall structure of this adding machine.

The circuit shown has a 6-bit address bus to address the memory for read and write operations. The 8-bit data bus of this machine is used for data in and out of the machine. Control signals, *reset*, *rd_mem*, and *wr_mem* are used for resetting, memory read, and memory write operations.

The machine starts reading its memory from location 0. An 8-bit word fetched from the memory consists of a 2-bit opcode and a 6-bit data or address. This field is either an immediate data or a memory address where the operand of the fetched instruction is. Figure 8.18 shows our machine's opcodes and instruction format.

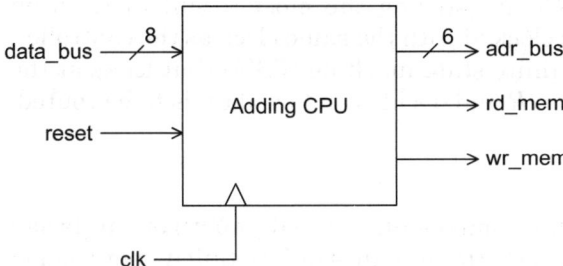

Figure 8.17   Interface of the Adding CPU Example

The Adding CPU has a main register called *AC* (accumulator). The **Load** instruction directs the machine to load the addressed data from the memory into *AC*. The **Store** instruction causes contents of *AC* to be written into the addressed location in the memory. The operand of the **Add** instruction is *immd* (immediate). This instruction adds *immd* to the present contents of *AC* and puts the result back into *AC*. The **Jump** instruction loads the 6-bit address into the program counter of our machine, causing the next instruction to be fetched from this address.

8.2.3 Designing the adding CPU

The first step in the design of our adding machine is to decide on its data and control partitioning and decide what goes into its data part and what behavior is expected from its controller.

The datapath of the design has the *AC* register for keeping data to operate on, the *PC* register to keep track of the address being fetched, and an adder unit to perform the addition. In addition, the datapath has

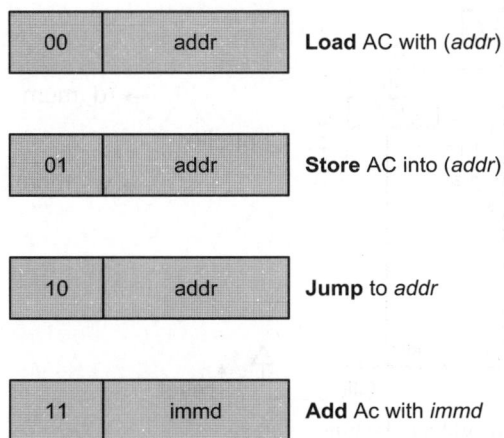

Figure 8.18   Instruction Format

an instruction register (*IR*) for storing the most recent instruction fetched. Data registers are clocked with the same clock as the controller.

The controller part is a finite state machine (FSM) that looks at the opcode of the instruction in *IR* and decides on how data is to be routed.

8.2.4   Design of datapath

As indicated above, the main components of the datapath of our design are *AC*, *PC* (program counter), *IR*, and an *ALU*. Detailed operation of these components will be decided once we decide on the architecture that incorporates them.

Given the Adding CPU description of Sec. 8.2.2, the bussing shown in Fig. 8.19 is appropriate for handling the necessary operations mentioned in this section. As shown here, the datapath has an internal *dbus* bus. The external bidirectional *data_bus* drives and is driven by *dbus*. This bus connects to the input of *IR* in order to bring instruction read from the memory into this register.

IR has a load input (*ld_ir*) that is activated to cause it load from *data_bus*. Similarly, this bus connects to *AC* to bring data read from the memory into this register. The control signal for loading *AC* is *ld_ac*. This control signal is issued when the **Load** instruction is being expected. *PC* has three control signals *ld_pc*, *inc_pc* and *clr_pc* to load, increment, and

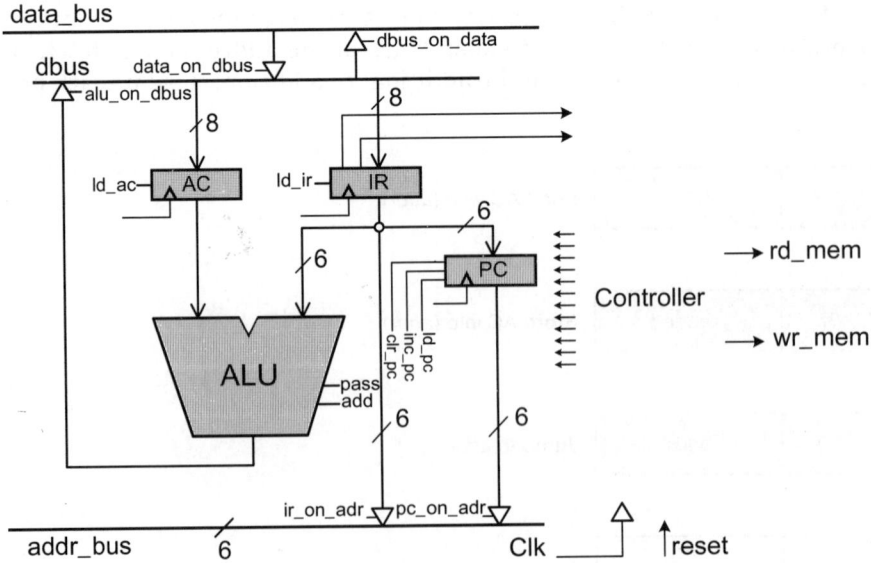

Figure 8.19 Architectural Design of our Adding Machine

clear it, respectively. The right most 6-bits of *IR* connect to the input of *PC* for execution of the **Jump** instruction.

For executing the **Store** instruction, *AC* is placed on the left input of *ALU* and from there to *dbus*, which eventually goes on *data_bus*. At the same time, *IR* is placed on *addr_bus* to specify the address in which *AC* data is to be stored. For this purpose, the adder unit (*ALU*) has a *pass* control input to make it pass its left input data to its output.

Execution of the **Add** instruction is done by taking one of the add operands from *AC* and the other from *IR*. For this instruction, activating the *add* control input of *ALU* causes it to perform addition.

The simple bussing structure described above facilitates execution of all four instructions of our simple Adding CPU. When a bus has more than one source driving it, e.g., *IR* and *PC* driving *addr_bus*, control signals from the controller select the source.

8.2.5 Control part design

After the design of the datapath and figuring control signals and their role in activities in the datapath, the design of the controller becomes a simple matter. The block diagram of this part is shown in Fig. 8.20.

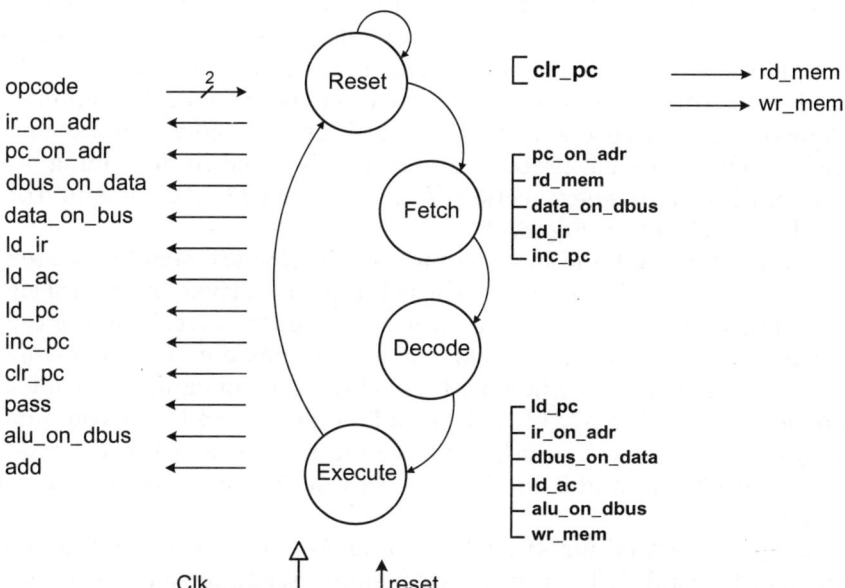

Figure 8.20 Controller of adding CPU

The controller of our simple von Neumann machine has four states, *Reset, Fetch, Decode,* and *Execute.* As the machine cycles through these states, various control signals are issued. In state *Reset,* for example, the *clr_pc* control signal is issued. State *Fetch* issues *pc_on_adr, rd_mem, data_on_dbus, ld_ir,* and *inc_pc,* to read memory from the present *PC* location, route it to *IR,* load it into *IR,* and increment *PC* for the next memory fetch. Depending on opcode bits, that are the controller inputs, the *Execute* state of the controller issues control signals for execution of **Load, Store, Add,** and **Jump** instructions. The next section discusses details of the controller signals and their role in execution of these instructions.

8.2.6 *AddingCPU* Verilog description

We develop the complete Verilog code of our simple adding machine by developing code for the blocks of Fig. 8.19. We first describe components of the datapath, and then will form the Verilog code of the datapath by instantiating and wiring these components. The controller will be described next, using a state machine coding style. At the end, the description of our small von Neumann example will be completed by wiring datapath and controller in a top-level Verilog module.

8.2.6.1 Data components.
Datapath components of *AddingCPU* could be described by **always** and **assign** statements directly in the datapath description of the machine. Recall that this coding technique was used for the multiplier example of the previous section. However, in this example we are taking a more general and extendable approach. We describe our components so that they can be independently simulated and tested. This is necessary for large designs with more complex components. The approach presented here will be used in describing our larger machine in the next section. Verilog code for *PC, AC, IR,* and *ALU* modules are shown in Fig. 8.21.

Accumulator (*AC*) and instruction register (*IR*) are simple registers with *load* enable control inputs. These inputs are driven by control signals coming from the controller through datapath ports. The program counter (*PC*) is a counter with parallel load, increment, and clear capabilities. As shown, this component has three control signals to control its functionality. The ALU (the last module in Fig. 8.21) is a combinational logic with *pass* and *add* control inputs. If *pass* is **1**, the *a* input goes on the output, and if *add* is **1**, the ALU output becomes the sum of *a* and *b*.

Codings presented for *AC, IR, PC,* and *ALU* are synthesizable and individually testable. These parts are instantiated in the datapath of our machine.

```
module AC ( input [7:0] data_in, input load, clk,
            output reg [7:0] data_out );
   always @( posedge clk )
      if( load ) data_out <= data_in;
endmodule
//
//
module PC ( input [5:0] data_in, input load, inc, clr, clk,
            output reg [5:0] data_out );
   always @( posedge clk )
      if( clr ) data_out <= 6'b000_000;
      else if( load ) data_out <= data_in;
      else if( inc ) data_out <= data_out + 1;
endmodule
//
//
module IR ( input [7:0] data_in, input load, clk,
            output reg [7:0] data_out );
   always @( posedge clk )
      if ( load ) data_out <= data_in;
endmodule
//
//
module ALU ( input [7:0] a, b, input pass, add,
             output reg[7:0] alu_out );
   always @(a or b or pass or add)
      if (pass) alu_out = a;
      else if (add) alu_out = a + b;
      else alu_out = 0;
endmodule
```

Figure 8.21 Datapath Components of Adding Machine

8.2.6.2 Datapath description. Figure 8.22 shows the datapath description of *AddingCPU*. The module name for this description is *DataPath* and it corresponds to the left part of Fig. 8.19. When studying the discussion below and the Verilog code of the datapath, the reader is encouraged to consider Fig. 8.19, and make correspondences between Verilog signals and constructs with graphical notations of this figure.

The inputs of the Verilog code of Fig. 8.22 are control signals coming from the controller, and the bidirectional *data_bus*. The outputs of this module are the opcode and address bus. The opcode goes out to the controller and the address bus goes to the memory for operand and instruction fetch.

```
module DataPath ( input ir_on_adr, pc_on_adr, dbus_on_data,
                        data_on_dbus, ld_ir, ld_ac, ld_pc,
                        inc_pc, clr_pc, pass, add, alu_on_dbus,
                        clk,
                  output [5:0] adr_bus,
                  output [1:0] op_code,
                  inout [7:0] data_bus );

    wire [7:0] dbus, ir_out, a_side, alu_out;
    wire [5:0] pc_out;

    IR ir ( dbus, ld_ir, clk, ir_out );
    PC pc ( ir_out[5:0], ld_pc, inc_pc, clr_pc, clk, pc_out );
    AC ac ( dbus, ld_ac, clk, a_side );
    ALU alu ( a_side, {2'b00,ir_out[5:0]}, pass, add, alu_out );

    assign adr_bus = ir_on_adr ? ir_out[5:0] : 6'bzz_zzzz;
    assign adr_bus = pc_on_adr ? pc_out : 6'bzz_zzzz;
    assign dbus = alu_on_dbus ? alu_out : 8'bzzzz_zzzz;
    assign data_bus = dbus_on_data ? dbus : 8'bzzzz_zzzz;
    assign dbus = data_on_dbus ? data_bus : 8'bzzzz_zzzz;

    assign op_code = ir_out[7:6];

endmodule
```

Figure 8.22 Datapath Description

Following the input and output declarations, the *DataPath* module declares internal datapath busses and signals. As shown, these declarations are followed by instantiation of data components, *IR*, *PC*, *AC*, and *ALU*. Interconnection of these components are done through wires and busses declared by **wire net** declarations. Control signals responsible for loading and incrementing registers and controlling the *ALU* function connect to the control inputs of *IR*, *PC*, *AC*, and *ALU*.

In the last part of *DataPath*, bus assignments take place. We use bus control signals coming from the controller to drive a left-hand side bus either with one of its sources or high-impedance. For example, *pc_on_adr* control signal either puts *PC* output (*pc_out*) or all **Z**s on *adr_bus*. The *dbus* bus is declared to connect to the external bidirectional *data_bus*. Two assignments are made to *dbus* using *alu_on_dbus* and *data_on_dbus* control signals. Placement of this intermediate bus to the external data bus of the datapath (*data_bus*) is controlled by *dbus_on_data* control signal.

The last **assign** statement shown in Fig. 8.22 places most significant *IR* bits on the *op_code* output of *DataPath* that goes out to the controller.

Although we have used tri-state busses, when synthesizing this circuit, we can direct our synthesis tool to use AND-OR logic or multiplexers to implement these busses.

8.2.6.3 Controller description.

The controller code for our adding machine example is shown in Fig. 8.23. This code corresponds to the right part of Fig. 8.19 which is shown in more details in Fig. 8.20. In addition to *clk* and *reset*, the controller has the *op_code* input that is driven by *IR* and comes to the controller from the DataPath module (see Fig. 8.19).

The sequencing of control states is implemented by a Huffman style Verilog code. In this style, an **always** block handles assignment of values to *present_state*, and another **always** statement uses this register output as the input of a combinational logic determining *next_state*. This combinational block also sets values to control signals that are outputs of the controller.

The former **always** block synthesizes as a register with active high *reset*, and the latter, (i.e., *combinational*) synthesizes to a combinational block. This block uses *present_state* and *reset* on its sensitivity list. For synthesis purposes and to avoid output latches, all outputs of this block, that are the control signals, are set to their inactive, **0**, values. In the body of the *combinational* **always** block, a **case** statement checks *present_state* against the states of the machine ('Reset, 'Fetch, 'Decode, and 'Execute), and activates the proper control signals.

The 'Reset state activates *clr_pc* to clear *PC* and sets 'Fetch as the next state of the machine. In the 'Fetch state, *pc_on_adr*, *rd_mem*, *data_on_dbus*, *ld_ir*, and *inc_pc* become active, and 'Decode is set to become the next state of the machine. By activating *pc_on_adr* and *rd_mem*, the *PC* output goes on the memory address and a read operation is issued. Assuming the memory responds in the same clock, contents of memory at the *PC* address will be put on *data_bus*. Issuance of *data_on_dbus* puts the contents of this bus on the internal *dbus* of *DataPath*. This bus is connected to the input of *IR* and issuance of *ld_ir* loads its contents into this register. The next state of the controller is 'Decode that makes the new contents of *IR* available for the controller. In the 'Execute state a newly fetched instruction in *IR* decides on control signals to issue to execute the instruction.

In the 'Execute state, *op_code* is used in a **case** expression to decide on control signals to issue depending on the opcode of the fetched instruction. The **case** alternatives in this statement are four *op_code* values of **00**, **01**, **10**, and **11** that correspond to **Load**, **Store**, **Jump**, and **Add** instructions.

For **load**, *ir_on_adr*, *rd_mem*, *data_on_dbus*, and *ld_ac* are issued. These control signals cause the address from *IR* to be placed on the

```verilog
`define Reset 2'b00
`define Fetch 2'b01
`define Decode 2'b10
`define Execute 2'b11
module Controller ( input reset, clk, input [1:0] op_code,
                    output reg rd_mem, wr_mem, ir_on_adr,
                               pc_on_adr, dbus_on_data,
                               data_on_dbus, ld_ir, ld_ac,
                               ld_pc, inc_pc, clr_pc,
                               pass, add, alu_on_dbus);
    reg [1:0] present_state, next_state;

    always @( posedge clk )
      if( reset ) present_state <= `Reset;
      else present_state <= next_state;

    always @( present_state or reset ) begin : Combinational
      rd_mem=1'b0; wr_mem=1'b0; ir_on_adr=1'b0; pc_on_adr=1'b0;
      dbus_on_data=1'b0; data_on_dbus=1'b0; ld_ir=1'b0;
      ld_ac=1'b0; ld_pc=1'b0; inc_pc=1'b0; clr_pc=1'b0;
      pass=0; add=0; alu_on_dbus=1'b0;

      case ( present_state )
      `Reset : begin next_state = reset ? `Reset : `Fetch;
        clr_pc = 1;
      end // End `Reset
      `Fetch : begin next_state = `Decode;
        pc_on_adr = 1; rd_mem = 1; data_on_dbus = 1;
        ld_ir = 1; inc_pc = 1;
      end // End `Fetch
      `Decode : next_state = `Execute; // End `Decode

      `Execute: begin next_state = `Fetch;
        case( op_code )
          2'b00: begin
            ir_on_adr = 1; rd_mem = 1;
            data_on_dbus = 1; ld_ac = 1;
          end
          2'b01: begin
            pass = 1;
            ir_on_adr = 1; alu_on_dbus = 1;
            dbus_on_data = 1; wr_mem = 1;
          end
```

(Continued)

Figure 8.23   Controller Verilog Code

```
            2'b10: ld_pc = 1;
            2'b11: begin
                add = 1; alu_on_dbus = 1; ld_ac = 1;
            end
          endcase
        end // End `Execute
        default : next_state = `Reset;
      endcase
    end
endmodule
```

Figure 8.23   Controller Verilog Code (*Continued*)

addr_bus address bus, memory read to take place, and data from memory to be loaded into *AC*. Data from the memory come through *data_bus* onto *dbus* of *DataPath* by the control signal *data_on_dbus*.

Controller executes the **Store** instruction by issuing *pass*, *ir_on_adr*, *alu_on_dbus*, *dbus_on_data*, and *wr_mem*. As shown in Fig. 8.19, these signals take contents of *AC* to the input bus of the memory (i.e., *data_bus*), and *wr_mem* causes the writing into the memory to take place. Note that *pass* causes *AC* to pass through *ALU* unchanged.

The **Jump** instruction is executed by enabling *PC* load input, which takes the jump address from *IR* (see Fig. 8.19).

The last instruction of this machine is **Add**, for execution of which, *add*, *alu_on_dbus*, and *ld_ac* are issued. This instruction adds data in the upper six bits of *IR* with *AC* and loads the result into *AC*. The *add* control signal instructs *ALU* to add its two inputs; the *alu_on_dbus* puts this output on the internal datapath *dbus*; and the *ld_ac* causes *AC* to be loaded with the result of addition.

8.2.6.4   The complete machine.   The top-level module for our adding machine example is shown in Fig. 8.24. In the *AddingCPU* module shown, *DataPath* and *Controller* modules are instantiated. Port connections of the *Controller* include its output control signals, the opcode input from *DataPath* and the reset external input. Port connections of *DataPath* consist of *adr_bus* and *data_bus* external busses, opcode output, and control signal inputs.

8.2.7   Testing adding CPU

In the testbench for the *AddingCPU* module, we model a simple memory with read and write operations. The memory is file-based and we will

```
module AddingCPU (input reset, clk,
                  output [5:0] adr_bus, output rd_mem, wr_mem,
                  inout [7:0] data_bus);

   wire  ir_on_adr, pc_on_adr, dbus_on_data, data_on_dbus,
         ld_ir,
         ld_ac, ld_pc, inc_pc, clr_pc, pass, add, alu_on_dbus;
   wire [1:0] op_code;

   Controller cu ( reset, clk, op_code, rd_mem, wr_mem, ir_on_adr,
                   pc_on_adr, dbus_on_data, data_on_dbus, ld_ir,
                   ld_ac, ld_pc, inc_pc, clr_pc, pass,
                   add, alu_on_dbus );

   DataPath dp ( ir_on_adr, pc_on_adr, dbus_on_data, data_on_dbus,
                 ld_ir, ld_ac, ld_pc, inc_pc, clr_pc, pass, add,
                 alu_on_dbus, clk, adr_bus, op_code, data_bus );
endmodule
```

Figure 8.24   Adding CPU Top-Level Module

use file I/O tasks for reading and writing from and to the memory. To make this testbench more complete, we use a task for converting instructions in mnemonic form from an external file to binary memory data. In general, memory modeling and file I/O are elaborated in the testbench of *AddingCPU*.

8.2.7.1   Testbench outline.   The outline of the testbench is shown in Fig. 8.25. This module reads the *InstructionFile.mem* file which contains instruction mnemonics and their addresses, converts them to hex and writes them to *HexadecimalFile.mem* file. After this conversion is done, every addressed memory read or write uses the hex file. Because the unit under test (UUT) does not have a large memory, all read and write operations are directly performed on the *HexadecimalFile.mem* file, and no image of its memory is kept in the testbench as a fast buffer as an array of **reg**. The processor of the next section uses such a memory buffering scheme for faster memory input-output operations.

As shown in Fig. 8.25, after declarations and instantiation of *AddingCPU*, an **initial** block calls the *Convert* **task** to read the instruction file (*InstructionFile.mem*), translate instruction mnemonics to hex, and write hex data in the *HexadecimalFile.mem* file. Following the invocation of *Convert*, the testbench opens the *HexadecimalFile.mem* file for

```
module Test_AddingCPU;
    reg reset=1, clk=0;
    wire [5:0] adr_bus;
    wire rd_mem, wr_mem;
    wire [7:0] data_bus;
    reg [7:0] mem_data=8'b0;
    reg control=0;
    integer HexFile, check;

    AddingCPU UUT (reset, clk, adr_bus, rd_mem, wr_mem, data_bus);

    always #10 clk = ~clk;

    initial begin
        Convert;
        HexFile = $fopen ("HexadecimalFile.mem", "r+");
        #25 reset=1'b0;
        #405 $fclose (HexFile);
        $stop;
    end

    always @(posedge clk) begin : Memory_Read_Write
        // . . .
    end

    // . . .

    task Convert;
        // . . .
    endtask
endmodule
```

Figure 8.25 Outline of Adding CPU Testbench

read and write, and sets the end of the simulation run time. The **$fopen** task opens this hex file and assigns the *HexFile* descriptor that is a declared integer to it. The **$fclose** task closes this file just before the simulation run is stopped.

This **initial** block is followed by the *Memory_Read_Write* **always** block. This block assumes 64 8-bit hex data are available in *HexadecimalFile.mem*. For accessing this file, its descriptor *HexFile*, will be used.

Figure 8.26 shows the details of *Memory_Read_Write* **always** block. After a short delay (1 ns) after the *posedge* of *clk*, *rd_mem* and *wr_mem* are expected to be stable. At this time, if *rd_mem* is **1**, data on *adr_bus*

```
always @(posedge clk) begin : Memory_Read_Write
   control = 0;
   #1;
   if (rd_mem) begin
      #1;
      check = $fseek (HexFile, 4 * adr_bus, 0);
      check = $fscanf (HexFile, "%h", mem_data);
      control = 1;
   end
   if (wr_mem) begin
      #1;
      check = $fseek (HexFile, 4 * adr_bus, 0);
      $fwrite (HexFile, "%h", data_bus);
      $fflush (HexFile);
   end
end

assign data_bus = (control) ? mem_data: 8'hZZ;
```

Figure 8.26   Memory Read and Write

is used to set the position of the next read from *HexFile*. Since data in *HexadecimalFile.mem* are in hex (2 bytes), a total of 4 bytes that include two "end of line" bytes are used for each memory entry. Therefore, **$fseek** of Fig. 8.26 positions the next reading from *4*adr_bus*. The **$fscanf** task that follows this task reads the hex data at the file position into *mem_data*. This variable is local to the testbench and is put on *data_bus* only when reading from the memory is to take place. The *control* variable is used to drive *data_bus* with *mem_data* or *8'hZZ*.

The next part of the **always** block of Fig. 8.26 handles writing into the memory. For this purpose, after file positioning, the **$fwrite** task writes contents of *data_bus* into *HexadecimalFile.mem*. After every writing **$fflush** writes any buffered output to this file.

The testbench outline of Fig. 8.25 shows the *convert* **task** that is used for converting instruction mnemonics of *InstructionFile.mem* to hex data in *HexadecimalFile.mem*. Figure 8.27 shows six lines of *InstructionFile.mem* and its corresponding hex translation in memory locations 0 to 15.

The *Convert* **task** reads a line of *InstructionFile.mem* that contains a memory location, its mnemonic, and the instruction operand. This task converts this line to an opcode and its data and writes it in its specified location in *HexadecimalFile.mem*. For example, the third line of the

```
InstructionFile                HexadecimalFile

Line:                          Location:
   1:  00  lda  0f                00:  0F
   2:  0f  :::  0f                01:  4A
   3:  01  sta  0a                02:  C1
   4:  02  add  01                03:  4B
   5:  03  sta  0b                04:  80
   6:  04  jmp  00                06:  00
                                 07:  00
                                 08:  00
                                 09:  00
                                 10:  00
                                 11:  00
                                 12:  00
                                 13:  00
                                 14:  00
                                 15:  0F
```

Figure 8.27   Instruction Mnemonics and Hex Memory Data

instruction file of Fig. 8.27 (*sta 0A*) is translated to *4A* and is put in location 1 of the hexadecimal file. For direct memory data, the instruction file uses the "*:::*" notation. *0F ::: 0f* shown in Fig. 8.27 is translated to data *0f* in location 15 of the hexadecimal file.

The *Convert* **task** is shown in Fig. 8.28. Initially all locations of *HexadecimalFile.mem* are initialized to "00". The *InstructionFile.mem* is opened for reading, (i.e., with *r* argument), and *HexadecimalFile.mem* is opened for reading and writing, (i.e., with *r+* argument) file descriptors for these two files are *InstFile* and *HexFile*, respectively.

Convert has a **while** loop that reads data from *InstFile*, converts it to hex and puts it in its corresponding location in *HexFile*. The **$fscanf** task, shown in this **while**, reads the first two hex digits of a line of instruction into *addr*. This variable is then used for setting the write position for the *HexFile* file. This file positioning is done by **$fseek**. This is followed by **$fgets** that reads the *opcode* string from the instruction file (*InstFile*). A **case** statement in *Convert* translates string opcodes to their hex equivalent, and an **$fwrite task** writes this hex data into the hex file (*HexFile*) at the location set by the **$fseek task**.

If the opcode string read from *InstFile* is "*:::*", the hex data that follows this string will be written into *HexFile* location specified by *addr*. The last part of *Convert* flushes *HexFile* and closes both instruction and hexadecimal files.

The testbench discussed above tests *AddingCPU* for all its instructions. This example shows the power and flexibility of Verilog file handling tasks for testbench development. In modeling larger memories,

```
task Convert;
   begin: block
      reg [5: 0] addr;
      reg [3 * 8: 1] opCode;
      reg [7: 0] data, writeData;
      reg JustData;
      integer i, HexFile, InstFile, check;
      HexFile = $fopen ("HexadecimalFile.mem");
      for (i = 0; i < 64; i = i + 1) $fwrite (HexFile, "00\n");
         $fflush (HexFile); $fclose (HexFile);

      InstFile = $fopen ("InstructionFile.mem", "r");
      HexFile = $fopen ("HexadecimalFile.mem", "r+");

      while ($fscanf (InstFile, "%h", addr) != -1) begin
         check = $fseek (HexFile, addr * 4, 0);
         check = $fgets (opCode, InstFile);
         JustData = 0;
         case (opCode)
            "lda": writeData[7: 6] = 0;
            "sta": writeData[7: 6] = 1;
            "jmp": writeData[7: 6] = 2;
            "add": writeData[7: 6] = 3;
            ":::": begin
               JustData = 1;
               check = $fscanf (InstFile, "%h", writeData);
            end
            default: begin
               JustData = 1;
               check = $fscanf (InstFile, "%h", writeData);
            end
         endcase

         if (JustData == 0) begin
            check = $fscanf (InstFile, "%h", data);
            writeData[5: 0] = data[5: 0];
         end
            $fwrite (HexFile, "%h", writeData);
         end

         $fflush (HexFile); $fclose (HexFile); $fclose (InstFile);
      end
   endtask
```

Figure 8.28 Converting Instructions to Hex

direct file read and write become inefficient and more elaborate memory
and related file handling should be done. The next section shows a
larger CPU with a more complete memory model.

8.3 CPU Design and Test

This section shows design, description, and test of a small processor in
Verilog. The CPU is simple architecture, yet enough hardware (SAYEH)
that has been designed for educational and benchmarking purposes.
The design is simple, and follows the design strategy used for the Adding
CPU of the previous section. For a better understanding of the material
presented here, the reader is expected to have a general understanding
of computer architectures.

8.3.1 Details of processor functionality

The simple CPU example discussed here has a register file that is used
for data processing instructions. The CPU has a 16-bit data bus and a
16-bit address bus. The processor has 8 and 16-bit instructions. Short
instructions contain shadow instructions, which effectively pack two
such instructions into a 16-bit word. Figure 8.29 shows SAYEH interface signals.

8.3.1.1 CPU components.   SAYEH uses its register file for most of its
data instructions. Addressing modes of this processor also take advantage of this structure. Because of this, the addressing hardware of
SAYEH is a simple one and the register file output is used in address
calculations.

 SAYEH components that are used by its instructions include the standard registers such as the Program Counter, Instruction Register, the
Arithmetic Logic Unit, and Status Register. In addition, this processor

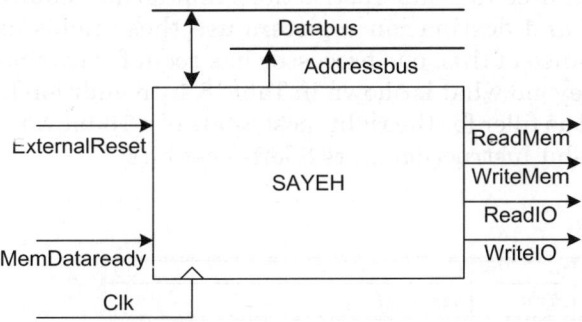

Figure 8.29   SAYEH Interface

has a register file forming registers *R0, R1, R2,* and *R3* as well as a Window Pointer that defines *R0, R1, R2,* and *R3* within the register file. CPU components and a brief description of each are shown below.

PC.   Program Counter, 16 bits

R0, R1, R2, and R3.   General purpose registers part of the register file, 16 bits

Reg file.   The general purpose registers form a window of 4 in a register file of 8 registers

WP.   Window Pointer points to the register file to define *R0, R1, R2,* and *R3*, 3 bits

IR.   Instruction Register that is loaded with a 16-bit, an 8-bit, or two 8-bit instructions, 16 bits

ALU.   The ALU that can AND, OR, NOT, Shift, Compare, Add, Subtract, and Multiply its inputs, 16 bit operands

Z flag.   Becomes **1** when the ALU output is **0**

C flag.   Becomes **1** when the ALU has a carry output

8.3.1.2   SAYEH instructions.   The general format of 8-bit and 16-bit SAYEH instructions is shown in Fig. 8.30. The 16-bit instructions have the *Immediate* field and the 8-bit instructions do not. The *OPCODE* filed is a 4-bit code that specifies the type of instruction. The *Left* and *Right* fields are two bit codes selecting *R0* through *R3* for source and/or destination of an instruction. Usually, *Left* is used for destination and *Right* for source. The *Immediate* field is used for immediate data, or if two 8-bit instructions are packed, it is used for the second instruction.

Our processor has a total of 29 instructions as shown in Table 8.1. Instructions with the "*I*" immediate field indicator are 16-bit instructions and the rest are 8-bit instructions. Instructions that use the *Destination* and *Source* fields (designated by *D* and *S* in the table of instruction set) have an opcode that is limited to 4 bits. Instructions that do not require specification of source and destination registers use these fields as opcode extensions. Because of this, our processor has room for extending its instruction set beyond what is shown in Table 8.1. In addition to *nop*, hex code 0F is used as filler for the right most 8-bits of a 16-bit word that only contains an 8-bit instruction in its 8 left-most bits.

15 12	11 10	09 08	07 00
OPCODE	*Left*	*Right*	*Immediate*

Figure 8.30   SAYEH Instruction Format

TABLE 8.1   Instruction Set of SAYEH

Instruction mnemonic and definition		Bits 15:0	RTL notation: comments or condition
nop	No operation	0000-00-00	No operation
hlt	Halt	0000-00-01	Halt, fetching stops
szf	Set zero flag	0000-00-10	Z <= '1'
czf	Clr zero flag	0000-00-11	Z <= '0'
scf	Set carry flag	0000-01-00	C <= '1'
ccf	Clr carry flag	0000-01-01	C <= '0'
cwp	Clr Window pointer	0000-01-10	WP <= "000"
mvr	Move Register	0001-D-S	$R_D <= R_S$
lda	Load Addressed	0010-D-S	$R_D <= (R_S)$
sta	Store Addressed	0011-D-S	$(R_D) <= R_S$
inp	Input from port	0100-D-S	In from port R_S and write to R_D
oup	Output to port	0101-D-S	Out to port R_D from R_S
and	AND Registers	0110-D-S	$R_D <= R_D \& R_S$
orr	OR Registers	0111-D-S	$R_D <= R_D \mid R_S$
not	NOT Register	1000-D-S	$R_D <= \sim R_S$
shl	Shift Left	1001-D-S	$R_D <= sla\ R_S$
shr	Shift Right	1010-D-S	$R_D <= sra\ R_S$
add	Add Registers	1011-D-S	$R_D <= R_D + R_S + C$
sub	Subtract Registers	1100-D-S	$R_D <= R_D - R_S - C$
mul	Multiply Registers	1101-D-S	$R_D <= R_D * R_S$:8-bit multiplication
cmp	Compare	1110-D-S	RD, RS (if equal:Z=1; if RD<RS: C=1)
mil	Move Immediate Low	1111-D-00-I	$R_{DL} <= \{8'bZ, I\}$
mih	Move Immediate High	1111-D-01-I	$R_{DH} <= \{I, 8'bZ\}$
spc	Save PC	1111-D-10-I	$R_D <= PC + I$
jpa	Jump Addressed	1111-D-11-I	$PC <= R_D + I$
jpr	Jump Relative	0000-01-11-I	$PC <= PC + I$
brz	Branch if Zero	0000-10-00-I	$PC <= PC + I$:if Z is 1
brc	Branch if Carry	0000-10-01-I	$PC <= PC + I$:if C is 1
awp	Add window pointer	0000-10-10-I	$WP <= WP + I$

In the instruction set, addressed locations in the memory are indicated by enclosing the address in a set of parenthesis. When these instructions are executed, the processor issues *ReadMem* or *WriteMem* signals to the memory. When input and output instructions (*inp, oup*) are executed, SAYEH issues *ReadIO* or *WriteIO* signals to its IO devices.

8.3.2   SAYEH datapath

The datapath of SAYEH is shown in Fig. 8.31. Main components and their lower-level structures are listed below.

1. Addressing Unit
   a. Program counter (PC)
   b. Address Logic
2. Instruction register (IR)
3. Window pointer (WP)

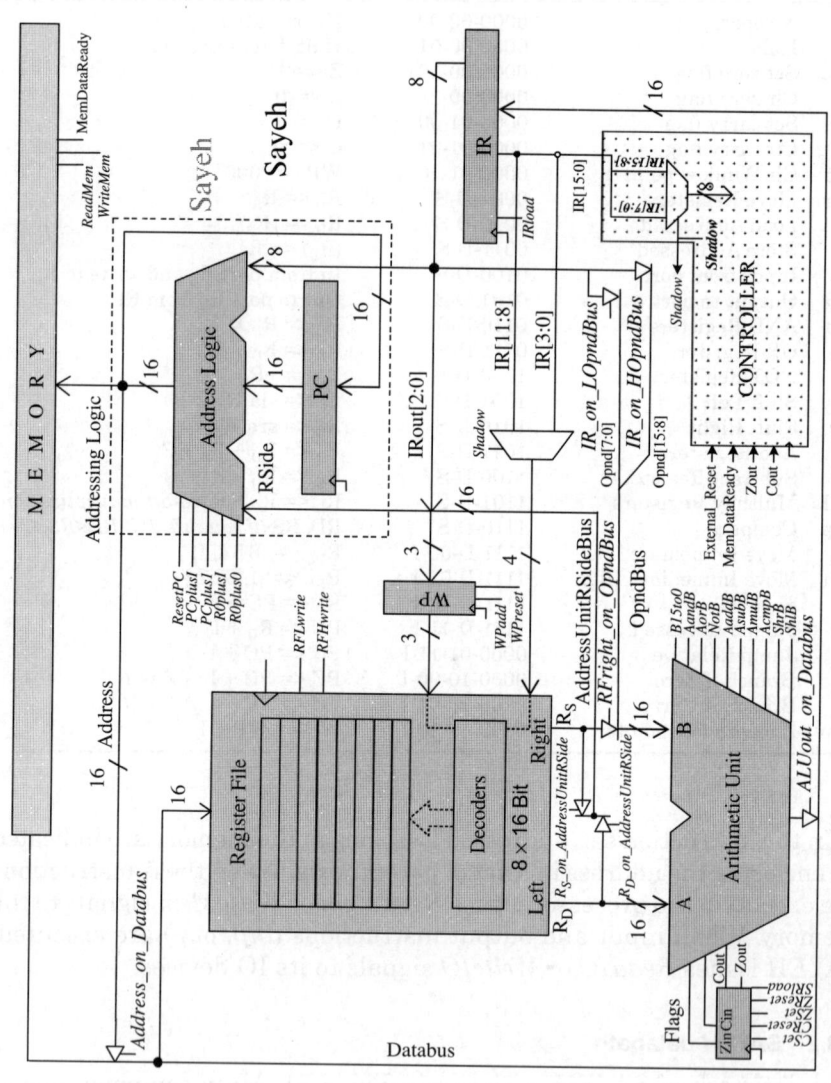

Figure 8.31 SAYEH datapath

284

4. Register File
   c. Decoder1 (Left)
   d. Decoder2 (Right)
5. Arithmetic logic unit (ALU)
6. Flags

As shown in Fig. 8.31, components are either hardwired or connected through three-state busses. Component inputs with multiple sources, such as the right-hand side input of ALU, use three-state buses. Three-state busses in this structure are *Dastabus* and *OpndBus*. Names shown on component interconnections are used in the Verilog description of the processor. In this figure, signals that are in italic are control signals issued by the controller. These signals control register clocking, logic unit operations and placement of data in busses.

8.3.2.1 Datapath components.   Figure 8.32 shows the hierarchical structure of SAYEH components. The processor has a *Datapath* and a *Controller*. *Datapath* components are *Addressing Unit, Instruction Register, Window Pointer, Register File, Arithmetic Logic Unit*, and the *Flags* register. The *Addressing Unit* is further partitioned into the *Program Counter* and *Address Logic*.

The *Addressing Logic* is a combinational circuit that is capable of adding its inputs to generate a 16-bit output that forms the address for the processor memory. The *Program Counter* and *Instruction Register* are 16-bit registers. The *Register File* is a 2-port memory and a file of 8, 16-bit registers. The *Window Pointer* is a 3-bit register that is used as the base of the *Register File*. Specific registers for read and write (*R0, R1, R2,* or *R3*) in the *Register File* are selected by its 4-bit input bus coming from the *Instruction Register*. 2 bits are used to select a source register and the other 2 bits select the destination register.

When the *Window Pointer* is enabled, it adds its 3-bit input to its current input. The *Flags* register is a 2-bit register that saves the flag outputs of the *Arithmetic Unit*. The *Arithmetic Unit* is a 16-bit arithmetic and logic unit that has the functions, as shown in Table 8.2. A 9-bit input selects the function of the ALU shown in this table. This code is provided by the processor controller.

Controller of SAYEH has eleven states for reset, fetch, decode, execute, and halt operations. Signals generated by the controller control logic unit operations and register clocking in the datapath.

SAYEH sequential data components and its controller are triggered on the falling edge of the main system clock. Control signals remain active after one falling edge through the next. This duration allows for propagation of signals through the busses and logic units in the datapath.

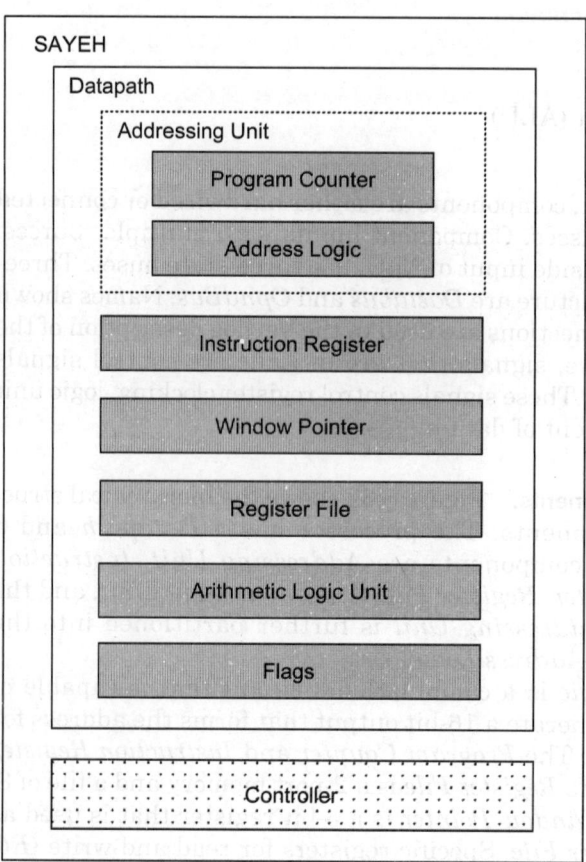

Figure 8.32   SAYEH Hierarchical Structure

TABLE 8.2   ALU Operations

Mnemonic	Description	Code
B15to0H	Place B on the output	1000000000
AandBH	Place A and B on the output	0100000000
AorBH	Place A or B on the output	0010000000
notBH	Place not B on the output	0001000000
shlBH	Shift B one bit to the left	0000100000
shrBH	Shift B one bit to the right	0000010000
AaddBH	Place A + B on the output	0000001000
AsubBH	Place A – B on the output	0000000100
AmulBH	Place A * B on the output	0000000010
AcmpBH	Z = 1 if A = B; C = 1 if A < B	0000000001

8.3.3 SAYEH Verilog description

SAYEH is described according to the hierarchical structure of Fig. 8.32. Data components are described separately, and then wired to form the datapath. The controller is described in a single Verilog module. In the complete SAYEH description, the datapath and controller are wired together.

8.3.3.1 Data components. Combinational and sequential SAYEH data components are described here. The combinational ones are like the ALU that perform arithmetic and logical operations. The function of such units is controlled by the controller. The sequential components are clocked with the negative edge of the main CPU clock. These components have functionalities like loading and resetting that are controlled by the controller.

Addressing unit. The Addressing Unit, shown in Fig. 8.33, consists of the PC and Address Logic. The PC is a simple register with enabling and resetting mechanisms, while the Address Logic is a small arithmetic unit that performs adding and incrementing for calculating PC or memory addresses.

This unit has a 16-bit input coming from the Register File, an 8-bit input from the Instruction Register, and a 16-bit address output. Control signals of the Addressing Unit are *ResetPC, PCplusI, PCplus1, RplusI, Rplus0,* and *PCenable.* These control signals select what goes on the output of this unit. Shown in Fig. 8.34 is the Verilog code of the PC. The Address Logic of Fig. 8.35 uses control signal inputs of the Addressing Unit to generate input data to the Program Counter via the *PCout* of Fig. 8.33.

Arithmetic unit. The ALU of SAYEH is shown in Fig. 8.36. For readability, control input codes of this unit are defined according to their function. For example, the select input that causes the ALU to perform the add operation is 0000001000, and it is defined as *AaddBH.* Control

```
module AddressingUnit (
    input [15:0], Rside, input [7:0] Iside, output [15:0] Address,
    input clk, ResetPC, PCplusI, PCplus1, RplusI, Rplus0, PCenable);

    wire [15:0] PCout;
    ProgramCounter PC (Address, PCenable, clk, PCout);
    AddressLogic AL (PCout, Rside, Iside, Address, ResetPC,
                     PCplusI, PCplus1, RplusI, Rplus0);
Endmodule
```

Figure 8.33 *AddressingUnit* Verilog Code

```
module ProgramCounter (
    input [15:0] in, input enable, clk, output reg [15:0] out);

    always @ (negedge clk) if (enable) out = in;

endmodule
```

Figure 8.34 *ProgramCounter* Verilog Code

inputs of this unit are *B15to0*, *AandB*, *AorB*, *notB*, *shlB*, *shrB*, *AaddB*, *AsubB*, *AmulB*, and *AcmpB* that select its various operations. In order to insure that no unwanted latches are made, all ALU outputs are set to their inactive values at the beginning of the **always** statement of its Verilog code. In a **case**-statement in this code, *aluout* and its flag outputs are set according to the selected control input of the ALU.

Instruction register.   SAYEH Instruction Register is shown in Fig. 8.37. This unit is a 16-bit register with an active high load-enable input. As shown the only control input of the *InstructionRegister* module is *IRload*.

Register file.   Figure 8.38 shows the Verilog code of SAYEH Register File. This is a 2-port memory with a moving window pointer. For reading

```
module AddressLogic ( input [15:0] PCside, Rside,
                      input [7:0] Iside,
                      input ResetPC, PCplusI, PCplus1, RplusI,
                            Rplus0,
                      output reg [15:0] ALout );

    always @ (PCside or Rside or Iside or ResetPC or
              PCplusI or PCplus1 or RplusI or Rplus0)
        case ({ResetPC, PCplusI, PCplus1, RplusI, Rplus0})
            5'b10000: ALout = 0;
            5'b01000: ALout = PCside + Iside;
            5'b00100: ALout = PCside + 1;
            5'b00010: ALout = Rside + Iside;
            5'b00001: ALout = Rside;
            default: ALout = PCside;
        endcase

endmodule
```

Figure 8.35 *AddressLogic* Verilog Code

```verilog
`define B15to0H 10'b1000000000
`define AandBH 10'b0100000000
`define AorBH  10'b0010000000
`define notBH  10'b0001000000
`define shlBH  10'b0000100000
`define shrBH  10'b0000010000
`define AaddBH 10'b0000001000
`define AsubBH 10'b0000000100
`define AmulBH 10'b0000000010
`define AcmpBH 10'b0000000001

module ArithmeticUnit ( A, B,
       B15to0, AandB, AorB, notB, shlB, shrB, AaddB, AsubB,
       AmulB, AcmpB, aluout, cin, zout, cout);
input [15:0] A, B;
input B15to0, AandB, AorB, notB, shlB, shrB,
      AaddB, AsubB, AmulB, AcmpB;
input cin;
output [15:0] aluout;
output zout, cout;
reg [15:0] aluout;
reg zout, cout;

   always @( A or B or B15to0 or AandB or AorB or notB or shlB or
            shrB or AaddB or AsubB or AmulB or AcmpB or cin)
   begin
     zout = 0; cout = 0; aluout = 0;
     case ({B15to0, AandB, AorB, notB, shlB,
        shrB, AaddB, AsubB, AmulB, AcmpB})
        `B15to0H:aluout = B;
        `AandBH: aluout = A & B;
        `AorBH: aluout = A | B;
        `notBH: aluout = ~B;
        `shlBH: aluout = {B[15:0], B[0]};
        `shrBH: aluout = {B[15], B[15:1]};
        `AaddBH: {cout, aluout} = A + B + cin;
        `AsubBH: {cout, aluout} = A - B - cin;
        `AmulBH: aluout = A[7:0] * B[7:0];
        `AcmpBH: begin
           aluout = A;
           if (A> B) cout = 1; else cout = 0;
           if (A==B) zout = 1; else zout = 0;
        end                              (Continued)
```

Figure 8.36   *ArithmeticUnit* Verilog Code

```
            default: aluout = 0;
        endcase
        if (aluout == 0) zout = 1'b1;
    end

endmodule
```

Figure 8.36 *ArithmeticUnit* Verilog Code (*Continued*)

```
module InstrunctionRegister (in, IRload, clk, out);
    input [15:0] in;
    input IRload, clk;
    output [15:0] out;
    reg [15:0] out;
        always @(negedge clk)  if (IRload == 1) out <= in;
endmodule
```

Figure 8.37 *InstructionRegister* Verilog Code

```
module RegisterFile ( input [15:0] in,
                      input clk, RFLwrite, RFHwrite,
                      input [1:0] Laddr, Raddr, input [2:0] Base,
                      output [15:0] Lout, Rout );

    reg [15:0] MemoryFile [0:7];
    wire [2:0] Laddress = Base + Laddr;
    wire [2:0] Raddress = Base + Raddr;

    assign Lout = MemoryFile [Laddress];
    assign Rout = MemoryFile [Raddress];

    reg [15:0] TempReg;

        always @(negedge clk) begin
            TempReg = MemoryFile [Laddress];
            if (RFLwrite) TempReg [7:0] = in [7:0];
            if (RFHwrite) TempReg [15:8] = in [15:8];
            MemoryFile [Laddress] = TempReg;
        end
endmodule
```

Figure 8.38 *RegisterFile* Verilog Code

from the memory, the base of the window pointer (*Base*) is added to the left and right addresses (*Laddress* and *Raddress*) and memory words are read on appropriate left and right outputs (*Lout* and *Rout*). Writing into the memory is done in the location pointed by its left address (left is used for instruction destinations). The *RFLwrite* and *RFHwrite* control signals decide whether a write is done to the low order or the high order bits of the Register File. If both these signals are active, writing is done in a 16-bit word addressed by *Laddress* plus *Base*.

8.3.3.2 SAYEH datapath. Figure 8.39 shows the datapath of SAYEH module. This module specifies component instantiations and bussing structure of the CPU according to the diagram of Fig. 8.31. Inputs of this module are the processor's data and address busses, as well as control signals that are provided by the controller of the CPU. Control signals shown in the *DataPath* module are routed to the instantiated data components or to the internal buses that are specified in this module.

Following the declarations, the *DataPath* module instantiates Addressing Unit, Arithmetic Unit, Register File, Instruction Register, Status Register, and the Window Pointer. Control signals that are inputs of the *DataPath* are passed from this module to the data components via their port connections. For example, *ResetPC* that is an input of *DataPath* and a control signal of *AddressingUnit* appears on the port list of *AddressingUnit* in its instantiation statement.

The part that follows module instantiations makes bus assignments to the internal buses of this module. For example, assignment of the output of *ArithmeticUnit* to *Databus* that is controlled by *ALU_on_Databus* is done by an **assign** statement with a right-hand side that is a conditional expression. Note the assignment of *16'bZZZZZZZZZZZZZZZZ* to *Databus* when none of the control signals of this bus are active.

In the last part of the *DataPath* module, bits of *IR* that indicate source and destination registers to the Register File are placed on *Laddr* and *Raddr* inputs of this register. The *Shadow* signal becomes 1 if a shadow instruction being executed is used to select appropriate bits of the *IR* as source and destination addresses.

8.3.3.3 SAYEH controller. The controller of SAYEH is a state machine with nine states that issues appropriate control signals to the Data Path. The controller uses the Huffman style of coding, in which the state machine has a large combinational part that is responsible for state transitions and issuing controller outputs. State transitions are done by setting next state values to the *Nstate* variable of **reg** type. Figure 8.40 shows a general outline of this controller. Various sections of this outline are discussed below.

```
module DataPath (
    clk, Databus, Addressbus,
    ResetPC, PCplusI, PCplus1, RplusI, Rplus0,
    Rs_on_AddressUnitRSide, Rd_on_AddressUnitRSide, EnablePC,
    B15to0, AandB, AorB, notB, shlB, shrB,
    AaddB, AsubB, AmulB, AcmpB,
    RFLwrite, RFHwrite,
    WPreset, WPadd, IRload, SRload, Address_on_Databus,
    ALU_on_Databus,
    IR_on_LOpndBus, IR_on_HOpndBus, RFright_on_OpndBus,
    Cset, Creset, Zset, Zreset, Shadow, Instruction, Cout, Zout );
input clk;
inout [15:0] Databus;
output [15:0] Addressbus, Instruction;
output Cout, Zout;
input
    ResetPC, PCplusI, PCplus1, RplusI, Rplus0,
    Rs_on_AddressUnitRSide, Rd_on_AddressUnitRSide, EnablePC,
    B15to0, AandB, AorB, notB, shlB, shrB,
    AaddB, AsubB, AmulB, AcmpB,
    RFLwrite, RFHwrite, WPreset, WPadd, IRload, SRload,
    Address_on_Databus, ALU_on_Databus, IR_on_LOpndBus,
    IR_on_HOpndBus, RFright_on_OpndBus,
    Cset, Creset, Zset, Zreset, Shadow;
wire [15:0] Right, Left, OpndBus, ALUout, IRout, Address,
        AddressUnitRSideBus;
wire SRCin, SRZin, SRZout, SRCout;
wire [2:0] WPout;
wire [1:0] Laddr, Raddr;
    AddressingUnit AU (AddressUnitRSideBus, IRout[7:0], Address,
                      clk, ResetPC, PCplusI, PCplus1, RplusI,
                      Rplus0, EnablePC);
    ArithmeticUnit AL (Left, OpndBus, B15to0, AandB, AorB, notB,
                      shlB, shrB, AaddB, AsubB, AmulB, AcmpB,
                      ALUout, SRCout, SRZin, SRCin);
    RegisterFile RF (Databus, clk, Laddr, Raddr, WPout, RFLwrite,
                      RFHwrite, Left, Right);
    InstrunctionRegister IR (Databus, IRload, clk, IRout);
    StatusRegister SR (SRCin, SRZin, SRload, clk, Cset, Creset,
                      Zset, Zreset, SRCout, SRZout);
    WindowPointer WP (IRout[2:0], clk, WPreset, WPadd, WPout);
                                                    (Continued)
```

Figure 8.39 SAYEH *DataPath* Module

```
    assign AddressUnitRSideBus = (Rs_on_AddressUnitRSide) ?
                                 right :
                                 (Rd_on_AddressUnitRSide) ?
                                 Left :
                                 16'bZZZZZZZZZZZZZZZZ;
    assign Addressbus = Address;
    assign Databus = (Address_on_Databus) ? Address :
                     (ALU_on_Databus) ? ALUout :
                     16'bZZZZZZZZZZZZZZZZ;
    assign OpndBus[07:0] = IR_on_LOpndBus == 1 ? IRout[7:0] :
                                                 8'bZZZZZZZZ;
    assign OpndBus[15:8] = IR_on_HOpndBus == 1 ? IRout[7:0] :
                                                 8'bZZZZZZZZ;
    assign OpndBus = RFright_on_OpndBus == 1 ? Right :
                     16'bZZZZZZZZZZZZZZZZ;

    assign Zout = SRZout;
    assign Cout = SRCout;
    assign Instruction = IRout[15:0];

    assign Laddr = (~Shadow) ? IRout[11:10] : IRout[3:2];
    assign Raddr = (~Shadow) ? IRout[09:08] : IRout[1:0];
endmodule
```

Figure 8.39   SAYEH *DataPath* Module (*Continued*)

Controller ports.   The instruction register output, ALU flags, and external control signals constitute the inputs of the controller. The outputs of the controller are 38 control signals going to the Data Path and a *Shadow* output that indicates that the controller is handling a shadow instruction. As shown in Fig. 8.40, controller outputs are declared as **reg** and are assigned values in the combinational **always** block of the controller module.

Control states.   A **parameter** declaration declares the nine states of the controller. States *reset* and *halt* are for the initial state of the machine and its halt state, respectively. In state *fetch* the machine begins fetching a 16-bit instruction that can include an 8-bit instruction and a shadow. State *memread* is entered while our controller is waiting for *memDataReady* signal from the memory indicating that its data is ready. Execution of instructions is performed in the *exec1* state. This state is entered from the *memread* state. The *lda* instruction is not completed by the *exec1* state and requires the additional state of *exec1lda* to complete its memory read. States *exec2* and *exec2lda* are like *exec1*

```verilog
module controller ( ExternalReset, clk, ResetPC, PCplusI,
                    PCplus1, RplusI, Rplus0, . . . );
    input  ExternalReset, clk, . . .
    output ResetPC, PCplusI, PCplus1, RplusI, Rplus0, . . .
    reg    ResetPC, PCplusI, PCplus1, RplusI, Rplus0, . . .
    parameter [3:0]
        reset = 0,  halt = 1,   fetch = 2,  memread = 3,
        exec1 = 4,  exec2 = 5,
        exec1lda = 6, exec2lda = 7, incpc = 8;
    parameter nop = 4'b0000;
    parameter hlt = 4'b0001;
    parameter szf = 4'b0010;
    . . .
    reg [3:0] Pstate, Nstate;

    wire ShadowEn = ~(Instruction[7:0] == 8'b000011111)
        always @(Instruction or Pstate or ExternalReset or
                 Cflag or Zflag or memDataReady) begin
            ResetPC            = 1'b0;
            PCplusI            = 1'b0;
            PCplus1            = 1'b0;
            RplusI             = 1'b0;
            Rplus0             = 1'b0;
            . . .
            case (Pstate)
                reset :
                . . .
                halt :
                . . .
                fetch :
                . . .
                memread :
                . . .
                exec1 :
                . . .
                exec1lda :
                . . .
                exec2 :
                . . .
                exec2lda :
                . . .
                incpc :
                . . .
                default: Nstate = reset;
            endcase
        end
    always @ (negedge clk) Pstate = Nstate;
endmodule
```

Figure 8.40   SAYEH Controller General Outline

294

and *exec1lda* except that they handle the shadow part of an instruction. The execute state of most instructions (*exec1* or *exec2*) increments the program counter while the instruction is being executed. However, certain instructions that use the address bus for their execution cannot increment *PC* while they are being executed. For these instructions, the *incpc* state increments the program counter.

Opcodes.   Referring to Fig. 8.40, instruction opcodes are declared as 4-bit parameters in the controller of SAYEH. These parameters are according to the processor's instruction set of Table 8.1.

State declarations.   As mentioned, the coding style of the controller is according to the Huffman style of coding discussed in Chap. 5. The next state and present states, required by this style of coding, are declared in the controller of SAYEH as 4-bit registers, *Nstate*, and *Pstate*.

Shadow instructions.   The *ShadowEn* signal that is internal to the controller is set when the hex code 0F (this code indicates that the right-most bits are not used) is not found in the right-most 8 bits of a 16-bit instruction. If this wire is **1** and execution of an 8-bit instruction is complete, the controller branches to *exec2* to execute the second half of the instruction before the next fetching begins.

Combinational block.   The combinational block of SAYEH controller has an **always** block that has a main **case** statement with case choices for every state of the machine. Transitions from one state to another and issuing control signals are performed in the **case** statement. At the beginning of the **always** statement, all control signals are set to their inactive values in order to avoid latches on these outputs.

Sequential block.   The last part of the code outline of Fig. 8.40 is the sequential **always** block for clocking *Pstate* into *Nstate*. The control state register of SAYEH and all its data registers are falling edge triggered. Control signals issues by the controller remain active through the next falling edge of the system clock.

Instruction execution.   Figure 8.41 zooms on the combinational **always** statement of the *controller* module and shows the details of execution of *mvr* in the *exec1* state of the controller. Signals issued for the execution of this instruction are shown in this figure. This instruction reads a word from the right address of the Register File and writes it into its left address. The right and left (source and destination) addresses are provided in the Data Path by connections made from *IR* to the Register File.

The *RFright_on_OpndBus* control signal is issued to read the source register from *RegisterFile* onto *OpndBus*. Since this bus is the input of

```
always @ (Instruction or Pstate or ExternalReset or Cflag or Zflag)
begin
    . . .
        case (Pstate)
        . . .
            exec1 :
                if (ExternalReset == 1'b1) Nstate = reset;
                else begin
                    case (Instruction[15:12])
                    . . .
                        mvr : begin
                            RFright_on_OpndBus = 1'b1;
                            B15to0 = 1'b1;
                            ALU_on_Databus = 1'b1;
                            RFLwrite = 1'b1;
                            RFHwrite = 1'b1;
                            SRload = 1'b1;
                            if (ShadowEn==1'b1)
                                Nstate = exec2;
                            else begin
                                PCplus1 = 1'b1;
                                EnablePC=1'b1;
                                Nstate = fetch;
                            end
                        end
                        lda : begin
                            Rplus0 = 1'b1;
                            Rs_on_AddressUnitRSide = 1'b1;
                            ReadMem = 1'b1;
                            Nstate = exec1lda;
                        end
                    . . .
                    endcase
                end
        endcase
end
. . .
```

Figure 8.41   Instruction Execution

the ALU, the data on the right input (*B*) of the ALU must pass through it to reach its output. For this purpose, the *B15to0* control input of ALU is issued. Once the data reaches the ALU output, it becomes available at the input of the Register File. Issuing *RFLwrite* and *RFHwrite* cause data to be written into the destination into *RegisterFile*.

The partial code of Fig. 8.41 shows the assignment of *exec2* to *Nstate* if the instruction we are executing has a shadow. Otherwise, signals for incrementing the Program Counter are issued and the next state is set to *fetch*.

The execution discussed here applies to most SAYEH instructions. However, instructions that require memory access, e.g., *lda*, require an extra clock for reading the memory. The first part of the execution of *lda* is shown in Fig. 8.41. As shown, for the execution of this instruction, the address is read from Register File and put on the address bus. At the same time, *ReadMem* is issued to initiate the memory read process.

The next state for execution of *lda* after *exec1* is *exec1lda* shown in the partial code of Fig. 8.42. In this state, *ReadMem* continues to be issued and state remains in *exec1lda* until *memDataReady* becomes **1**. In this

```verilog
always @ (Instruction, Pstate, ExternalReset, Cflag, Zflag) begin
   . . .
     case (Pstate)
         . . .
        exec1lda :
             if (ExternalReset == 1'b1)
                 Nstate = reset;
             else begin
             if (memDataReady == 1'b0) begin
                 Rplus0 = 1'b1;
                 Rs_on_AddressUnitRSide = 1'b1;
                 ReadMem = 1'b1;
                 Nstate = exec1lda;
             end
             else begin
                 RFLwrite = 1'b1;
                 RFHwrite = 1'b1;
                 if (ShadowEn==1'b1)
                     Nstate = exec2;
                 else begin
                     PCplus1 = 1'b1;
                     EnablePC=1'b1;
                     Nstate = fetch;
                 end
             end
         end
         . . .
     endcase
 end
 . . .
```

Figure 8.42 Memory Handshaking for *exec1lda*

case, memory data that is available on *Databus* will be clocked into *RegisterFile* by issuing *RFLwrite* and *RFHwrite*. Executions of other SAYEH instructions are similar to the examples we discussed. The complete Verilog code of SAYEH controller is over 800 lines and is included in the CD that accompanies this book.

8.3.3.4 Complete SAYEH processor. The top-level Verilog code of SAYEH that is shown in Fig. 8.43 consists of instantiation of *DataPath* and *controller* modules. In the *Sayeh* module, control signal outputs of *controller* are wired to the similarly named signals of *DataPath*. The ports of the processor are according to the block diagram of Fig. 8.29.

8.3.4 SAYEH top-level testbench

The complete Verilog description of SAYEH consists of component descriptions like registers, counters, logic units, and a state machine for its controller. Chapter 6 has shown how such components can be tested with testbenches for data application and assertion monitoring. Obviously testing SAYEH begins with testing its components using such techniques. On the other hand, a complete test of the processor when all its tested components are put together is still necessary. This section discusses top-level testing of SAYEH.

```
module Sayeh ( clk, ReadMem, WriteMem, ReadIO, WriteIO,
               Databus, Addressbus, ExternalReset, MemDataready);
input clk;
output ReadMem, WriteMem, ReadIO, WriteIO;
inout [15: 0] Databus;
output [15: 0] Addressbus;
input ExternalReset, MemDataready;

wire [15:0] Instruction;
wire esetPC, PCplusI, PCplus1, RplusI, Rplus0,
   . . .
DataPath dp (clk, Databus, Addressbus,
             ResetPC, PCplusI, PCplus1, RplusI, Rplus0, . . . );

controller ctrl (ExternalReset, clk,
                 ResetPC, PCplusI, PCplus1, RplusI, Rplus0,
                 . . . );

endmodule
```

Figure 8.43 SAYEH Top-Level Description

In a testbench, we instantiate SAYEH, and through a memory model, we apply instructions to the CPU and watch its response to these test instructions. For developing such a top-level testbench that is easy for the design engineer to work with, two issues must be considered: Test data format and memory size handling.

Test data format must be at a high level so that the designer testing the CPU can apply large volumes of instructions and data to the CPU. For this purpose, our testbench takes test data in the form of SAYEH instructions and translates them to binary data for the processor to be tested. This scheme was used for our simple *AddingCPU* and was discussed in the previous section. SAYEH testbench has such a translation program that is, of course, much larger than that of *AddingCPU*.

The other issue that must be considered for a testbench for this design is memory size handling. Recall that our *AddingCPU* example did all its reading and writings directly into an external file representing its complete memory. Having the complete memory image in one file is not practical for the relatively large size of SAYEH memory. Furthermore, moving the complete memory image of the processor being tested into its testbench and declaring it as a two dimensional **reg** requires too much memory of the computer performing the simulation. In a large design, the actual memory of a design being tested may be larger than the computer it is being tested on.

In developing a testbench for SAYEH, we focus on the issue of memory size handling discussed above. Instead of having all memory image in one file, or all of it declared as a **reg**, we take an in-between approach. We partition the memory of the circuit being tested (SAYEH CPU, in our case) into several pages, and use one file for each page. The file corresponding to a page is named according to the page number it represents. Then, the actual testbench declares a **reg** of the size of only one such page. This **reg** is regarded as a buffer. When the CPU model addresses a memory location, the testbench checks to see if that is available in the buffer. If so, data from the buffer will be read or written into according to the CPU request. On the other hand, if a memory location is addressed that is not in the buffer, the testbench writes the contents of the buffer into its corresponding memory file, and loads the page that has the addressed location into the buffer.

Figure 8.44 shows the overall structure of our testbench. The sections that follow discuss the details of the Verilog code of this testbench. The complete code of this testbench is included in the CD that accompanies this book.

8.3.4.1 Testbench Verilog outline. The outline of the Verilog code of SAYEH testbench is shown in Fig. 8.45. This code corresponds to the

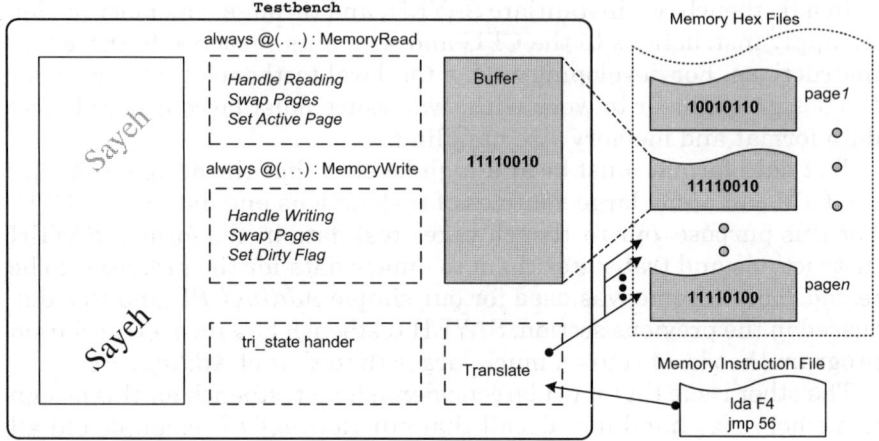

Figure 8.44   Graphical Representation of SAYEH Testbench

diagram of Fig. 8.44. The *Sayeh Testbench* module declares *totalAddrLen* and *pageLen* parameters for the total number of address lines, and address lines that are only considered for page addressing. With these parameters, the total memory being handled is 2^{16}, there are 2^4 pages, and each page is $2^{16-4} = 4096$ words. The Verilog code of Fig. 8.45 shows *SayehRAM* declared as a memory of 4096 16-bit words. This **reg** is the memory buffer that contains one of the 16 pages of the memory.

The **initial** statement shown in the testbench of SAYEH is responsible for reading an instruction file, initializing test signals, and controlling the simulation start and stop times.

8.3.4.2   Instruction translation.   The *Translate* **task** called in the **initial** block of Fig. 8.45 handles translation of instructions to hex. The instruction input file is *inst.mem* and hex files are *memFileXX.mem*, where *XX* is 00 to 15 representing memory file pages. Instruction and data read from *inst.mem* are translated to hex and placed in appropriate locations in the corresponding memory pages. At the end, the *Translate* **task** generates 16 files representing the complete image of the initial memory of SAYEH.

Except for the page handling, the *Translate* **task** of SAYEH testbench is similar to *Convert* of *AddingCPU* testbench. *Translate* handles more instructions (all of SAYEH instructions) and writes 16-bit data into its hexadecimal files. As in *Convert*, if instead of a mnemonic, "*:::*" appears in a line of *inst.mem*, *Translate* treats it as a directive for writing data directly into the specified memory location.

```verilog
module SayehTestbench();
    parameter totalAddrLen = 16;
    parameter pageLen = 4;
    parameter pageNumberLen = 2;

    reg clk, ExternalReset, MemDataready;
    reg [15:0] MemoryData;
    wire [15:0] Databus;
    wire [totalAddrLen - 1: 0] Addressbus;
    wire ReadMem, WriteMem, ReadIO, WriteIO;
    wire [totalAddrLen - pageLen - 1: 0] physicalAddr;

    reg [15:0] SayehRAM [0:(1<<(totalAddrLen-pageLen))-1];

    reg dirty;
    reg [pageLen - 1: 0] prePage;
    reg [pageNumberLen * 8: 1] pageNumber;
    integer i, file;

    always #20 clk = ~clk;

    initial begin
        Translate; //convert file
        clk = 0;
        ExternalReset = 0;
        MemDataready = 0;
        MemoryData = 16'bZ;
        dirty = 0;
        prePage = 15;

        #05 ExternalReset = 1;
        #81 ExternalReset = 0;  //run CPU
        #370000;
        $stop;
    end

    always @(negedge clk) begin : MemoryRead
        // . . .
    end

    always @(negedge clk) begin : MemoryWrite
        // . . .
    end                                          (Continued)
```

Figure 8.45   Testbench Verilog Outline

```
    assign Databus = MemoryData;
    assign physicalAddr = Addressbus[totalAddrLen-pageLen-1 : 0];

    Sayeh U1 (clk, ReadMem, WriteMem, ReadIO, WriteIO,
            Databus, Addressbus, ExternalReset, MemDataready);

    task Translate;
     // . . .
    endtask

endmodule
```

Figure 8.45   Testbench Verilog Outline (*Continued*)

8.3.4.3 Memory read procedure. The *MemoryRead* **always** block of Fig. 8.46 handles reading data requested by SAYEH from its memory image. If *ReadMem* is **1**, this block performs file and buffer handling and completes the read operation. If this signal is **0** (see the last part of the code), *MemDataReady* is set to **0**, and *MemoryData* is set to high-impedance.

If *ReadMem* is **1**, the **always** block of Fig. 8.46 checks to see if the addressed word is in a page that is in the *SayehRAM* buffer. If it is not there, and the existing buffer has data written into it (*dirty* == **1**), then the present page image from *SayehRAM* buffer is written to the page it belongs. If *dirty* is not **1**, the page image in the buffer can simply be over-written with the contents of the requested page. The **$sformat** shown in this part of the code, generates a string corresponding to the dirty page for creating the file name to open for writing.

When reading from a page whose image is not in *SayehRAM*, the **$readmemh task** shown in Fig. 8.46 reads the file that corresponds to this page, and loads it into *SayehRAM* buffer. For generating the proper file name, the **$sformat task** converts *prePage* integer page number to a two-character string. The argument of **$readmemh** concatenates this string with the rest of the file name and reads its contents.

When *SayehRAM* has the data of the addressed location of memory, *physicalAddress* that is the least significant 12 bits of SAYEH address bus is used for reading data from *SayehRAM*. This data is placed on *MemoryData*, and *MemDataReady* is issued. *MemoryData* is assigned to the CPU *Databus* with an **assign** statement.

8.3.4.4 Memory write procedure. Figure 8.47 shows the *MemoryWrite* **always** block. Handling nonexisting pages, dirty pages, address calculations, and file name generation of this block are similar to that discussed

```
always @(negedge clk) begin : MemoryRead
    if (ReadMem) begin
        #1
        if (prePage != Addressbus[totalAddrLen-1:totalAddrLen-
                                   pageLen])
        begin
            if (dirty) begin
                $sformat (pageNumber, "%0d", prePage);
                file = $fopen ({"memfile", pageNumber, ".mem"}, "r+");
                // opens the corresponding memory file
                for (i = 0; i < (1<<(totalAddrLen - pageLen)); i=i+1)
                    $fwrite (file, "%h\n", SayehRAM[i]);
                dirty = 0;
                $fclose (file);
            end
            prePage = Addressbus[totalAddrLen-1:totalAddrLen-
                                 pageLen];
            $sformat (pageNumber, "%0d", prePage);
            $readmemh ({"memfile", pageNumber, ".mem"}, SayehRAM);
            // reads the corresponding memory file
        end
        MemDataready = 1;
        MemoryData = SayehRAM [physicalAddr];
    end else begin
        #1
        MemDataready = 0;
        MemoryData = 16'hZZZZ;
    end
end
```

Figure 8.46   Reading from the Memory

in conjunction with the *MemoryRead* block. As shown near the end of code of Fig. 8.47, when writing, the 12-bit *physicalAddress* corresponding to the least significant bits of *Addressbus* is used to address the *SayehRAM* buffer. After a write, *dirty* is set to indicate that page-data in *SayehRAM* is different from its corresponding file. As discussed above, this flag is used to indicate if an unwanted page can simply be ignored or its file image needs to be updated.

The last part of the outline Verilog code of Fig. 8.45 shows instantiation of SAYEH. This processor issues *ReadMem* and *WriteMem* signals that activate *MemoryRead* and *MemoryWrite* **always** blocks discussed above. In the next section a program written for SAYEH in its assembly language is discussed. This program is initially entered

```
always @(negedge clk) begin : MemoryWrite
    #1
    if (WriteMem) begin
        if (prePage !=
            Addressbus [totalAddrLen-1 : totalAddrLen - pageLen] )
        begin
            if (dirty)  begin
                $sformat (pageNumber, "%0d", prePage);
                file = $fopen ({"memfile", pageNumber, ".mem"}, "r+");
                // opens the corresponding memory file
                for (i=0; i < (1<<(totalAddrLen-pageLen)); i=i+1)
                    $fwrite (file, "%h\n", SayehRAM[i]);
                dirty = 0;
                $fclose (file);
            end
            prePage =
            Addressbus[totalAddrLen-1:totalAddrLen-pageLen];
            $sformat (pageNumber, "%0d", prePage);
            $readmemh ({"memfile", pageNumber, ".mem"}, SayehRAM);
            // reads the corresponding memory file
        end
        #1 SayehRAM [physicalAddr] = Databus;
        dirty = 1;
    end
end
```

Figure 8.47   Writing into the Memory

in *inst.mem* file and is translated and applied to SAYEH by the testbench discussed here.

8.3.5 Sorting test program

Figure 8.48 shows a sorting program for SAYEH. This program reads data starting from the CPU memory and sorts them in descending order. The number of data item to sort is in location 768 and data begins in the next memory location. This sorting program uses two loops for the sorting to be done. When completed, the CPU is put into the halt state.

    The program shown in Fig. 8.48 is translated into its hexadecimal equivalent and is put in *SayehRAM.hex* file. As discussed in the previous section, SAYEH testbench reads instructions from this file and applies to the CPU.

8.3.6 SAYEH hardware realization

The SAYEH CPU described in this chapter has been synthesized and programmed into a number of FPGAs and tested on Altera development boards.

```
0000 mil r0 00 :r0=768      starting address in memory
0001 mih r0 03 :
0002 lda r1 r0 :r1=         total number of elements
0003 awp 5 :
0004 mil r0 01 :r5=1        for adding with index each time
0005 mih r0 00 :
0006 cwp    :
0006 add r1 r0 :r1=         limit for final r4
0007 mvr r2 r1 :
0008 awp 2 :
0009 sub r0 r3 :r2=         limit for index r3
0009 cwp    :
000A mvr r3 r0 :r3=         outer index
000A nop    :
000B cwp    :
000B cmp r3 r2 :            is the outer index is equal to its limit
000C brz 19:                branch to 0025 if zero
000D awp 3 :
000E add r0 r2 :r3=r3+1     increment outer index
000E mvr r1 r0 :r4=r3       set inner index to outer index as
                            initial
000F cwp    :
0010 awp 1 :
0011 cmp r3 r0 :            check if inner index reaches its limit
0012 brz 10:                branch to 0022 if zero
0013 awp 2 :
0014 lda r3 r0 :r6=(r3)
0015 awp 1 :
0016 add r0 r1 :r4=r4+r5    increment inner index
0016 lda r3 r0 :r7=(r4)
0017 cmp r2 r3 :            check if r6 is greater than r7
0018 brc 07:                branch to 001F if carry
0019 lda r1 r0 :r5=(r4)     r5 as an temporary register
0019 sta r0 r2 :(r4)=r6
001A cwp    :
001B awp 3 :
001C sta r0 r2 :(r3)=r5
001D mil r2 01 :
001E mih r2 00 :r5=1        for adding with index each time
001F cwp    :
0020 awp 5 :
0021 jpa r0 0E :            jump to 000F
0022 cwp    :
0023 awp 5 :
0024 jpa r0 0A :            jump to 000B
0025 hlt    :
```

Figure 8.48 Sorting Program for SAYEH

One implementation has been on Altera's FLEX device of an Altera UP2. We used a RAM from Altera's megafunctions and configured it as a memory of 1024 16-bit words. The number of logic elements used by this CPU was 1,125, which is 30% of the available LEs. Memory bits used was 16,384, which is 44% of the available memory bits. This usage indicates that we can form a complete system with a keyboard and VGA output on a FLEX 10K of UP2.

8.4 Summary

This chapter showed design, testing, and implementation of several RT level designs of a complete CPU. This design put all that we have covered in this book into one package. The design is complete and typical of any large system with a complex controller and data path. Use of the synthesizable subset of Verilog to develop for development of a design for FPGA programming was shown. On the other hand, utilization of behavioral constructs of Verilog was demonstrated in developing a testbench for our processor.

Problems

8.1 Design a 2's complement multiplier using *Booth*'s algorithm. Your design should consist of two units, a *controller* and a *datapath*. The multiplier has an 8-bit A and B inputs and a 16-bit result.

8.2 Design a multiplier that performs its 4 × 4 multiplications by a memory lookup. Your design should consist of a *controller* and a *datapath*. The multiplier has 8-bit A and B inputs and a 16-bit result. The addresses in the memory can at most be 8-bit.

8.3 In this problem you are to design a *Single Cycle* processor. The processor has a 16 bit external data bus, and a 12 bit address bus for memory or read and write. The processor has a 16 bit register (*acc*), which is used as an operand in arithmetic instructions and as an accumulator for holding data read from memory in load instructions. The processor can address 4048 16-bit words of the memory. Memory *read* and *write* operations are synchronous with the system clock and they are completed in one cycle. In reading from the memory, the address bus containing the address of the desired memory cell and the *read* signal should be issued to the data memory block. In writing into memory, the address and the *write* signal should be issued. The machine has 16-bit instructions consisting of a 4-bit opcode and a 12-bit address. The processor has three types of instructions: arithmetic and logical instructions (**ADD, CPL, AND, INC**), data-transfer instructions (**STA, LDA**), and control-flow instructions (**JNZ, JMP**). JNZ is jump and **acc** is zero. Table shown below, shows processor instructions.

Opcode	Instruction	Instruction class	Description
0000	ADD *adr*	Arithmetic-logical	*acc*<= *acc*+Mem[*adr*]
0001	CPL *adr*	Arithmetic-logical	*acc*<= Mem[*adr*]
0010	AND *adr*	Arithmetic-logical	*acc*<=*acc* & Mem[*adr*]
0011	INC *adr*	Arithmetic-logical	*acc*<= (Mem[*adr*]+1)
0100	LDA *adr*	Data-transfer	*acc*<= Mem[*adr*]
0101	STA *adr*	Data-transfer	Mem[*adr*] <= *acc*
0110	JMP *adr*	Control-flow	Unconditional Jump to *adr*
0111	JNZ *adr*	Control-flow	Conditional Jump to *adr*

8.4 In this problem, you are to design a stack based multi-cycle processor. This processor has an 8 bit data bus and a 5 bit address bus. It's memory is 32×8 (32 bytes). Processor instructions are 8 bits wide with a 3-bit opcode. All the processor instructions use the processor's internal stack. For example, the ADD instruction takes its operands from the top two stack locations. For this instruction, two top operands on the stack are popped and then the result of their addition is pushed back into the stack. The instructions and their opcodes are shown in the following table. Processor has eight different instructions that are categorized into three types: arithmetic and logical instructions (ADD, SUB, AND, NOT), memory-access instructions (PUSH, POP), and control-flow instructions (JMP).

Mnemonic	Instruction description	Bits [7:5]
ADD	Pop two operands, add, push result	000
SUB	Pop two operands, subtract, push result	001
AND	Pop two operands, AND, push result	010
NOT	Pop operand, complement, push result	011
PUSH	Load Address	100
POP	Store Address	101
JMP	Jump Address	110
JZ	Jump Address if top of stack is zero	111

Suggested Reading

IEEE Std 1364-2001, *IEEE Standard Verilog Language Reference Manual*, SH94921-TBR (print) SS94921-TBR (electronic), ISBN 0-7381-2827-9 (print and electronic), 2001.

Navabi, Z., *Digital Design and Implementation with Field Programmable Devices*, Kluwer Academic Publishers, Boston, 2005, ISBN: 1-4020-8011-5.

Navabi, Z., *Verilog Computer-Based Training Course*, CBT CD with hardcopy User's manual, McGraw-Hill, New York, 2002, ISBN 0-07-137473-6.

Navabi, Z., *VHDL: Analysis and Modeling of Digital Systems (Series in Electrical and Computer Engineering)*, McGraw-Hill College Division, New York, 1992, ISBN: 0070464723.

Patterson, D.A., J.L. Hennessy, P.J. Ashenden, et al., *Computer Organization and Design: The Hardware/Software Interface, Third Edition*, 3rd ed, Morgan Kaufmann, San Fransisco, 2004, ISBN: 1558606041.

List of Keywords

Verilog keywords are predefined identifiers used in the Verilog language constructs.

always	endcase	include
and	endconfig	initial
assign	endfunction	inout
automatic	endgenerate	input
begin	endmodule	instance
buf	endprimitive	integer
bufif0	endspecify	join
bufif1	endtable	large
case	endtask	liblist
casex	event	library
casez	for	localparam
cell	force	macromodule
cmos	forever	medium
config	fork	module
deassign	function	nand
default	generate	negedge
defparam	genvar	nmos
design	highz0	nor
disable	highz1	noshowcancelled
edge	if	not
else	ifnone	notif0
end	incdir	notif1

or	rpmos	tranif1
output	rtran	tri
parameter	rtranif0	tri0
pmos	rtranif1	tri1
posedge	scalared	triand
primitive	showcancelled	trior
pull0	signed	trireg
pull1	small	unsigned
pulldown	specify	use
pullup	specparam	vectored
pulsestyle_onevent	strong0	wait
pulsestyle_ondetect	strong1	wand
rcmos	supply0	weak0
real	supply1	weak1
realtime	table	while
reg	task	wire
release	time	wor
repeat	tran	xnor
rnmos	tranif0	xor

Frequently Used
System Tasks and Functions

This appendix includes a list of frequently used Verilog system tasks and functions for reference. For each such utility a brief description is provided. Examples for these and other system tasks are included in the chapters. The details of system tasks not discussed here and corresponding examples can be found in Chap. 7.

B.1 Display Tasks

$display The **$displayb** task displays its arguments in the order that they appear. Display will be done to the standard output device. When invoked, it always inserts a newline character at the end of its output string. Strings to be displayed as well as format specifications must appear in double quotes as an argument of this task. Task invocation,

```
$display ( "Counter value is: %d", cnt);
```

prints value of *cnt* variable in decimal format. Decimal format is assumed if no format specification exists for a variable or expression.

$displayb The **$displayb** task displays its arguments in the order that they appear. Display will be done to the standard output device. When invoked, it always inserts a newline character at the end of its output string. Strings to be displayed as well as format specifications must appear in double quotes as an argument of this task. Task invocation,

```
$displayb ( "Counter value is: %o", cnt);
```

prints value of *cnt* variable in octal format. Binary format is assumed if no format specification exists for a variable or expression.

$displayh   The **$displayh** task displays its arguments in the order that they appear. Display will be done to the standard output device. When invoked, it always inserts a newline character at the end of its output string. Strings to be displayed as well as format specifications must appear in double quotes as an argument of this task. Task invocation,

    $displayh ("Counter value is: %b", cnt);

prints value of *cnt* variable in binary format. Hexadecimal format is assumed if no format specification exists for a variable or expression.

$displayo   The **$displayo** task displays its arguments in the order that they appear. Display will be done to the standard output device. When invoked, it always inserts a newline character at the end of its output string. Strings to be displayed, as well as, format specifications must appear in double quotes as an argument of this task. Task invocation,

    $displayo ("Counter value is: %h", cnt);

prints value of *cnt* variable in hexadecimal format. Octal format is assumed if no format specification exists for a variable or expression.

$monitoron   Turns on the monitor flag used by various forms of the **$monitor** system task. Monitoring will be enabled.

$monitoroff   Turns off the monitor flag used by various forms of the **$monitor** system task. Monitoring will be disabled.

$monitor   While monitor flag is on, when a variable or an expression on the argument list changes value, the entire argument list is displayed as in the **$display** system task.

$monitorb   While monitor flag is on, when a variable or an expression on the argument list changes value, the entire argument list is displayed as in the **$displayb** system task.

$monitorh   While monitor flag is on, when a variable or an expression on the argument list changes value, the entire argument list is displayed as in the **$displayh** system task.

$monitoro   While monitor flag is on, when a variable or an expression on the argument list changes value, the entire argument list is displayed as in the **$displayo** system task.

$strobe   Using the same format as in **$display**, the **$strobe** system task displays its arguments in a simulation cycle after all events have expired.

$strobeb   Using the same format as in **$displayb**, the **$strobeb** system task displays its arguments in a simulation cycle after all events have expired.

$strobeh   Using the same format as in **$displayh**, the **$strobeh** system task displays its arguments in a simulation cycle after all events have expired.

$strobeo   Using the same format as in **$displayo**, the **$strobeo** system task displays its arguments in a simulation cycle after all events have expired.

$write   The **$write** task displays its arguments in the order that they appear using the same format as in **$display**. Unlike the **$display** task a newline character is not added to the end of its output, and consecutive outputs continue on the same line.

$writeb   The **$writeb** task displays its arguments in the order that they appear using the same format as in **$displayb**. Unlike the **$displayb** task a newline character is not added to the end of its output, and consecutive outputs continue on the same line.

$writeh   The **$writeh** task displays its arguments in the order that they appear using the same format as in **$displayh**. Unlike the **$displayh** task a newline character is not added to the end of its output, and consecutive outputs continue on the same line.

$writeo   The **$writeo** task displays its arguments in the order that they appear using the same format as in **$displayo**. Unlike the **$displayo** task a newline character is not added to the end of its output, and consecutive outputs continue on the same line.

B.2   File I/O Tasks

$fopen   The **$fopen** system function returns a file descriptor for the physical file specified as a string in the function argument. The following example makes *desc* a descriptor for the physical file *dataset.dat*.

```
integer desc = $fopen ("dataset.dat");
```

$fclose   The **$fclose** task closes an open file. The only argument of this task is a file descriptor for an open file.

$fdisplay   The **$fdisplay** task outputs its arguments in the order that they appear. Writing will be done to a file specified by its descriptor. The file descriptor must appear first in the task argument list. This task uses the same formatting as in the **$display** task. Task invocation,

```
$fdisplay (desc, "Counter value is: %d", cnt);
```

prints value of *cnt* variable in decimal format. Decimal format is assumed if no format specification exists for a variable or expression.

$fdisplayb   The **$fdisplay** task outputs its arguments in the order that they appear. Writing will be done to a file specified by its descriptor. The file descriptor must appear first in the task argument list. This task uses the same formatting as in the **$displayb** task. Task invocation,

```
$fdisplayb (desc, "Counter value is: %d", cnt);
```

prints value of *cnt* variable in decimal format. Binary format is assumed if no format specification exists for a variable or expression.

$fdisplayh   The **$fdisplayh** task outputs its arguments in the order that they appear. Writing will be done to a file specified by its descriptor. The file descriptor must appear first in the task argument list. This task uses the same formatting as in the **$displayh** task. Task invocation,

```
$fdisplayh (desc, "Counter value is: %d", cnt);
```

prints value of *cnt* variable in decimal format. Hexadecimal format is assumed if no format specification exists for a variable or expression.

$fdisplayo   The **$fdisplayo** task outputs its arguments in the order that they appear. Writing will be done to a file specified by its descriptor. The file descriptor must appear first in the task argument list. This task uses the same formatting as in the **$displayo** task. Task invocation,

```
$fdisplayo (desc, "Counter value is: %d", cnt);
```

prints value of *cnt* variable in decimal format. Octal format is assumed if no format specification exists for a variable or expression.

$fmonitor   While monitor flag is on, when a variable or an expression on the argument list changes value, the entire argument list is written into a file. The file is specified by its descriptor which is the first argument in the task argument list. This task uses the same formatting as in the **$display** task.

$fmonitorb   While monitor flag is on, when a variable or an expression on the argument list changes value, the entire argument list is written into a file. The file is specified by its descriptor which is the first argument in the task argument list. This task uses the same formatting as in the **$displayb** task.

$fmonitorh   While monitor flag is on, when a variable or an expression on the argument list changes value, the entire argument list is written into a file. The file is specified by its descriptor which is the first argument in the task argument list. This task uses the same formatting as in the **$displayh** task.

$fmonitoro   While monitor flag is on, when a variable or an expression on the argument list changes value, the entire argument list is written into a file. The file is specified by its descriptor which is the first argument in the task argument list. This task uses the same formatting as in the **$displayo** task.

$fstrobe   Using the same format as in **$display**, the **$fstrobe** system task writes its arguments into a file specified by its descriptor as the first argument of the task. Writing will be done in a simulation cycle after all events have expired.

$fstrobeb   Using the same format as in **$displayb**, the **$fstrobeb** system task writes its arguments into a file specified by its descriptor as the first argument of the task. Writing will be done in a simulation cycle after all events have expired.

$fstrobeh   Using the same format as in **$displayh**, the **$fstrobeh** system task writes its arguments into a file specified by its descriptor as the first argument of the task. Writing will be done in a simulation cycle after all events have expired.

$fstrobeo   Using the same format as in **$displayo**, the **$fstrobeo** system task writes its arguments into a file specified by its descriptor as the first argument of the task. Writing will be done in a simulation cycle after all events have expired.

$fwrite   The **$fwrite** task is similar to the **$fdisplay** task except that it does not insert a newline character at the end of its output string. Descriptor for the file into which writing is done appears first in the argument list of this task.

$fwriteb   The **$fwriteb** task is similar to the **$fdisplayb** task except that it does not insert a newline character at the end of its output string. Descriptor for the file into which writing is done appears first in the argument list of this task.

$fwriteh   The **$fwriteh** task is similar to the **$fdisplayh** task except that it does not insert a newline character at the end of its output string. Descriptor for the file into which writing is done appears first in the argument list of this task.

$fwriteo   The **$fwriteo** task is similar to the **$fdisplayo** task except that it does not insert a newline character at the end of its output string. Descriptor for the file into which writing is done appears first in the argument list of this task.

$readmemb   A physical file name and a memory name are required arguments of the **$readmemb** task. When invoked, this task reads binary data from file specified in its argument and loads this data into the memory specified as its second parameter. Optionally, invocation of this task may contain range of memory words to fill.

If *mem* is declared as

```
reg [15:0] mem [0:511],
```

then, invocation shown below reads 16 bit words in binary from *memdata.dat* file and loads this data into memory locations 12 to 412.

```
$readmemb("memdata.dat", mem, 12, 412);
```

$readmemh   A physical file name and a memory name are required arguments of the **#readmemh** task. When invoked, this task reads hexadecimal data from file specified in its argument and loads this data into the memory specified as its second parameter. Optionally, invocation of this task may contain range of memory words to fill.

If *mem* is declared as

```
reg [15:0] mem [0:511],
```

then, invocation shown below reads 16 bit words in hexadecimal from *memdata.dat* file and loads this data into memory locations 12 to 412.

$readmemh("memdata.dat", mem, 12, 412);

$swrite   The **$swrite** string output system task is similar to **$fwrite** but instead of writing to a file it writes its arguments into a register. This task can be used to convert data to string.

$sformat   The **$sformat** task is another string output system task. This task is similar to **$swrite** with one major difference; it always interprets its arguments as a format string.

$fgetc   The **$fgetc** reads a byte (character) from a specified file, i.e., **$fgetc** (*fd*), reads from *fd*. The specified file must be opened with either **r** or **r+** type values to make them available for reading.

$ungetc   The **$ungetc** inserts a character into the buffer of a specified file. The character shall be returned by the next **$fgetc** call of the specified file. The file itself remains unchanged.

$fgets   It reads characters from a specified file into a string register until either the register is filled or a newline character is read from the file. An example usage is:

$fgets (string_name, file_descriptor);

$fscanf   It reads data from a specified file into a register. It reads characters and interprets them according to the format specified by it arguments.

$sscanf   The **$sscanf** is similar to the **$fscanf**. The difference is that **$sscanf** reads from a string register.

$fread   It reads binary data from a specified file and writes into a memory. The address of the first data to be written and the number of them can be specified in this task.

$ftell   The **$ftell** returns the offset from the beginning of a specified file. This task is useful for file positioning.

$fseek   The **$fseek** task sets the position of the next input or output operation on a specified file. The positioning can be specified by offset bytes from beginning of the file or current position, or from the end of the file. General format of use of **$fseek** is shown below. The value of operation (0, 1, or 2) determines if position is set to *offset*, *offset* plus current location, or EOF plus *offset*.

Code = **$fseek** (fd, offset, operation);

$rewind   The **$rewind** task sets the position of a specified file to the beginning of the file.

$fflush   The **$fflush** task writes any buffered output to a specified file. If **$fflush** is invoked with no arguments, it writes any buffered output to all open files.

$ferror   It creates a string description of the type of the error encountered by the most recent file I/O operation and writes it into a specified register.

B.3   Timescale Tasks

$printtimescale   The **$printtimescale** task prints time unit and time precision specified by `timescale directive. If used without an argument this task considers `timescale of the module within which it is invoked. If used with an argument, the argument must be the hierarchical name of the module considered.

$timeformat   The **$timeformat** task specifies how %t format specification reports time information for various forms of display tasks. Arguments of this task are unit number, precision number, suffix string, and field width. Unit number is an integer between 0 and 15 specifying time units 1s to 1fs respectively. Precision number argument specifies number of fractional digits of time reported. Suffix string argument is a string for textual representation of time unit. The last argument specifies the width of time information output string.

B.3.1   Simulation control tasks

$finish   When encountered in a procedural flow the **$finish** system task terminates and exits simulation. An integer between 0 and 2 passed to this task as an argument specifies the type of message printed when task is invoked.

$stop   The **$stop** system task suspends simulation when invoked. An integer between 0 and 2 passed to this task as an argument specifies the type of message printed when task is invoked.

B.4   Timing Check Tasks

$hold   The **$hold** system task reports a violation when a reference event occurs too close to a data event. A time limit specifying hold time is the allowed time distance between the reference and data events. The first argument specifies the reference event such as the clock edge. The second argument specifies the data signal. The third argument specifies the hold time.

```
$hold (posedge clk, data, holdtime);
```

$period   Time distance between consecutive events of the same kind (positive or negative) is monitored by the **$period** task. The first argument is the reference event and the second event is the time specifying the period.

$setup   The **$setup** system task reports a violation when a data event occurs too close to a reference event. A time limit, which is the setup time, specifies allowed time distance between the data and reference events. The first argument specifies the name of data signal. The second argument specifies the reference event such as the clock edge. The third argument is the setup time. An example is:

$setup (data, **posedge** clk, setuptime);

$skew   The **$skew** system task reports a violation when a reference event and a data event are too far a part in time. As in the **$hold** task, the reference event is the first argument, the data event is the second argument and the skew time is the third argument of this task.

$nochange   The **$nochange** task reports a violation if during a level specified by transition on reference event of its first argument, its second argument changes value. Offset time values specified by third and fourth arguments expand or shrink the time within which data events are monitored. The following statement report a violation if *go* changes while *start* is 0.

$nochange (**negedge** start, go, 0, 0);

$recovery   The **$recovery** task is similar to the **$setup** task except that the **$recovery** task reports a violation if the data event and reference event occur at the same simulation time.

$setuphold   The **$setuphold** task is invoked with arguments specifying a reference event, a data event, setup time, and hold time in this order. This task performs both **$setup** and **$hold** tasks.

$width   The **$width** system task reports a violation when a reference event, specified by its first argument, occurs too close to an opposite event on this argument. The second argument of this task is the allowed pulse width.

Compiler Directives

This appendix briefly describes Verilog HDL compiler directives. Use and examples of such a language utility where described in Chaps. 7 and 9.

`celldefine` `endcelldefine`	Bracketing modules between `celldefine` and `endcelldefine` tags the modules as cells.
`default_nettype`	The `default_nettype` directive sets the type of implicit nets. The default is **wire**.
`define` `undef`	The `define` directive aliases an expression with a name. The `undef` directive turns of aliases set by `define`.

`ifdef`
`else`
`endif`
	Directives `ifdef`, `else`, and `endif` are if-then-else type bracketing for optional compilation of a Verilog code.

`include`	The `include` directive inserts text from an external file.
`unconnected_drive` `nounconnected_drive`	The `unconnected_drive` and `nounconnected_drive` directives bracket a portion of code for which unconnected input ports will be treated pulled up or pulled down instead of normal default.
`resetall`	The `resetall` directive resets all directives to their default values.
`timescale`	For setting time scale and time precision, `timescale is used.

Verilog Formal
Syntax Definition

The formal syntax of Verilog HDL is described using Backus-Naur Form (BNF).

D.1   Source text

D.1.1   Library Source Text

```
library_text ::= { library_descriptions }
library_descriptions ::=
          library_declaration
        | include_statement
        | config_declaration
library_declaration ::=
          library library_identifier file_path_spec [ { , file_path_spec } ]
          [ -incdir file_path_spec [ { , file_path_spec } ] ] ;
file_path_spec ::= file_path
include_statement ::= include <file_path_spec> ;
```

D.1.2   Configuration Source Text

```
config_declaration ::=
        config config_identifier ;
        design_statement
        {config_rule_statement}
        endconfig
design_statement ::= design { [library_identifier.]cell_identifier } ;
config_rule_statement ::=
```

```
          default_clause liblist_clause
         | inst_clause liblist_clause
         | inst_clause use_clause
         | cell_clause liblist_clause
         | cell_clauseuse_clause
default_clause ::= default
inst_clause ::= instance inst_name
inst_name ::= topmodule_identifier{.instance_identifier}
cell_clause ::= cell [ library_identifier.]cell_identifier
liblist_clause ::= liblist [{library_identifier}]
use_clause ::= use [library_identifier.]cell_identifier[:config]
```

D.1.3   Module and Primitive Source Text

```
source_text ::= { description }
description ::=
          module_declaration
         | udp_declaration
module_declaration ::=
          { attribute_instance } module_keyword module_identifier [ module_
             param eter_port_list ]
                  [ list_of_ports ] ; { module_item }
                  endmodule
         | { attribute_instance } module_keyword module_identifier [ module_
             parameter_port_list ]
                  [ list_of_port_declarations ] ; { non_port_module_item }
                  endmodule
module_keyword ::= module  | macromodule
```

D.1.4   Module Parameters and Ports

```
module_parameter_port_list ::= # ( parameter_declaration { , parameter_declaration } )
list_of_ports ::= ( port { , port } )
list_of_port_declarations ::=
          ( port_declaration { , port_declaration } )
         | ( )
port ::=
          [ port_expression ]
         | . port_identifier ( [ port_expression ] )
port_expression ::=
          port_reference
         | { port_reference { , port_reference } }
port_reference ::=
          port_identifier
         | port_identifier [ constant_expression ]
         | port_identifier [ range_expression ]
port_declaration ::=
          {attribute_instance} inout_declaration
         | {attribute_instance} input_declaration
         | {attribute_instance} output_declaration
```

D.1.5   Module Items

module_item ::=
> module_or_generate_item
> | port_declaration ;
> | { attribute_instance } generated_instantiation
> | { attribute_instance } local_parameter_declaration
> | { attribute_instance } parameter_declaration
> | { attribute_instance } specify_block
> | { attribute_instance } specparam_declaration

module_or_generate_item ::=
> { attribute_instance } module_or_generate_item_declaration
> | { attribute_instance } parameter_override
> | { attribute_instance } continuous_assign
> | { attribute_instance } gate_instantiation
> | { attribute_instance } udp_instantiation
> | { attribute_instance } module_instantiation
> | { attribute_instance } initial_construct
> | { attribute_instance } always_construct

module_or_generate_item_declaration ::=
> net_declaration
> | reg_declaration
> | integer_declaration
> | real_declaration
> | time_declaration
> | realtime_declaration
> | event_declaration
> | genvar_declaration
> | task_declaration
> | function_declaration

non_port_module_item ::=
> { attribute_instance } generated_instantiation
> | { attribute_instance } local_parameter_declaration
> | { attribute_instance } module_or_generate_item
> | { attribute_instance } parameter_declaration
> | { attribute_instance } specify_block
> | { attribute_instance } specparam_declaration

parameter_override ::= **defparam** list_of_param_assignments ;

D.2   Declarations

D.2.1   Declaration Types

D.2.1.1   Module Parameter Declarations

local_parameter_declaration ::=
> **localparam** [**signed**] [range] list_of_param_assignments ;
> | **localparam integer** list_of_param_assignments ;
> | **localparam real** list_of_param_assignments ;
> | **localparam realtime** list_of_param_assignments ;
> | **localparam time** list_of_param_assignments ;

parameter_declaration ::=

        parameter [**signed**] [range] list_of_param_assignments ;

        | **parameter integer** list_of_param_assignments ;

        | **parameter real** list_of_param_assignments ;

        | **parameter realtime** list_of_param_assignments ;

        | **parameter time** list_of_param_assignments ;

specparam_declaration ::= **specparam** [range] list_of_specparam_assignments ;

D.2.1.2 Port Declarations

inout_declaration ::= **inout** [net_type] [**signed**] [range]

               list_of_port_identifiers

input_declaration ::= **input** [net_type] [**signed**] [range]

               list_of_port_identifiers

output_declaration ::=

        output [net_type] [**signed**] [range]

            list_of_port_identifiers

        | **output** [**reg**] [**signed**] [range]

            list_of_port_identifiers

        | **output reg** [**signed**] [range]

            list_of_variable_port_identifiers

        | **output** [output_variable_type]

            list_of_port_identifiers

        | **output** output_variable_type

            list_of_variable_port_identifiers

D.2.1.3 Type Declarations

event_declaration ::= **event** list_of_event_identifiers ;

genvar_declaration ::= **genvar** list_of_genvar_identifiers ;

integer_declaration ::= **integer** list_of_variable_identifiers ;

net_declaration ::=

        net_type [**signed**]

            [delay3] list_of_net_identifiers ;

        | net_type [drive_strength] [**signed**]

            [delay3] list_of_net_decl_assignments ;

        | net_type [**vectored** | **scalared**] [**signed**]

        range [delay3] list_of_net_identifiers ;

        | net_type [drive_strength] [**vectored** | **scalared**] [**signed**]

        range [delay3] list_of_net_decl_assignments ;

        | **trireg** [charge_strength] [**signed**]

            [delay3] list_of_net_identifiers ;

        | **trireg** [drive_strength] [**signed**]

            [delay3] list_of_net_decl_assignments ;

        | **trireg** [charge_strength] [**vectored** | **scalared**] [**signed**]

            range [delay3] list_of_net_identifiers ;

        | **trireg** [drive_strength] [**vectored** | **scalared**] [**signed**]

            range [delay3] list_of_net_decl_assignments ;

real_declaration ::= **real** list_of_real_identifiers ;

realtime_declaration ::= **realtime** list_of_real_identifiers ;

```
reg_declaration ::= reg [ signed ] [ range ]
              list_of_variable_identifiers ;
time_declaration ::= time list_of_variable_identifiers ;
```

D.2.2 Declaration Data Types

D.2.2.1 Net and Variable Types

```
net_type ::=
          supply0 | supply1
         | tri | triand | trior | tri0 | tri1
         | wire | wand | wor
output_variable_type ::= integer | time
real_type ::=
          real_identifier [ = constant_expression ]
         | real_identifier dimension { dimension }
variable_type ::=
          variable_identifier [ = constant_expression ]
         | variable_identifier dimension { dimension }
```

D.2.2.2 Strengths

```
drive_strength ::=
          ( strength0 , strength1 )
         | ( strength1 , strength0 )
         | ( strength0 , highz1 )
         | ( strength1 , highz0 )
         | ( highz0 , strength1 )
         | ( highz1 , strength0 )
strength0 ::= supply0 | strong0 | pull0 | weak0
strength1 ::= supply1 | strong1 | pull1 | weak1
charge_strength ::= ( small ) | ( medium ) | ( large )
```

D.2.2.3 Delays

```
delay3 ::= # delay_value | # ( delay_value [ , delay_value [ , delay_value ] ] )
delay2 ::= # delay_value | # ( delay_value [ , delay_value ] )
delay_value ::=
          unsigned_number
         | parameter_identifier
         | specparam_identifier
         | mintypmax_expression
```

D.2.3 Declaration Lists

```
list_of_event_identifiers ::= event_identifier [ dimension { dimension }]
                { , event_identifier [ dimension { dimension }] }
list_of_genvar_identifiers ::= genvar_identifier { , genvar_identifier }
list_of_net_decl_assignments ::= net_decl_assignment { , net_decl_assignment }
```

list_of_net_identifiers ::= net_identifier [dimension { dimension }]
                    { , net_identifier [dimension { dimension }] }
list_of_param_assignments ::= param_assignment { , param_assignment }
list_of_port_identifiers ::= port_identifier { , port_identifier }
list_of_real_identifiers ::= real_type { , real_type }
list_of_specparam_assignments ::= specparam_assignment { , specparam_assignment }
list_of_variable_identifiers ::= variable_type { , variable_type }
list_of_variable_port_identifiers ::= port_identifier [= constant_expression]
                    { , port_identifier [= constant_expression] }

D.2.4 Declaration Assignments

net_decl_assignment ::= net_identifier = expression
param_assignment ::= parameter_identifier = constant_expression
specparam_assignment ::=
        specparam_identifier = constant_mintypmax_expression
        | pulse_control_specparam
pulse_control_specparam ::=
        PATHPULSE$ = (reject_limit_value [, error_limit_value]) ;
        | **PATHPULSE$**specify_input_terminal_descriptor$specify_output_
        terminal_descriptor
                    = (reject_limit_value [, error_limit_value]) ;
error_limit_value ::= limit_value
reject_limit_value ::= limit_value
limit_value ::= constant_mintypmax_expression

D.2.5 Declaration Ranges

dimension ::= [dimension_constant_expression : dimension_constant_expression]
range ::= [msb_constant_expression : lsb_constant_expression]

D.2.6 Function Declarations

function_declaration ::=
        function [**automatic**] [**signed**] [range_or_type] function_identifier ;
        function_item_declaration { function_item_declaration }
        function_statement
        endfunction
        | **function** [**automatic**] [**signed**] [range_or_type] function_identifier
        (function_port_list) ;
        block_item_declaration { block_item_declaration }
        function_statement
        endfunction
function_item_declaration ::=
        block_item_declaration
        | tf_input_declaration ;
function_port_list ::= { attribute_instance } tf_input_declaration { , { attribute_instance }
                    tf_input_declaration }
range_or_type ::= range | **integer** | **real** | **realtime** | **time**

D.2.7   Task Declarations

task_declaration ::=
       task [**automatic**] task_identifier **;**
       { task_item_declaration }
       statement
       endtask
       | **task** [**automatic**] task_identifier **(** task_port_list **) ;**
       { block_item_declaration }
       statement
       endtask
task_item_declaration ::=
       block_item_declaration
       | { attribute_instance } tf_input_declaration ;
       | { attribute_instance } tf_output_declaration ;
       | { attribute_instance } tf_inout_declaration ;
task_port_list ::= task_port_item { , task_port_item }
task_port_item ::=
       { attribute_instance } tf_input_declaration
       | { attribute_instance } tf_output_declaration
       | { attribute_instance } tf_inout_declaration
tf_input_declaration ::=
       input [**reg**] [**signed**] [range] list_of_port_identifiers
       | **input** [task_port_type] list_of_port_identifiers
tf_output_declaration ::=
       output [**reg**] [**signed**] [range] list_of_port_identifiers
       | **output** [task_port_type] list_of_port_identifiers
tf_inout_declaration ::=
       inout [**reg**] [**signed**] [range] list_of_port_identifiers
       | **inout [** task_port_type] list_of_port_identifiers
task_port_type ::=
       time | **real** | **realtime** | **integer**

D.2.8   Block Item Declarations

block_item_declaration ::=
       { attribute_instance } block_reg_declaration
       | { attribute_instance } event_declaration
       | { attribute_instance } integer_declaration
       | { attribute_instance } local_parameter_declaration
       | { attribute_instance } parameter_declaration
       | { attribute_instance } real_declaration
       | { attribute_instance } realtime_declaration
       | { attribute_instance } time_declaration
block_reg_declaration ::= **reg** [**signed**] [range]
          list_of_block_variable_identifiers ;
list_of_block_variable_identifiers ::=
          block_variable_type { , block_variable_type }
block_variable_type ::=
       variable_identifier
       | variable_identifier dimension { dimension }

D.3   Primitive instances

D.3.1   Primitive Instantiation and Instances

gate_instantiation ::=
>   cmos_switchtype [delay3]
>>   cmos_switch_instance { , cmos_switch_instance } ;
> | enable_gatetype [drive_strength] [delay3]
>>   enable_gate_instance { , enable_gate_instance } ;
> | mos_switchtype [delay3]
>>   mos_switch_instance { , mos_switch_instance } ;
> | n_input_gatetype [drive_strength] [delay2]
>>   n_input_gate_instance { , n_input_gate_instance } ;
> | n_output_gatetype [drive_strength] [delay2]
>>   n_output_gate_instance { , n_output_gate_instance } ;
> | pass_en_switchtype [delay2]
>>   pass_enable_switch_instance { , pass_enable_switch_instance } ;
> | pass_switchtype
>>   pass_switch_instance { , pass_switch_instance } ;
> | **pulldown** [pulldown_strength]
>>   pull_gate_instance { , pull_gate_instance } ;
> | **pullup** [pullup_strength]
>>   pull_gate_instance { , pull_gate_instance } ;

cmos_switch_instance ::= [name_of_gate_instance] (output_terminal , input_terminal ,
>   ncontrol_terminal , pcontrol_terminal)

enable_gate_instance ::= [name_of_gate_instance] (output_terminal , input_terminal ,
>   enable_terminal)

mos_switch_instance ::= [name_of_gate_instance] (output_terminal , input_terminal ,
>   enable_terminal)

n_input_gate_instance ::= [name_of_gate_instance] (output_terminal , input_terminal
>   {, input_terminal })

n_output_gate_instance ::= [name_of_gate_instance] (output_terminal
>   { , output_terminal } , input_terminal)

pass_switch_instance ::= [name_of_gate_instance] (inout_terminal , inout_terminal)

pass_enable_switch_instance ::= [name_of_gate_instance]
>   (inout_terminal , inout_terminal , enable_terminal)

pull_gate_instance ::= [name_of_gate_instance] (output_terminal)

name_of_gate_instance ::= gate_instance_identifier [range]

D.3.2   Primitive Strengths

pulldown_strength ::=
>   (strength0 , strength1)
> | (strength1 , strength0)
> | (strength0)

pullup_strength ::=
>   (strength0 , strength1)
> | (strength1 , strength0)
> | (strength1)

D.3.3 Primitive Terminals

```
enable_terminal ::= expression
inout_terminal ::= net_lvalue
input_terminal ::= expression
ncontrol_terminal ::= expression
output_terminal ::= net_lvalue
pcontrol_terminal ::= expression
```

D.3.4 Primitive Gate and Switch Types

```
cmos_switchtype ::= cmos | rcmos
enable_gatetype ::= bufif0 | bufif1 | notif0 | notif1
mos_switchtype ::= nmos | pmos | rnmos | rpmos
n_input_gatetype ::= and | nand | or | nor | xor | xnor
n_output_gatetype ::= buf | not
pass_en_switchtype ::= tranif0 | tranif1 | rtranif1 | rtranif0
pass_switchtype ::= tran | rtran
```

D.4 Module and Generated Instantiation

D.4.1 Module Instantiation

```
module_instantiation ::=
        module_identifier [ parameter_value_assignment ]
                module_instance { , module_instance } ;
parameter_value_assignment ::= # ( list_of_parameter_assignments )
list_of_parameter_assignments ::=
        ordered_parameter_assignment { , ordered_parameter_assignment } |
        named_parameter_assignment { , named_parameter_assignment }
ordered_parameter_assignment ::= expression
named_parameter_assignment ::= . parameter_identifier ( [ expression ] )
module_instance ::= name_of_instance ( [ list_of_port_connections ] )
name_of_instance ::= module_instance_identifier [ range ]
list_of_port_connections ::=
        ordered_port_connection { , ordered_port_connection }
        | named_port_connection { , named_port_connection }
ordered_port_connection ::= { attribute_instance } [ expression ]
named_port_connection ::= { attribute_instance } .port_identifier ( [ expression ] )
```

D.4.2 Generated Instantiation

```
generated_instantiation ::= generate { generate_item } endgenerate
generate_item_or_null ::= generate_item | ;
generate_item ::=
        generate_conditional_statement
        | generate_case_statement
        | generate_loop_statement
        | generate_block
        | module_or_generate_item
```

generate_conditional_statement ::=
  if (constant_expression) generate_item_or_null [**else** generate_item_or_null]
generate_case_statement ::= **case** (constant_expression)
    genvar_case_item { genvar_case_item } **endcase**
genvar_case_item ::= constant_expression { , constant_expression } :
    generate_item_or_null | **default** [:] generate_item_or_null
generate_loop_statement ::= **for** (genvar_assignment ; constant_expression ;
    genvar_ assignment)
   begin : generate_block_identifier { generate_item } **end**
genvar_assignment ::= genvar_identifier = constant_expression
generate_block ::= **begin** [: generate_block_identifier] { generate_item } **end**

D.5   UDP Declaration and Instantiation

D.5.1   UDP Declaration

udp_declaration ::=
  { attribute_instance } **primitive** udp_identifier (udp_port_list) ;
  udp_port_declaration { udp_port_declaration }
  udp_body
  endprimitive
 | { attribute_instance } **primitive** udp_identifier (udp_declaration_port_list) ;
  udp_body
  endprimitive

D.5.2   UDP Ports

udp_port_list ::= output_port_identifier , input_port_identifier { , input_port_identifier }
udp_declaration_port_list ::=
  udp_output_declaration , udp_input_declaration { , udp_input_declaration }
udp_port_declaration ::=
  udp_output_declaration ;
  | udp_input_declaration ;
  | udp_reg_declaration ;
udp_output_declaration ::=
  { attribute_instance } **output** port_identifier
  | { attribute_instance } **output reg** port_identifier [= constant_expression]
udp_input_declaration ::= { attribute_instance } **input** list_of_port_identifiers
udp_reg_declaration ::= { attribute_instance } **reg** variable_identifier

D.5.3   UDP Body

udp_body ::= combinational_body | sequential_body
combinational_body ::= **table** combinational_entry { combinational_entry } **endtable**
combinational_entry ::= level_input_list : output_symbol ;
sequential_body ::= [udp_initial_statement] **table** sequential_entry { sequential_entry }
endtable
udp_initial_statement ::= **initial** output_port_identifier = init_val ;
init_val ::= **1'b0** | **1'b1** | **1'bx** | **1'bX** | **1'B0** | **1'B1** | **1'Bx** | **1'BX** | **1** | **0**
sequential_entry ::= seq_input_list : current_state : next_state ;

seq_input_list ::= level_input_list | edge_input_list
level_input_list ::= level_symbol { level_symbol }
edge_input_list ::= { level_symbol } edge_indicator { level_symbol }
edge_indicator ::= **(** level_symbol level_symbol **)** | edge_symbol
current_state ::= level_symbol
next_state ::= output_symbol | -
output_symbol ::= **0** | **1** | **x** | **X**
level_symbol ::= **0** | **1** | **x** | **X** | **?** | **b** | **B**
edge_symbol ::= **r** | **R** | **f** | **F** | **p** | **P** | **n** | **N** | *****

D.5.4 UDP Instantiation

udp_instantiation ::= udp_identifier [drive_strength] [delay2]
                    udp_instance { , udp_instance } **;**
udp_instance ::= [name_of_udp_instance] **(** output_terminal **,** input_terminal
                    { **,** input_terminal } **)**
name_of_udp_instance ::= udp_instance_identifier [range]

D.6 Behavioral Statements

D.6.1 Continuous Assignment Statements

continuous_assign ::= **assign** [drive_strength] [delay3] list_of_net_assignments **;**
list_of_net_assignments ::= net_assignment { **,** net_assignment }
net_assignment ::= net_lvalue = expression

D.6.2 Procedural Blocks and Assignments

initial_construct ::= **initial** statement
always_construct ::= **always** statement
blocking_assignment ::= variable_lvalue = [delay_or_event_control] expression
nonblocking_assignment ::= variable_lvalue **<=** [delay_or_event_control] expression
procedural_continuous_assignments ::=
        assign variable_assignment
        | **deassign** variable_lvalue
        | **force** variable_assignment
        | **force** net_assignment
        | **release** variable_lvalue
        | **release** net_lvalue
function_blocking_assignment ::= variable_lvalue = expression
function_statement_or_null ::=
        function_statement
        | { attribute_instance } **;**

D.6.3 Parallel and Sequential Blocks

function_seq_block ::= **begin** [**:** block_identifier
                { block_item_declaration }] { function_statement } **end**
variable_assignment ::= variable_lvalue = expression

```
par_block ::= fork [ : block_identifier
                    { block_item_declaration } ] { statement } join
seq_block ::= begin [ : block_identifier
                    { block_item_declaration } ] { statement } end
```

D.6.4   Statements

```
statement ::=
            { attribute_instance } blocking_assignment ;
          | { attribute_instance } case_statement
          | { attribute_instance } conditional_statement
          | { attribute_instance } disable_statement
          | { attribute_instance } event_trigger
          | { attribute_instance } loop_statement
          | { attribute_instance } nonblocking_assignment ;
          | { attribute_instance } par_block
          | { attribute_instance } procedural_continuous_assignments ;
          | { attribute_instance } procedural_timing_control_statement
          | { attribute_instance } seq_block
          | { attribute_instance } system_task_enable
          | { attribute_instance } task_enable
          | { attribute_instance } wait_statement
statement_or_null ::=
            statement
          | { attribute_instance } ;
function_statement ::=
            { attribute_instance } function_blocking_assignment ;
          | { attribute_instance } function_case_statement
          | { attribute_instance } function_conditional_statement
          | { attribute_instance } function_loop_statement
          | { attribute_instance } function_seq_block
          | { attribute_instance } disable_statement
          | { attribute_instance } system_task_enable
```

D.6.5   Timing Control Statements

```
delay_control ::=
            # delay_value
          | # ( mintypmax_expression )
delay_or_event_control ::=
            delay_control
          | event_control
          | repeat ( expression ) event_control
disable_statement ::=
            disable hierarchical_task_identifier ;
          | disable hierarchical_block_identifier ;
event_control ::=
            @ event_identifier
          | @ ( event_expression )
          | @*
          | @ (*)
```

event_trigger ::=
       –> hierarchical_event_identifier ;
event_expression ::=
     expression
     | hierarchical_identifier
     | **posedge** expression
     | **negedge** expression
     | event_expression **or** event_expression
     | event_expression , event_expression
procedural_timing_control_statement ::=
     delay_or_event_control statement_or_null
wait_statement ::=
     wait (expression) statement_or_null

D.6.6   Conditional Statements

conditional_statement ::=
     if (expression)
statement_or_null [**else** statement_or_null]
     | if_else_if_statement
if_else_if_statement ::=
     if (expression) statement_or_null
     { **else if** (expression) statement_or_null }
     [**else** statement_or_null]
function_conditional_statement ::=
     if (expression) function_statement_or_null
     [**else** function_statement_or_null]
     | function_if_else_if_statement
function_if_else_if_statement ::=
     if (expression) function_statement_or_null
     { **else if** (expression) function_statement_or_null }
     [**else** function_statement_or_null]

D.6.7   Case Statements

case_statement ::=
     case (expression)
        case_item { case_item } **endcase**
     | **casez** (expression)
        case_item { case_item } **endcase**
     | **casex** (expression)
        case_item { case_item } **endcase**
case_item ::=
     expression { , expression } : statement_or_null
     | **default** [:] statement_or_null
function_case_statement ::=
     case (expression)
        function_case_item { function_case_item } **endcase**
     | **casez** (expression)
        function_case_item { function_case_item } **endcase**

```
                | casex ( expression )
                        function_case_item { function_case_item } endcase
function_case_item ::=
                expression { , expression } : function_statement_or_null
                | default [ : ] function_statement_or_null
```

D.6.8   Looping Statements

```
function_loop_statement ::=
                forever function_statement
                | repeat ( expression ) function_statement
                | while ( expression ) function_statement
                | for ( variable_assignment ; expression ; variable_assignment )
                        function_statement
loop_statement ::=
                forever statement
                | repeat ( expression ) statement
                | while ( expression ) statement
                | for ( variable_assignment ; expression ; variable_assignment )
                        statement
```

D.6.9   Task Enable Statements

```
system_task_enable ::= system_task_identifier [ ( expression { , expression } ) ] ;
task_enable ::= hierarchical_task_identifier [ ( expression { , expression } ) ] ;
```

D.7   Specify Section

D.7.1   Specify Block Declaration

```
specify_block ::= specify { specify_item } endspecify
specify_item ::=
                specparam_declaration
                | pulsestyle_declaration
                | showcancelled_declaration
                | path_declaration
                | system_timing_check
pulsestyle_declaration ::=
                pulsestyle_onevent list_of_path_outputs ;
                | pulsestyle_ondetect list_of_path_outputs ;
showcancelled_declaration ::=
                showcancelled list_of_path_outputs ;
                | noshowcancelled list_of_path_outputs ;
```

D.7.2   Specify Path Declarations

```
path_declaration ::=
                simple_path_declaration ;
```

      | edge_sensitive_path_declaration ;
      | state_dependent_path_declaration ;
simple_path_declaration ::=
      parallel_path_description = path_delay_value
      | full_path_description = path_delay_value
parallel_path_description ::=
      (specify_input_terminal_descriptor [polarity_operator] => specify_output_
      terminal_descriptor)
full_path_description ::=
      (list_of_path_inputs [polarity_operator] *> list_of_path_outputs)
list_of_path_inputs ::=
      specify_input_terminal_descriptor { , specify_input_terminal_descriptor }
list_of_path_outputs ::=
      specify_output_terminal_descriptor { , specify_output_terminal_descriptor }

D.7.3 Specify Block Terminals

specify_input_terminal_descriptor ::=
      input_identifier
      | input_identifier [constant_expression]
      | input_identifier [range_expression]
specify_output_terminal_descriptor ::=
      output_identifier
      | output_identifier [constant_expression]
      | output_identifier [range_expression]
input_identifier ::= input_port_identifier | inout_port_identifier
output_identifier ::= output_port_identifier | inout_port_identifier

D.7.4 Specify Path Delays

path_delay_value ::=
      list_of_path_delay_expressions
      | (list_of_path_delay_expressions)
list_of_path_delay_expressions ::=
      t_path_delay_expression
      | trise_path_delay_expression , tfall_path_delay_expression
      | trise_path_delay_expression , tfall_path_delay_expression , tz_path_delay_
      expression
      | t01_path_delay_expression , t10_path_delay_expression , t0z_path_delay_
      expression , tz1_path_delay_expression , t1z_path_delay_expression , tz0_
      path_delay_expression
      | t01_path_delay_expression , t10_path_delay_expression , t0z_path_delay_
      expression , tz1_path_delay_expression , t1z_path_delay_expression , tz0_
      path_delay_ expression
      t0x_path_delay_expression , tx1_path_delay_expression , t1x_path_delay_
      expression , tx0_path_delay_expression , txz_path_delay_expression , tzx_
      path_delay_ expression
t_path_delay_expression ::= path_delay_expression
trise_path_delay_expression ::= path_delay_expression
tfall_path_delay_expression ::= path_delay_expression

tz_path_delay_expression ::= path_delay_expression
t01_path_delay_expression ::= path_delay_expression
t10_path_delay_expression ::= path_delay_expression
t0z_path_delay_expression ::= path_delay_expression
tz1_path_delay_expression ::= path_delay_expression
t1z_path_delay_expression ::= path_delay_expression
tz0_path_delay_expression ::= path_delay_expression
t0x_path_delay_expression ::= path_delay_expression
tx1_path_delay_expression ::= path_delay_expression
t1x_path_delay_expression ::= path_delay_expression
tx0_path_delay_expression ::= path_delay_expression
txz_path_delay_expression ::= path_delay_expression
tzx_path_delay_expression ::= path_delay_expression
path_delay_expression ::= constant_mintypmax_expression
edge_sensitive_path_declaration ::=
              parallel_edge_sensitive_path_description = path_delay_value
           | full_edge_sensitive_path_description = path_delay_value
parallel_edge_sensitive_path_description ::=
           ([edge_identifier] specify_input_terminal_descriptor =>
                   specify_output_terminal_descriptor [polarity_operator] : data_
                   source_expression)
full_edge_sensitive_path_description ::=
           ([edge_identifier] list_of_path_inputs *>
                   list_of_path_outputs [polarity_operator] : data_source_expression)
data_source_expression ::= expression
edge_identifier ::= **posedge** | **negedge**
state_dependent_path_declaration ::=
              if (module_path_expression) simple_path_declaration
           | **if** (module_path_expression) edge_sensitive_path_declaration
           | **ifnone** simple_path_declaration
polarity_operator ::= **+** | **–**

D.7.5   System Timing Checks

D.7.5.1   System Timing Check Commands

system_timing_check ::=
              $setup_timing_check
           | $hold _timing_check
           | $setuphold_timing_check
           | $recovery_timing_check
           | $removal_timing_check
           | $recrem_timing_check
           | $skew_timing_check
           | $timeskew_timing_check
           | $fullskew_timing_check
           | $period_timing_check
           | $width_timing_check
           | $nochange_timing_check
$setup_timing_check ::=
           $setup (data_event , reference_event , timing_check_limit [, [notify_reg]]) ;

```
$hold _timing_check ::=
        $hold ( reference_event , data_event , timing_check_limit [ , [ notify_reg ] ] ) ;
$setuphold_timing_check ::=
        $setuphold ( reference_event , data_event , timing_check_limit , timing_
                check_limit
                        [ , [ notify_reg ] [ , [ stamptime_condition ] [ , [ checktime_
                                condition ]
                        [ , [ delayed_reference ] [ , [ delayed_data ] ] ] ] ] ] ) ;
$recovery_timing_check ::=
        $recovery ( reference_event , data_event , timing_check_limit [ , [ notify_reg ] ] ) ;
$removal_timing_check ::=
        $removal ( reference_event , data_event , timing_check_limit [ , [ notify_reg ] ] ) ;
$recrem_timing_check ::=
        $recrem ( reference_event , data_event , timing_check_limit , timing_check_limit
                        [ , [ notify_reg ] [ , [ stamptime_condition ] [ , [ checktime_
                                condition ]
                        [ , [ delayed_reference ] [ , [ delayed_data ] ] ] ] ] ] ) ;
        $skew_timing_check ::=
        $skew ( reference_event , data_event , timing_check_limit [ , [ notify_reg ] ] ) ;
$timeskew_timing_check ::=
        $timeskew ( reference_event , data_event , timing_check_limit
                [ , [ notify_reg ] [ , [ event_based_flag ] [ , [ remain_active_flag ] ] ] ] ) ;
$fullskew_timing_check ::=
        $fullskew ( reference_event , data_event , timing_check_limit , timing_check_limit
                [ , [ notify_reg ] [ , [ event_based_flag ] [ , [ remain_active_flag ] ] ] ] ) ;
$period_timing_check ::=
        $period ( controlled_reference_event , timing_check_limit [ , [ notify_reg ] ] ) ;
$width_timing_check ::=
        $width ( controlled_reference_event , timing_check_limit ,
        threshold [ , [ notify_reg ] ] ) ;
$nochange_timing_check ::=
        $nochange ( reference_event , data_event , start_edge_offset ,
                        end_edge_offset [ , [ notify_reg ] ] ) ;
```

D.7.5.2 System Timing Check Command Arguments

```
checktime_condition ::= mintypmax_expression
controlled_reference_event ::= controlled_timing_check_event
data_event ::= timing_check_event
delayed_data ::=
        terminal_identifier
        | terminal_identifier [ constant_mintypmax_expression ]
delayed_reference ::=
        terminal_identifier
        | terminal_identifier [ constant_mintypmax_expression ]
end_edge_offset ::= mintypmax_expression
event_based_flag ::= constant_expression
notify_reg ::= variable_identifier
reference_event ::= timing_check_event
remain_active_flag ::= constant_mintypmax_expression
stamptime_condition ::= mintypmax_expression
start_edge_offset ::= mintypmax_expression
```

threshold ::=constant_expression
timing_check_limit ::= expression

D.7.5.3   System Timing Check Event Definitions

timing_check_event ::=
      [timing_check_event_control]   specify_terminal_descriptor   [**&&&**
      timing_check_condition]
Controlled_timing_check_event ::=
      timing_check_event_control specify_terminal_descriptor [**&&&** timing_check_
      condition]
timing_check_event_control ::=
      posedge
      | **negedge**
      | edge_control_specifier
specify_terminal_descriptor ::=
      specify_input_terminal_descriptor
      | specify_output_terminal_descriptor
edge_control_specifier ::= **edge** [edge_descriptor [, edge_descriptor]]
edge_descriptor1 ::=
      01
      | **10**
      | z_or_x zero_or_one
      | zero_or_one z_or_x
zero_or_one ::= **0** | **1**
z_or_x ::= **x** | **X** | **z** | **Z**
timing_check_condition ::=
      scalar_timing_check_condition
      | (scalar_timing_check_condition)
scalar_timing_check_condition ::=
      expression
      | ~ expression
      | expression == scalar_constant
      | expression === scalar_constant
      | expression != scalar_constant
      | expression !== scalar_constant
scalar_constant ::=
      1'b0 | **1'b1** | **1'B0** | **1'B1** | **'b0** | **'b1** | **'B0** | **'B1** | **1** | **0**

D.8   Expressions

D.8.1   Concatenations

concatenation ::= { expression { , expression } }
constant_concatenation ::= { constant_expression { , constant_expression } }
constant_multiple_concatenation ::= { constant_expression constant_concatenation }
module_path_concatenation ::= { module_path_expression { , module_path_expression } }
module_path_multiple_concatenation ::= { constant_expression module_path_
                                   concatenation }
multiple_concatenation ::= { constant_expression concatenation }

net_concatenation ::= { net_concatenation_value { , net_concatenation_value } }
net_concatenation_value ::=
        hierarchical_net_identifier
        | hierarchical_net_identifier [expression] { [expression] }
        | hierarchical_net_identifier [expression] { [expression] } [range_expression]
        | hierarchical_net_identifier [range_expression]
        | net_concatenation
variable_concatenation ::= { variable_concatenation_value { , variable_concatenation_value } }
variable_concatenation_value ::=
        hierarchical_variable_identifier
        | hierarchical_variable_identifier [expression] { [expression] }
        | hierarchical_variable_identifier [expression] { [expression] } [range_
        expression]
        | hierarchical_variable_identifier [range_expression]
        | variable_concatenation

D.8.2 Function Calls

constant_function_call ::= function_identifier { attribute_instance }
        (constant_expression { , constant_expression })
function_call ::= hierarchical_function_identifier { attribute_instance }
        (expression { , expression })
genvar_function_call ::= genvar_function_identifier { attribute_instance }
        (constant_expression { , constant_expression })
system_function_call ::= system_function_identifier
        [(expression { , expression })]

D.8.3 Expressions

base_expression ::= expression
conditional_expression ::= expression1 ? { attribute_instance } expression2 : expression3
constant_base_expression ::= constant_expression
constant_expression ::=
        constant_primary
        | unary_operator { attribute_instance } constant_primary
        | constant_expression binary_operator { attribute_instance } constant_expression
        | constant_expression ? { attribute_instance } constant_expression : constant_
        expression
        | string
constant_mintypmax_expression ::=
        constant_expression
        | constant_expression : constant_expression : constant_expression
constant_range_expression ::=
        constant_expression
        | msb_constant_expression : lsb_constant_expression
        | constant_base_expression +: width_constant_expression
        | constant_base_expression -: width_constant_expression
dimension_constant_expression ::= constant_expression
expression1 ::= expression
expression2 ::= expression

expression3 ::= expression
expression ::=
       primary
      | unary_operator { attribute_instance } primary
      | expression binary_operator { attribute_instance } expression
      | conditional_expression
      | string
lsb_constant_expression ::= constant_expression
mintypmax_expression ::=
       expression
      | expression : expression : expression
module_path_conditional_expression ::= module_path_expression **?** { attribute_instance }
       module_path_expression : module_path_expression
module_path_expression ::=
       module_path_primary
      | unary_module_path_operator { attribute_instance } module_path_primary
      | module_path_expression binary_module_path_operator { attribute_instance }
       module_path_expression
      | module_path_conditional_expression
module_path_mintypmax_expression ::=
       module_path_expression
      | module_path_expression : module_path_expression : module_path_expression
msb_constant_expression ::= constant_expression
range_expression ::=
       expression
      | msb_constant_expression : lsb_constant_expression
      | base_expression **+:** width_constant_expression
      | base_expression **-:** width_constant_expression
width_constant_expression ::= constant_expression

D.8.4 Primaries

constant_primary ::=
       constant_concatenation
      | constant_function_call
      | **(** constant_mintypmax_expression **)**
      | constant_multiple_concatenation
      | genvar_identifier
      | number
      | parameter_identifier
      | specparam_identifier
module_path_primary ::=
       number
      | identifier
      | module_path_concatenation
      | module_path_multiple_concatenation
      | function_call
      | system_function_call
      | constant_function_call
      | **(** module_path_mintypmax_expression **)**
primary ::=
       number

      | hierarchical_identifier
      | hierarchical_identifier [expression] { [expression] }
      | hierarchical_identifier [expression] { [expression] } [range_expression]
      | hierarchical_identifier [range_expression]
      | concatenation
      | multiple_concatenation
      | function_call
      | system_function_call
      | constant_function_call
      | (mintypmax_expression)

D.8.5   Expression Left-Side Values

net_lvalue ::=
      hierarchical_net_identifier
      | hierarchical_net_identifier [constant_expression] { [constant_expression] }
      | hierarchical_net_identifier [constant_expression] { [constant_expression] } [
           constant_range_expression]
      | hierarchical_net_identifier [constant_range_expression]
      | net_concatenation
variable_lvalue ::=
      hierarchical_variable_identifier
      | hierarchical_variable_identifier [expression] { [expression] }
      | hierarchical_variable_identifier [expression] { [expression] } [range_
        expression]
      | hierarchical_variable_identifier [range_expression]
      | variable_concatenation

D.8.6   Operators

unary_operator ::=
      + | – | ! | ~ | & | ~& | | | ~| | ^ | ~^ | ^~
binary_operator ::=
      + | – | * | / | % | == | != | === | !== | && | || | **
      | < | <= | > | >= | & | | | ^ | ^~ | ~^ | >> | << | >>> | <<<
unary_module_path_operator ::=
      ! | ~ | & | ~& | | | ~| | ^ | ~^ | ^~
binary_module_path_operator ::=
      == | != | && | || | & | | | ^ | ^~ | ~^

D.8.7   Numbers

number ::=
      decimal_number
      | octal_number
      | binary_number
      | hex_number
      | real_number
real_number1 ::=
      unsigned_number . unsigned_number

```
            | unsigned_number [ . unsigned_number ] exp [ sign ] unsigned_number
exp ::= e | E
decimal_number ::=
            unsigned_number
            | [ size ] decimal_base unsigned_number
            | [ size ] decimal_base x_digit { _ }
            | [ size ] decimal_base z_digit { _ }
binary_number ::= [ size ] binary_base binary_value
octal_number ::= [ size ] octal_base octal_value
hex_number ::= [ size ] hex_base hex_value
sign ::= + | −
size ::= non_zero_unsigned_number
non_zero_unsigned_number¹ ::= non_zero_decimal_digit { _ | decimal_digit}
unsigned_number¹ ::= decimal_digit { _ | decimal_digit }
binary_value¹ ::= binary_digit { _ | binary_digit }
octal_value¹ ::= octal_digit { _ | octal_digit }
hex_value¹ ::= hex_digit { _ | hex_digit }
decimal_base¹ ::= '[s|S]d |'[s|S]D
binary_base¹ ::= '[s|S]b |'[s|S]B
octal_base¹ ::= '[s|S]o |'[s|S]O
hex_base¹ ::= '[s|S]h |'[s|S]H
non_zero_decimal_digit ::= 1 |2 |3 |4 |5 |6 |7 |8 |9
decimal_digit ::= 0 |1 |2 |3 |4 |5 |6 |7 |8 |9
binary_digit ::= x_digit |z_digit |0 |1
octal_digit ::= x_digit |z_digit |0 |1 |2 |3 |4 |5 |6 |7
hex_digit ::=
            x_digit |z_digit |0 |1 |2 |3 |4 |5 |6 |7 |8 |9
         |a |b |c |d |e |f |A |B |C |D |E |F
x_digit ::= x |X
z_digit ::= z |Z |?
```

D.8.8   Strings

```
string ::= " { Any_ASCII_Characters_except_new_line } "
```

D.9   General

D.9.1   Attributes

```
attribute_instance ::= (* attr_spec { , attr_spec } *)
attr_spec ::=
         attr_name = constant_expression
         | attr_name
attr_name ::= identifier
```

D.9.2   Comments

```
comment ::=
         one_line_comment
         | block_comment
```

one_line_comment ::= **//** comment_text \n
block_comment ::= **/*** comment_text ***/**
comment_text ::= { Any_ASCII_character }

D.9.3   Identifiers

arrayed_identifier ::=
          simple_arrayed_identifier
      | escaped_arrayed_identifier
block_identifier ::= identifier
cell_identifier ::= identifier
config_identifier ::= identifier
escaped_arrayed_identifier ::= escaped_identifier [range]
escaped_hierarchical_identifier[4] ::=
              escaped_hierarchical_branch
                  { .simple_hierarchical_branch | .escaped_hierarchical_branch }
escaped_identifier ::= \ {Any_ASCII_character_except_white_space} white_space
event_identifier ::= identifier
function_identifier ::= identifier
gate_instance_identifier ::= arrayed_identifier
generate_block_identifier ::= identifier
genvar_function_identifier ::= identifier **/*** Hierarchy disallowed ***/**
genvar_identifier ::= identifier
hierarchical_block_identifier ::= hierarchical_identifier
hierarchical_event_identifier ::= hierarchical_identifier
hierarchical_function_identifier ::= hierarchical_identifier
hierarchical_identifier ::=
         simple_hierarchical_identifier
      | escaped_hierarchical_identifier
hierarchical_net_identifier ::= hierarchical_identifier
hierarchical_variable_identifier ::= hierarchical_identifier
hierarchical_task_identifier ::= hierarchical_identifier
identifier ::=
         simple_identifier
      | escaped_identifier
inout_port_identifier ::= identifier
input_port_identifier ::= identifier
instance_identifier ::= identifier
library_identifier ::= identifier
memory_identifier ::= identifier
module_identifier ::= identifier
module_instance_identifier ::= arrayed_identifier
net_identifier ::= identifier
output_port_identifier ::= identifier
parameter_identifier ::= identifier
port_identifier ::= identifier
real_identifier ::= identifier
simple_arrayed_identifier ::= simple_identifier [range]
simple_hierarchical_identifier[3] ::=
         simple_hierarchical_branch [.escaped_identifier]
simple_identifier[2] ::= **[a-zA-Z_]** { **[a-zA-Z0-9_$]** }
specparam_identifier ::= identifier

system_function_identifier[5] ::= $[a-zA-Z0-9_$]{ [a-zA-Z0-9_$] }
system_task_identifier5 ::= $[a-zA-Z0-9_$]{ [a-zA-Z0-9_$] }
task_identifier ::= identifier
terminal_identifier ::= identifier
text_macro_identifier ::= simple_identifier
topmodule_identifier ::= identifier
udp_identifier ::= identifier
udp_instance_identifier ::= arrayed_identifier
variable_identifier ::= identifier

D.9.4 Identifier Branches

simple_hierarchical_branch[3] ::=
       simple_identifier [[unsigned_number]]
           [{ .simple_identifier [[unsigned_number]] }]
escaped_hierarchical_branch[4] ::=
       escaped_identifier [[unsigned_number]]
           [{ .escaped_identifier [[unsigned_number]] }]

D.9.5 White Space

white_space ::= space | tab | newline | eof[6]

Notes

1. Embedded spaces are illegal.

2. A simple_identifier and arrayed_reference shall start with an alpha or underscore (_) character, shall have at least one character, and shall not have any spaces.

3. The period (.) in simple_hierarchical_identifier and simple_hierarchical_branch shall not be preceded or followed by white_space.

4. The period in escaped_hierarchical_identifier and escaped_hierarchical_branch shall be preceded by white_space, but shall not be followed by white_space.

5. The $ character in a system_function_identifier or system_task_identifier shall not be followed by white_space. A system_function_identifier or system_task_identifier shall not be escaped.

6. End of file.

Verilog Assertion Monitors

OVL assertion checkers are Verilog modules with parameters and ports that can be used to check the occurrence of certain conditions in a design. The assertion library must be installed in order to be used with a Verilog simulation program. The Accelera Standard OVL V1.0 document in the accompanying CD of this book has details of OVL library installation and usage. Timing diagrams and examples in this document provide a comprehensive document for proper use of this library by Verilog designers. This appendix serves as a quick reference for OVL assertion checkers and does not include installation and detailed analysis contained in the Accelera document.

An OVL assertion checker is invoked by a set of optional parameters and a set of ports (arguments) that provide conditions that lead to checking what is referred to as the checker's test expression. A set of integer values have been defined for the checkers' optional parameters. When the OVL library is installed, the Verilog `define construct is used for associating integer constants with meaningful variable names. Variables used for assertion parameters and their corresponding values are as follows:

```
// active   edges
`define OVL_NOEDGE  0
`define OVL_POSEDGE  1
`define OVL_NEGEDGE  2
`define OVL_ANYEDGE  3

// severity   levels
`define OVL_FATAL 0
`define OVL_ERROR 1
`define OVL_WARNING 2
`define OVL_INFO 3
```

```
// c o v e r a g e   l e v e l s
`define OVL_COVER_NONE 0
`define OVL_COVER_ALL 1

// p r o p e r t y   t y p e
`define OVL_ASSERT 0
`define OVL_ASSUME 1

// n e c e s s a r y   c o n d i t i o n
`define OVL_TRIGGER_ON_MOST_PIPE 0
`define OVL_TRIGGER_ON_FIRST_PIPE 1
`define OVL_TRIGGER_ON_FIRST_NOPIPE 2

// a c t i o n   o n   n e w s t a r t
`define OVL_IGNORE_NEW_START 0
`define OVL_RESET_ON_NEW_START 1
`define OVL_ERROR_ON_NEW_START 2

// i n a c t i v e   l e v e l s
`define OVL_ALL_ZEROS 0
`define OVL_ALL_ONES 1
`define OVL_ONE_COLD 2
```

In the sections that follow we will give a brief description of OVL assertion checkers. Parameters and ports (arguments) of each are discussed here.

E.1   assert_always

The **assert_always** assertion checker evaluates the expression *test_expr* at every positive edge of the triggering event or clock *clk*. This assertion contends that the expression always evaluates to TRUE. Should the expression evaluate to FALSE, the message *msg* will be reported with severity *severity_level*.

```
assert_always
        #( severity_level, property_type, msg, coverage_level )
        instance_name ( clk, reset_n, test_expr )
```

- *severity_level* is for handling an assertion violation. The values are OVL_FATAL, OVL_ERROR, OVL_WARNING, or OVL_INFO. The default value is OVL_ERROR.
- *property_type* determines whether to use the assertion as an assert property or an assume property. The default value is OVL_ASSERT.

- *msg* is a string expression displayed whenever the assertion fails. It has a language-dependent default value.

- *coverage_level* is for enabling or disabling coverage monitoring for the checker. The default value is OVL_COVERALL.

- *clk* is a signal whose positive edge triggers the checking of the assertion.

- *reset_n* is a signal, that when deasserted, indicates that the assertion is to be monitored.

- *test_expr* is the expression being monitored by this instance of **assert_always**.

E.2   assert_always_on_edge

The **assert_always_on_edge** assertion checker monitors the *test_expr* at every specified edge of the sampling event and positive edge of clock *clk*. It contends that a specified expression will always evaluate to TRUE on the edge of a sampling event. Whenever *test_expr* evaluates to FALSE, the message *msg* will be reported with the *severity_level*.

```
assert_always_on_edge
    #( severity_level, edge_type, property_type,
       msg, coverage_level )
    instance_name (clk, reset_n, sampling_event, test_expr )
```

- *severity_level* is for handling an assertion violation. The values are OVL_FATAL, OVL_ERROR, OVL_WARNING, or OVL_INFO. The default value is OVL_ERROR.

- *edge_type* determines which type of transition of *sampling_event* initiates the check.

  - `OVL_POSEDGE, initiates the check if *sampling_event* transitions to **1**.

  - `OVL_NEGEDGE, initiates the check if *sampling_event* transitions to **0**.

  - `OVL_POSEDGE, initiates the check if *sampling_event* transitions to **1** or to **0**.

  - `OVL_NOEDGE (default), always initiates the check and *sampling_event* is never sampled.

- *property_type* determines whether to use the assertion as an assert property or an assume property. The default value is OVL_ASSERT.

- *msg* is a string expression displayed whenever the assertion fails. It has a language-dependent default value.

- *coverage_level* is for enabling or disabling coverage monitoring for the checker. The default value is OVL_COVERALL.

- *clk* is a signal whose positive edge triggers the checking of the assertion, provided that an edge of the type indicated by edge type simultaneously occurs on the sampling event.

- *reset_n* is a signal, that when deasserted, indicates that the assertion is to be monitored.

- *sampling_event* is an boolean expression, that when TRUE, enables the monitoring of *test_expr*.

- *test_expr* is the expression being monitored by this instance of **assert_always_on_edge**.

E.3   assert_change

The **assert_change** assertion monitors the *start_event* at every positive edge of the start event or positive edge of clock *clk*. It contends that the test expression will change value within *num_cks* clock cycles. If this condition does not hold, the message *msg* will be reported with *severity_level*.

```
assert_change
        #( severity_level, width, num_cks, action_on_new_start,
           property_type, msg, coverage_level )
        instance_name (clk, reset_n, start_event, test_expr)
```

- *severity_level* is for handling an assertion violation. The values are OVL_FATAL, OVL_ERROR, OVL_WARNING, or OVL_INFO. The default value is OVL_ERROR.

- *width* is the width of *test_expr*, with a default of 1.

- *num_cks* is the number of clock cycles within which *test_expr* is to change.

- *action_on_new_start* is the method of handling a new start that occurs before *test_expr* changes value or *num_cks* cycles transpire without a change.

  - `OVL_IGNORE_NEW_START (default), ignores all assertions of *start_event* after the first assertion has been detected.

  - `OVL_RESET_ON_NEW_START, restart monitoring of *test_expr* if *start_event* is asserted in any subsequent clock assertion during the monitoring of *test_expr*.

- `OVL_ERROR_ON_NEW_START, while monitoring, issue an error if *start_event* is asserted in any clock cycle subsequent to the cycle in which monitoring began.

- *property_type* determines whether to use the assertion as an assert property or an assume property. The default value is OVL_ASSERT.

- *msg* is a string expression displayed whenever the assertion fails. It has a language-dependent default value.

- *coverage_level* is for enabling or disabling coverage monitoring for the checker. The default value is OVL_COVERALL.

- *clk* is a signal whose positive edge triggers the checking of the assertion.

- *reset_n* is a signal, that when deasserted, indicates that the assertion is to be monitored.

- *start_event* Starting event that triggers the monitoring of *test_expr*.

- *test_expr* the expression being monitored by this instance of **assert_change**.

E.4 assert_cycle_sequence

The **assert_cycle_sequence** assertion contends that a specified sequence of events occurs. It checks sequences in three ways:

- If *necessary_condition* is `OVL_TRIGGER_ON_MOST_PIPE (default), it contends that, if the first *num_cks*-1 events of *event_sequence* occur in sequence, then the final one (*event_sequence*(0)) also occurs.

- If *necessary_condition* is `OVL_TRIGGER_ON_FIRST_PIPE, it contends that once the first event (*event_sequence*(*num_cks*-1)) occurs, all the remaining events also occur in sequence. This allows simultaneous checking to run, i.e., overlapping of correct sequences is allowed.

- If *necessary_condition* is `OVL_TRIGGER_ON_FIRST_NOPIPE, it contends that once the first event (*event_sequence*(*num_cks*-1)) occurs, all the remaining events also occur in sequence. This does not allow a new checking to begin so long as another checking is in progress, i.e., no overlapping is allowed and one sequence has to be completed before the next checking begins.

```
assert_cycle_sequence
    #( severity_level, num_cks, necessary_condition,
       property_type, msg, coverage_level )
    instance_name (clk, reset_n, event_sequence)
```

- *severity_level* is for handling an assertion violation. The values are OVL_FATAL, OVL_ERROR, OVL_WARNING, or OVL_INFO. The default value is OVL_ERROR.

- *num_cks* The number of events in the *event_sequence* (and the length of the *event_sequence* array).

- *necessary_condition*, either 0 or 1, with a default of 0.

- *property_type* determines whether to use the assertion as an assert property or an assume property. The default value is OVL_ASSERT.

- *msg* is a string expression displayed whenever the assertion fails. It has a language-dependent default value.

- *coverage_level* is for enabling or disabling coverage monitoring for the checker. The default value is OVL_COVERALL.

- *clk* is a signal whose positive edge triggers the checking of the assertion.

- *reset_n* is a signal, that when deasserted, indicates that the assertion is to be monitored.

- *event_sequence* A Verilog or VHDL concatenation expression, where each bit represents an event.

E.5   assert_decrement

The **assert_decrement** assertion checker monitors the expression *test_expr* at every positive edge of the clock *clk*. This assertion contends that the expression never decreases in value except in steps of *value*. The expression *test_expr* can be any valid integral expression. The check begins at the first rising edge of *clk* after reset_n is deasserted.

```
assert_decrement
      #( severity_level, width, value, property_type,
         msg, coverage_level )
      instance_name ( clk, reset_n, test_expr )
```

- *severity_level* is for handling an assertion violation. The values are OVL_FATAL, OVL_ERROR, OVL_WARNING, or OVL_INFO. The default value is OVL_ERROR.

- *width* is the width of the test expression (default value is 1).

- *value* permitted decrement value (default value is 1).

- *property_type* determines whether to use the assertion as an assert property or an assume property. The default value is OVL_ASSERT.

- *msg* is a string expression displayed whenever the assertion fails. It has a language-dependent default value.

- *coverage_level* is for enabling or disabling coverage monitoring for the checker. The default value is OVL_COVERALL.

- *clk* is a signal whose positive edge triggers the checking of the assertion.

- *reset_n* is a signal, that when deasserted, indicates that the assertion is to be monitored.

- *test_expr[width-1:0]* is the expression being monitored by this instance of **assert_decrement**.

E.6 assert_delta

The **assert_delta** assertion monitors the expression *test_expr* at every positive edge of the triggering event or clock *clk*. This assertion contends that the expression never changes value by less than the *min* value or more than the *max* value.

```
assert_delta
      #( severity_level, width, min, max, property_type,
         msg, coverage_level )
      instance_name ( clk, reset_n, test_expr )
```

- *severity_level* is for handling an assertion violation. The values are OVL_FATAL, OVL_ERROR, OVL_WARNING, or OVL_INFO. The default value is OVL_ERROR.

- *width* is the width of the test expression (default value is 1).

- *min* is the minimum value by which the test expression can change in value in one step (default value is 1).

- *max* is the maximum value by which the test expression can change in value in one step (default value is 1).

- *property_type* determines whether to use the assertion as an assert property or an assume property. The default value is OVL_ASSERT.

- *msg* is a string expression displayed whenever the assertion fails. It has a language-dependent default value.

- *coverage_level* is for enabling or disabling coverage monitoring for the checker. The default value is OVL_COVERALL.

- *clk* is a signal whose positive edge triggers the checking of the assertion.

- *reset_n* is a signal, that when deasserted, indicates that the assertion is to be monitored.

- *test_expr[width-1:0]* is the expression being monitored by this instance of **assert_delta**.

E.7   assert_even_parity

The **assert_even_parity** assertion monitors the expression *test_expr* at every positive edge of the triggering event or clock *clk*. This assertion contends that the expression always has an even number of bits asserted.

```
assert_even_parity
        #( severity_level, width, property_type,
          msg, coverage_level )
        instance_name ( clk, reset_n, test_expr )
```

- *severity_level* is for handling an assertion violation. The values are OVL_FATAL, OVL_ERROR, OVL_WARNING, or OVL_INFO. The default value is OVL_ERROR.

- *width* is the width of the test expression (default value is 1).

- *property_type* determines whether to use the assertion as an assert property or an assume property. The default value is OVL_ASSERT.

- *msg* is a string expression displayed whenever the assertion fails. It has a language-dependent default value.

- *coverage_level* is for enabling or disabling coverage monitoring for the checker. The default value is OVL_COVERALL.

- *clk* is a signal whose positive edge triggers the checking of the assertion.

- *reset_n* is a signal, that when deasserted, indicates that the assertion is to be monitored.

- *test_expr[width-1:0]* is the expression being monitored by this instance of **assert_even_parity**.

E.8   assert_fifo_index

The **assert_fifo_index** assertion monitors the index of a FIFO-like structure to assure that it never either overflows or underflows. The assertion supports multiple pushes (writes) into and pops (reads) from a FIFO within a given clock cycle.

assert_fifo_index
```
#( severity_level, depth, push_width, pop_width, property_type,
   msg, coverage_level, simultaneous_push_pop )
instance_name ( clk, reset_n, push, pop )
```

- *severity_level* is for handling an assertion violation. The values are OVL_FATAL, OVL_ERROR, OVL_WARNING, or OVL_INFO. The default value is OVL_ERROR.

- *depth* is an integral expression specifying the maximum number of elements the queue or FIFO can hold.

- *push_width* is an integral expression defining the maximum number of pushes (or writes) that is possible in a single clock cycle. The default value is 1.

- *pop_width* is an integral expression defining the maximum number of pops (or reads) that is possible in a single clock cycle. The default value is 1.

- *property_type* determines whether to use the assertion as an assert property or an assume property. The default value is OVL_ASSERT.

- *msg* is a string expression displayed whenever the assertion fails. It has a language-dependent default value.

- *coverage_level* is for enabling or disabling coverage monitoring for the checker. The default value is OVL_COVERALL.

- *clk* is a signal whose positive edge triggers the checking of the assertion.

- *reset_n* is a signal, that when deasserted, indicates that the assertion is to be monitored.

- *Push[push_width-1:0]* is an integral expression indicating the number of pushes (writes) occurring in the current clock cycle, with a default value of 1.

- *Pop[pop_width-1:0]* is an integral expression indicating the number of pops (reads) occurring in the current clock cycle, with a default value of 1.

E.9   assert_frame

The **assert_frame** assertion monitors the timing relationships between two events in a sequence of events. When *start_event* is asserted, *test_expr* must evaluate to TRUE within a given range of clock cycles.

assert_frame
```
#( severity_level, min_cks, max_cks, action_on_new_start,
   property_type, msg, coverage_level)
instance_name ( clk, reset_n, start_event, test_expr )
```

- *severity_level* is for handling an assertion violation. The values are OVL_FATAL, OVL_ERROR, OVL_WARNING, or OVL_INFO. The default value is OVL_ERROR.

- *min_cks* is the minimum number of clock cycles to wait after *start_event* is asserted (the first event in the sequence) for the *test_expr* to be asserted (the second event in the sequence). If 0, there is no minimum number of cycles to wait prior to checking for the assertion of *test_expr*.

- *max_cks* is the maximum number of clock cycles to wait after *start_event* is asserted for *test_expr* to be asserted.

- *action_on_new_start* is the method of handling a new start that occurs before *test_expr* changes value or *num_cks* cycles transpire without a change.

  - `OVL_IGNORE_NEW_START (default), ignores all assertions of *start_event* after the first assertion has been detected.

  - `OVL_RESET_ON_NEW_START, restart monitoring of *test_expr* if *start_event* is asserted in any subsequent clock assertion during the monitoring of *test_expr*.

  - `OVL_ERROR_ON_NEW_START, while monitoring, issue an error if *start_event* is asserted in any clock cycle subsequent to the cycle in which monitoring began.

- *property_type* determines whether to use the assertion as an assert property or an assume property. The default value is OVL_ASSERT.

- *msg* is a string expression displayed whenever the assertion fails. It has a language-dependent default value.

- *coverage_level* is for enabling or disabling coverage monitoring for the checker. The default value is OVL_COVERALL.

- *clk* is a signal whose positive edge triggers the checking of the assertion.

- *reset_n* is a signal, that when deasserted, indicates that the assertion is to be monitored.

- *start_event* is an expression defining the first event in the sequence of events.

- *test_expr* is an expression defining the second event in the sequence of events.

E.10   assert_handshake

The **assert_handshake** assertion monitors the handshake signals *req* and *ack* at every positive edge of either the triggering event or *clk*.

assert_handshake
```
    #( severity_level, min_ack_cycle, max_ack_cycle,
       req_drop, deassert_count, max_ack_length, property_type,
       msg, coverage_level)
    instance_name ( clk, reset_n, req, ack )
```

- *severity_level* is for handling an assertion violation. The values are OVL_FATAL, OVL_ERROR, OVL_WARNING, or OVL_INFO. The default value is OVL_ERROR.

- *min_ack_cycle* is an integer-valued expression that is nonnegative. Specifies that *ack* occurs no sooner than *min_ack_cycle* clock cycles after the clock cycle in which *req* occurs.

- *max_ack_cycle* is an integer-valued expression that is nonnegative. Specifies that *ack* does not occur after *max_ack_cycle* clock cycles after the clock cycle in which *req* occurs.

- *req_drop* is an integer-valued expression. If equal to one, **assert_handshake** ensures that *req* remains asserted until *ack* is asserted.

- *deassert_count* is a nonnegative, integer-valued expression. If greater than 0, **assert_handshake** ensures that *req* becomes deasserted within *deassert_count* clock cycles after the deasertion of *ack*. (This check guards against a stuck-active *req*.)

- *max_ack_length* is a nonnegative, integer-valued expression. If greater than 0, **assert_handshake** verifies that *ack* is not asserted for more than *max_ack_length* clock cycles without being deasserted. (This check guards against a stuck-active *ack*.)

- *property_type* determines whether to use the assertion as an assert property or an assume property. The default value is OVL_ASSERT.

- *msg* is a string expression displayed whenever the assertion fails. It has a language-dependent default value.

- *coverage_level* is for enabling or disabling coverage monitoring for the checker. The default value is OVL_COVERALL.

- *clk* is a signal whose positive edge triggers the checking of the assertion.

- *reset_n* is a signal, that when deasserted, indicates that the assertion is to be monitored.

- *req* is the signal defining the start of the handshake.

- *ack* is the signal defining the end of the handshake.

E.11   assert_implication

The **assert_implication** assertion monitors the *antecedent* expression. Whenever it evaluates TRUE, the assertion contends that the *consequence* is also TRUE.

```
assert_implication
    #( severity_level, property_type, msg, coverage_level )
    instance_name ( clk, reset_n, antecedent_expr,
                        consequence_expr )
```

- *severity_level* is for handling an assertion violation. The values are OVL_FATAL, OVL_ERROR, OVL_WARNING, or OVL_INFO. The default value is OVL_ERROR.
- *property_type* determines whether to use the assertion as an assert property or an assume property. The default value is OVL_ASSERT.
- *msg* is a string expression displayed whenever the assertion fails. It has a language-dependent default value.
- *coverage_level* is for enabling or disabling coverage monitoring for the checker. The default value is OVL_COVERALL.
- *clk* is a signal whose positive edge triggers the checking of the assertion.
- *reset_n* is a signal, that when deasserted, indicates that the assertion is to be monitored.
- *antecedent_expr* is an expression that, when TRUE, triggers the monitoring of the *consequence* expression.
- *consequence_expr* is the expression that is monitored when the *antecedent* expression is TRUE.

E.12   assert_increment

The **assert_increment** assertion monitors the expression *test_expr* at every positive edge of the triggering event or clock *clk*. This assertion contends that the expression never increases in value except by *value*.

```
assert_increment
    #( severity_level, width, value, property_type,
        msg, coverage_level )
    instance_name ( clk, reset_n, test_expr )
```

- *severity_level* is for handling an assertion violation. The values are OVL_FATAL, OVL_ERROR, OVL_WARNING, or OVL_INFO. The default value is OVL_ERROR.

- *width* is the width of the test expression (default value is 1).

- *value* is permitted increment value (default value is 1).

- *property_type* determines whether to use the assertion as an assert property or an assume property. The default value is OVL_ASSERT.

- *msg* is a string expression displayed whenever the assertion fails. It has a language-dependent default value.

- *coverage_level* is for enabling or disabling coverage monitoring for the checker. The default value is OVL_COVERALL.

- *clk* is a signal whose positive edge triggers the checking of the assertion.

- *reset_n* is a signal, that when deasserted, indicates that the assertion is to be monitored.

- *test_expr[width-1:0]* is the expression being monitored by this instance of **assert_increment**.

E.13 assert_never

The **assert_never** assertion evaluates the expression *test_expr* at every positive edge of the triggering event or clock *clk*. This assertion contends that the expression never evaluates to TRUE. Should *text_expr* evaluate to TRUE, the message *msg* will be reported with *severity_level*.

```
assert_never
      #( severity_level, property_type, msg, coverage_level )
      instance_name ( clk, reset_n, test_expr )
```

- *severity_level* is for handling an assertion violation. The values are OVL_FATAL, OVL_ERROR, OVL_WARNING, or OVL_INFO. The default value is OVL_ERROR.

- *property_type* determines whether to use the assertion as an assert property or an assume property. The default value is OVL_ASSERT.

- *msg* is a string expression displayed whenever the assertion fails. It has a language-dependent default value.

- *coverage_level* is for enabling or disabling coverage monitoring for the checker. The default value is OVL_COVERALL.

- *clk* is a signal whose positive edge triggers the checking of the assertion.

- *reset_n* is a signal, that when deasserted, indicates that the assertion is to be monitored.

- *test_expr* is the expression being monitored by this instance of **assert_never**.

E.14   assert_next

The **assert_next** assertion verifies that a specified expression is TRUE a specified number of cycles after a start event. When *start_event* asserts, then *test_expr* must assert exactly *num_cks* clock cycles later. This assertion supports overlapping sequences of events. The reassertion of *start_event* fewer than *num_cks* cycles after a given assertion of *start_event* defines an overlapping sequence of events. The sequences must terminate by assertions of *test_expr* as defined by the values of *start_event* and *num_cks* that initiated each sequence.

```
assert_next
    #( severity_level, num_cks, check_overlapping,
       check_missing_start, property_type,
       msg, coverage_level )
    instance_name ( clk, reset_n, start_event, test_expr )
```

- *severity_level* is for handling an assertion violation. The values are OVL_FATAL, OVL_ERROR, OVL_WARNING, or OVL_INFO. The default value is OVL_ERROR.

- *num_cks* is an integer-valued, nonnegative expression that specifies the number of clock cycles that must elapse between the assertion of the *start_event* and the *test_expr*.

- *check_overlapping*, if **0**, overlap checking is performed. After a *start* and during the wait time for *test_expr*, another *start_event* will not cause a new search for *test_expr* to be initiated. If *check_overlapping* is **1**, then these multiple overlapping sequences will be independently verified.

- *check_missing_start*, if **1**, verifies that the exact sequence of the assertion of *start_event* followed exact *num_cks* clock periods later, are then followed by the assertion of *test_expr*. If *test_expr* is asserted out of this sequence, **assert_next** fails. If *check_missing_start* is **0**, an out of sequence *test_expr* is allowed, and the only requirement is that after *start_event* and *num_cks* clocks, *test_expr* also becomes **1**.

- *property_type* determines whether to use the assertion as an assert property or an assume property. The default value is OVL_ASSERT.

- *msg* is a string expression displayed whenever the assertion fails. It has a language-dependent default value.

- *coverage_level* is for enabling or disabling coverage monitoring for the checker. The default value is OVL_COVERALL.

- *clk* is a signal whose positive edge triggers the checking of the assertion.

- *reset_n* is a signal, that when deasserted, indicates that the assertion is to be monitored.
- *start_event* is the expression whose assertion defines the start of the event sequence.
- *test_expr* is the expression whose assertion defines the end of the event sequence.

E.15   assert_no_overflow

The **assert_no_overflow** assertion evaluates the expression *test_expr* at every positive edge of the triggering event or clock *clk*. This assertion contends that the expression never takes on a value outside of the range *min* to *max*, inclusive.

```
assert_no_overflow
      #( severity_level, width, min, max, property_type,
         msg, coverage_level )
      instance_name ( clk, reset_n, test_expr )
```

- *severity_level* is for handling an assertion violation. The values are OVL_FATAL, OVL_ERROR, OVL_WARNING, or OVL_INFO. The default value is OVL_ERROR.
- *width* is an integer-valued, nonnegative expression describing the width of *test_expr*. Implementations are allowed, but not required to restrict the maximum value of *width*.
- *min* The permissible minimum value of *test_expr*. The default value is 0.
- *property_type* determines whether to use the assertion as an assert property or an assume property. The default value is OVL_ASSERT.
- *max* The permissible maximum value of *test_expr*.
- *msg* is a string expression displayed whenever the assertion fails. It has a language-dependent default value.
- *coverage_level* is for enabling or disabling coverage monitoring for the checker. The default value is OVL_COVERALL.
- *clk* is a signal whose positive edge triggers the checking of the assertion.
- *reset_n* is a signal, that when deasserted, indicates that the assertion is to be monitored.
- *test_expr[width-1:0]* the expression being monitored by this instance of **assert_no_overflow**.

E.16   assert_no_transition

The **assert_no_transition** assertion checks the expression *test_expr* at every positive edge of the triggering event or clock *clk*. When the value of *text_expr* is equal to *start_state* (i.e., the machine is in the start state) this assertion contends that *test_expr* never transits to the value of *next_state* (i.e., the machine stays in the start state). The *width* parameter defines the size of *test_expr*.

```
assert_no_transition
      #( severity_level, width, property_type, msg, coverage_level )
         instance_name ( clk, reset_n, test_expr, start_state,
         next_state )
```

- *severity_level* is for handling an assertion violation. The values are OVL_FATAL, OVL_ERROR, OVL_WARNING, or OVL_INFO. The default value is OVL_ERROR.

- *width* is an integer-valued, nonnegative expression describing the width of *test_expr*. The default value is 1.

- *property_type* determines whether to use the assertion as an assert property or an assume property. The default value is OVL_ASSERT.

- *msg* is a string expression displayed whenever the assertion fails. It has a language-dependent default value.

- *coverage_level* is for enabling or disabling coverage monitoring for the checker. The default value is OVL_COVERALL.

- *clk* is a signal whose positive edge triggers the checking of the assertion.

- *reset_n* is a signal, that when deasserted, indicates that the assertion is to be monitored.

- *test_expr[width-1:0]* is the expression being monitored by this instance of **assert_no_transition**.

- *start_state[width-1:0]* is the value of *test_expr* that triggers the assertion.

- *next_state[width-1:0]* is the value of *test_expr* being guarded against.

E.17   assert_no_underflow

The **assert_no_underflow** assertion evaluates the expression *test_expr* at every positive edge of the triggering event or clock *clk*. This assertion contends that the expression never takes on a value outside of the range *min* to *max*, exclusive.

assert_no_underflow
```
    #( severity_level, width, min, max, property_type,
       msg, coverage_level )
    instance_name ( clk, reset_n, test_expr )
```

- *severity_level* is for handling an assertion violation. The values are OVL_FATAL, OVL_ERROR, OVL_WARNING, or OVL_INFO. The default value is OVL_ERROR.

- *width* is an integer-valued, nonnegative expression describing the width of *test_expr*. Implementations are allowed, but not required to restrict the maximum value of *width*.

- *min* is the permissible minimum value of *test_expr*. The default value is 0.

- *max* is the permissible maximum value of *test_expr*.

- *property_type* determines whether to use the assertion as an assert property or an assume property. The default value is OVL_ASSERT.

- *msg* is a string expression displayed whenever the assertion fails. It has a language-dependent default value.

- *coverage_level* is for enabling or disabling coverage monitoring for the checker. The default value is OVL_COVERALL.

- *clk* is a signal whose positive edge triggers the checking of the assertion.

- *reset_n* is a signal, that when deasserted, indicates that the assertion is to be monitored.

- *test_expr[width-1:0]* is the expression being monitored by this instance of **assert_no_underflow**.

E.18   assert_odd_parity

The **assert_odd_parit**y assertion monitors the expression *test_expr* at every positive edge of the triggering event or clock *clk*. This assertion contends that the expression always has an odd number of bits asserted.

assert_odd_parity
```
    #( severity_level, width, property_type, msg, coverage_level )
    instance_name ( clk, reset_n, test_expr )
```

- *severity_level* is for handling an assertion violation. The values are OVL_FATAL, OVL_ERROR, OVL_WARNING, or OVL_INFO. The default value is OVL_ERROR.

- *width* is the width of the test expression (default value is 1).

- *property_type* determines whether to use the assertion as an assert property or an assume property. The default value is OVL_ASSERT.
- *msg* is a string expression displayed whenever the assertion fails. It has a language-dependent default value.
- *coverage_level* is for enabling or disabling coverage monitoring for the checker. The default value is OVL_COVERALL.
- *clk* is a signal whose positive edge triggers the checking of the assertion.
- *reset_n* is a signal, that when deasserted, indicates that the assertion is to be monitored.
- *test_expr[width-1:0]* is the expression being monitored by this instance of **assert_odd_parity**.

E.19   assert_one_cold

The **assert_one_cold** assertion monitors the expression *test_expr* at every positive edge of the triggering event or clock *clk*. This assertion contends that the expression always has either exactly one bit deasserted or has the appropriate inactive state (either all zero bits or all one bits), depending on the value of the *inactive* parameter.

```
assert_one_cold
      #( severity_level, width, inactive, property_type,
        msg, coverage_level )
      instance_name ( clk, reset_n, test_expr )
```

- *severity_level* is for handling an assertion violation. The values are OVL_FATAL, OVL_ERROR, OVL_WARNING, or OVL_INFO. The default value is OVL_ERROR.
- *width* is the width of the test expression (default value is 1).
- *inactive* specifies the inactive state of *test_expr*. It has three possible values:
  - `OVL_ALL_ZEROS, allows an all-zero value of *test_expr* without the generation of a report.
  - `OVL_ALL_ONES, allows an all-one value of *test_expr* without the generation of a report.
  - `OVL_ONE_COLD (default) allows no inactive state. At all times, *test_expr* must have exactly one bit deasserted.
- *property_type* determines whether to use the assertion as an assert property or an assume property. The default value is OVL_ASSERT.

- *msg* is a string expression displayed whenever the assertion fails. It has a language-dependent default value.

- *coverage_level* is for enabling or disabling coverage monitoring for the checker. The default value is OVL_COVERALL.

- *clk* is a signal whose positive edge triggers the checking of the assertion.

- *reset_n* is a signal, that when deasserted, indicates that the assertion is to be monitored.

- *test_expr[width-1:0]* is the expression being monitored by this instance of **assert_one_cold**.

E.20   assert_one_hot

The **assert_one_hot** assertion monitors the expression *test_expr* at every positive edge of the triggering event or clock *clk*. This assertion contends that the expression always has exactly one bit asserted.

```
assert_one_hot
      #( severity_level, width, property_type, msg, coverage_level )
      instance_name ( clk, reset_n, test_expr )
```

- *severity_level* is for handling an assertion violation. The values are OVL_FATAL, OVL_ERROR, OVL_WARNING, or OVL_INFO. The default value is OVL_ERROR.

- *width* is the width of the test expression (default value is 1).

- *property_type* determines whether to use the assertion as an assert property or an assume property. The default value is OVL_ASSERT.

- *msg* is a string expression displayed whenever the assertion fails. It has a language-dependent default value.

- *coverage_level* is for enabling or disabling coverage monitoring for the checker. The default value is OVL_COVERALL.

- *clk* is a signal whose positive edge triggers the checking of the assertion.

- *reset_n* is a signal, that when deasserted, indicates that the assertion is to be monitored.

- *test_expr[width-1:0]* is the expression being monitored by this instance of **assert_one_hot**.

E.21   assert_proposition

The **assert_proposition** assertion continually monitors the expression *test_expr*; unlike **assert_always**, *test_expr* is not sampled by a clock. This assertion contends that the expression always evaluates to

TRUE. Should the expression evaluate to FALSE, the message *msg* will be reported with *severity_level*.

```
assert_proposition
      #( severity_level, property_type, msg, coverage_level )
      instance_name ( reset_n, test_expr )
```

- *severity_level* is for handling an assertion violation. The values are OVL_FATAL, OVL_ERROR, OVL_WARNING, or OVL_INFO. The default value is OVL_ERROR.

- *property_type* determines whether to use the assertion as an assert property or an assume property. The default value is OVL_ASSERT.

- *msg* is a string expression displayed whenever the assertion fails. It has a language-dependent default value.

- *coverage_level* is for enabling or disabling coverage monitoring for the checker. The default value is OVL_COVERALL.

- *reset_n* is a signal, that when deasserted, indicates that the assertion is to be monitored.

- *test_expr* the expression being monitored by this instance of **assert_proposition**.

E.22   assert_quiescent_state

The **assert_quiescent_state** assertion continually monitors the sample event on every rising edge of *clk*. This assertion contends that *state_expr* is equal to *check_value* whenever *sample_event* is asserted, and optionally when the entire system is quiescent.

```
assert_quiescent_state
      #( severity_level, width, property_type, msg, coverage_level )
      instance_name ( clk, reset_n, state_expr, check_value,
                      sample_event )
```

- *severity_level* is for handling an assertion violation. The values are OVL_FATAL, OVL_ERROR, OVL_WARNING, or OVL_INFO. The default value is OVL_ERROR.

- *width* is the width of the test expression (default value is 1).

- *property_type* determines whether to use the assertion as an assert property or an assume property. The default value is OVL_ASSERT.

- *msg* is a string expression displayed whenever the assertion fails. It has a language-dependent default value.

- *coverage_level* is for enabling or disabling coverage monitoring for the checker. The default value is OVL_COVERALL.
- *clk* is a signal whose positive edge triggers the checking of the assertion.
- *reset_n* is a signal, that when deasserted, indicates that the assertion is to be monitored.
- *state_expr[width-1:0]* is the expression being monitored by this instance of **assert_quiescent_state**.
- *check_value[width-1:0]* is the value of *state_expr* that indicates quiescence.
- *sample_event* when TRUE, causes the assertion to check *state_expr*.

E.23   assert_range

The **assert_range** assertion evaluates the expression *test_expr* at every positive edge of the triggering event or clock *clk*. This assertion contends that the expression never takes on a value outside of the range *min* to *max*, inclusive.

```
assert_range
    #( severity_level, width, min, max, property_type,
       msg, coverage_level )
    instance_name ( clk, reset_n, test_expr )
```

- *severity_level* is for handling an assertion violation. The values are OVL_FATAL, OVL_ERROR, OVL_WARNING, or OVL_INFO. The default value is OVL_ERROR.
- *width* is an integer-valued, nonnegative expression describing the width of *test_expr*.
- *min* is the permissible minimum value of *test_expr*. The default value is 0.
- *max* is the permissible maximum value of *test_expr*. The default value is $2**width - 1$. *max* must be greater than or equal to *min*.
- *property_type* determines whether to use the assertion as an assert property or an assume property. The default value is OVL_ASSERT.
- *msg* is a string expression displayed whenever the assertion fails. It has a language-dependent default value.
- *coverage_level* is for enabling or disabling coverage monitoring for the checker. The default value is OVL_COVERALL.
- *clk* is a signal whose positive edge triggers the checking of the assertion.

- *reset_n* is a signal, that when deasserted, indicates that the assertion is to be monitored.

- *test_expr[width-1:0]* is the expression being monitored by this instance of **assert_range**.

E.24   assert_time

The **assert_time** assertion continually evaluates the expression *test_expr*. This assertion contends that the expression remains TRUE for at least a specified number of clock periods.

```
assert_time
    #( severity_level, num_cks, action_on_new_start,
       property_type, msg, coverage_level )
    instance_name ( clk, reset_n, start_event, test_expr )
```

- *severity_level* is for handling an assertion violation. The values are OVL_FATAL, OVL_ERROR, OVL_WARNING, or OVL_INFO. The default value is OVL_ERROR.

- *num_cks* specifies the minimum number of clock cycles that *test_expr* must hold each time it asserts.

- *action_on_new_start* is the method of handling a new start that occurs before *test_expr* changes value or *num_cks* cycles transpire without a change.
  - `OVL_IGNORE_NEW_START (default), ignores all assertions of *start_event* after the first assertion has been detected.
  - `OVL_RESET_ON_NEW_START, restart monitoring of *test_expr* if *start_event* is asserted in any subsequent clock assertion during the monitoring of *test_expr*.
  - `OVL_ERROR_ON_NEW_START, while monitoring, issue an error if *start_event* is asserted in any clock cycle subsequent to the cycle in which monitoring began.

- *property_type* determines whether to use the assertion as an assert property or an assume property. The default value is OVL_ASSERT.

- *msg* is a string expression displayed whenever the assertion fails. It has a language-dependent default value.

- *coverage_level* is for enabling or disabling coverage monitoring for the checker. The default value is OVL_COVERALL.

- *clk* is a signal whose positive edge triggers the checking of the assertion.

- *reset_n* is a signal, that when deasserted, indicates that the assertion is to be monitored.

- *start_event* is the event that, when asserted, triggers the monitoring of *test_expr*.

- *test_expr* is the expression being monitored by this instance of **assert_time**.

E.25 assert_transition

The **assert_transition** assertion monitors the expression *test_expr* at every positive edge of the triggering event or clock *clk*. When the value of *test_expr* is equal to *start_state* this assertion contends that, if *test_expr* changes value, it will change to the value of *end_state*. The *width* parameter defines the size of *test_expr*.

```
assert_transition
      #( severity_level, width, property_type, msg, coverage_level )
      instance_name ( clk, reset_n, test_expr, start_state,
                      next_state )
```

- *severity_level* is for handling an assertion violation. The values are OVL_FATAL, OVL_ERROR, OVL_WARNING, or OVL_INFO. The default value is OVL_ERROR.

- *width* is an integer-valued, nonnegative expression describing the width of *test_expr*. The default value is 1.

- *property_type* determines whether to use the assertion as an assert property or an assume property. The default value is OVL_ASSERT.

- *msg* is a string expression displayed whenever the assertion fails. It has a language-dependent default value.

- *coverage_level* is for enabling or disabling coverage monitoring for the checker. The default value is OVL_COVERALL.

- *clk* is a signal whose positive edge triggers the checking of the assertion.

- *reset_n* is a signal, that when deasserted, indicates that the assertion is to be monitored.

- *test_expr[width-1:0]* is the expression being monitored by this instance of **assert_transition**.

- *start_state[width-1:0]* is the value of *test_expr* that triggers the assertion.

- *next_state[width-1:0]* is the value that *test_expr* should assume in the next clock cycle.

E.26 assert_unchange

The **assert_unchange** assertion evaluates the expression *test_expr* at every positive edge of the triggering event or clock *clk*. This assertion contends that, when triggered, *test_expr* remains stable for at least a specified number of clock periods.

```
assert_unchange
    #( severity_level, num_cks, action_on_new_start,
        property_type, msg, coverage_level )
    instance_name ( clk, reset_n, start_event, test_expr )
```

- *severity_level* is for handling an assertion violation. The values are OVL_FATAL, OVL_ERROR, OVL_WARNING, or OVL_INFO. The default value is OVL_ERROR.

- *num_cks* specifies the minimum number of clock cycles that *test_expr* must remain unchanged each time *start_event* is asserted.

- *action_on_new_start* is the method of handling a new start that occurs before *test_expr* changes value or *num_cks* cycles transpire without a change.

  - `OVL_IGNORE_NEW_START (default), ignores all assertions of *start_event* after the first assertion has been detected.

  - `OVL_RESET_ON_NEW_START, restart monitoring of *test_expr* if *start_event* is asserted in any subsequent clock assertion during the monitoring of *test_expr*.

  - `OVL_ERROR_ON_NEW_START, while monitoring, issue an error if *start_event* is asserted in any clock cycle subsequent to the cycle in which monitoring began.

- *property_type* determines whether to use the assertion as an assert property or an assume property. The default value is OVL_ASSERT.

- *msg* is a string expression displayed whenever the assertion fails. It has a language-dependent default value.

- *coverage_level* is for enabling or disabling coverage monitoring for the checker. The default value is OVL_COVERALL.

- *clk* is a signal whose positive edge triggers the checking of the assertion.

- *reset_n* is a signal, that when deasserted, indicates that the assertion is to be monitored.

- *start_event* is the event that, when asserted, triggers the monitoring of *test_expr*.

- *test_expr[width-1:0]* is the expression being monitored by this instance of **assert_unchange**.

E.27 assert_width

The **assert_width** assertion continually evaluates the expression *test_expr*. This assertion contends that the expression remains TRUE for at least a specified minimum number of clock periods and no longer than a specified maximum number of clock periods.

```
assert_width
        #( severity_level, min_cks, max_cks, property_type,
           msg, coverage_level )
        instance_name ( clk, reset_n, test_expr )
```

- *severity_level* is for handling an assertion violation. The values are OVL_FATAL, OVL_ERROR, OVL_WARNING, or OVL_INFO. The default value is OVL_ERROR.

- *min_cks* specifies the minimum number of clock cycles that *test_expr* must hold each time it asserts. If 0, there is no minimum number of clocks; *test_expr* may be asserted at *start_event*.

- *max_cks* specifies the maximum number of clock cycles that *test_expr* must hold each time it asserts. If 0, there is no maximum check; *test_expr* may deassert at any time after it asserts.

- *property_type* determines whether to use the assertion as an assert property or an assume property. The default value is OVL_ASSERT.

- *msg* is a string expression displayed whenever the assertion fails. It has a language-dependent default value.

- *coverage_level* is for enabling or disabling coverage monitoring for the checker. The default value is OVL_COVERALL.

- *clk* is a signal whose positive edge triggers the checking of the assertion.

- *reset_n* is a signal, that when deasserted, indicates that the assertion is to be monitored.

- *start_event* is the event that, when asserted, triggers the monitoring of *test_expr*.

- *test_expr* is the expression being monitored by this instance of **assert_width**.

E.28 assert_win_change

The **assert_win_change** assertion continually monitors *start_event* at every positive edge of the triggering event or clock *clk*. When *start_event* is asserted, this assertion monitors the expression *test_expr* to assure that it changes value prior to the assertion of the *end_event*. When *start_event* asserts, this assertion verifies that the *test_expr* changes value on a clock edge that occurs before or coincident with the next assertion of *end_event*. Once *test_expr* changes value, it need not remain stable until *end_event* asserts.

assert_win_change
```
#( severity_level, width, property_type, msg, coverage_level )
instance_name ( clk, reset_n, start_event, test_expr,
                end_event )
```

- *severity_level* is for handling an assertion violation. The values are OVL_FATAL, OVL_ERROR, OVL_WARNING, or OVL_INFO. The default value is OVL_ERROR.
- *width* is an integer-valued, nonnegative expression describing the width of *test_expr*. The default value is 1.
- *property_type* determines whether to use the assertion as an assert property or an assume property. The default value is OVL_ASSERT.
- *msg* is a string expression displayed whenever the assertion fails. It has a language-dependent default value.
- *coverage_level* is for enabling or disabling coverage monitoring for the checker. The default value is OVL_COVERALL.
- *clk* is a signal whose positive edge triggers the checking of the assertion.
- *reset_n* is a signal, that when deasserted, indicates that the assertion is to be monitored.
- *start_event* is the event that, when asserted, triggers the monitoring of *test_expr*.
- *test_expr[width-1:0]* is the expression being monitored by this instance of **assert_win_change**.
- *end_event* is the event that, when asserted, terminates the monitoring of *test_expr*.

E.29   assert_win_unchange

The **assert_win_unchange** assertion continually monitors *start_event* at every positive edge of the triggering event or clock *clk*. When *start_event* is asserted, this assertion monitors the expression *test_expr* to assure that it does not change value prior to the assertion of *end_event*.

assert_win_unchange
```
        #( severity_level, width, property_type, msg, coverage_level )
        instance_name ( clk, reset_n, start_event, test_expr,
                        end_event )
```

- *severity_level* is for handling an assertion violation. The values are OVL_FATAL, OVL_ERROR, OVL_WARNING, or OVL_INFO. The default value is OVL_ERROR.

- *width* is an integer-valued, nonnegative expression describing the width of *test_expr*. The default value is 1.

- *property_type* determines whether to use the assertion as an assert property or an assume property. The default value is OVL_ASSERT.

- *msg* is a string expression displayed whenever the assertion fails. It has a language-dependent default value.

- *coverage_level* is for enabling or disabling coverage monitoring for the checker. The default value is OVL_COVERALL.

- *clk* is a signal whose positive edge triggers the checking of the assertion.

- *reset_n* is a signal, that when deasserted, indicates that the assertion is to be monitored.

- *start_event* is the event that, when asserted, triggers the monitoring of *test_expr*.

- *test_expr[width-1:0]* is the expression being monitored by this instance of **assert_win_unchange**.

- *end_event* is the event that, when asserted, terminates the monitoring of *test_expr*.

E.30 assert_window

The **assert_window** assertion continually monitors the *start_event* at every positive edge of the triggering event or clock *clk*. When *start_event* is asserted, this assertion contends that the expression *test_expr* remains TRUE at every subsequent positive clock edge until the assertion of the *end_event*.

assert_window
```
        #( severity_level, property_type, msg, coverage_level )
        instance_name ( clk, reset_n, start_event, test_expr,
                        end_event )
```

- *severity_level* is for handling an assertion violation. The values are OVL_FATAL, OVL_ERROR, OVL_WARNING, or OVL_INFO. The default value is OVL_ERROR.

- *property_type* determines whether to use the assertion as an assert property or an assume property. The default value is OVL_ASSERT.

- *msg* is a string expression displayed whenever the assertion fails. It has a language-dependent default value.

- *coverage_level* is for enabling or disabling coverage monitoring for the checker. The default value is OVL_COVERALL.

- *clk* is a signal whose positive edge triggers the checking of the assertion.

- *reset_n* is a signal, that when deasserted, indicates that the assertion is to be monitored.

- *start_event* is the event that, when asserted, triggers the monitoring of *test_expr*.

- *test_expr* is the expression being monitored by this instance of **assert_window**.

- *end_event* is the event that, when asserted, terminates the monitoring of *test_expr*.

E.31   assert_zero_one_hot

The **assert_zero_one_hot** assertion checker checks the expression *test_expr* at every positive edge of the triggering event or clock *clk*. This assertion contends that the expression always has at most one bit asserted.

```
assert_zero_one_hot
        #( severity_level, width, property_type, msg, coverage_level )
        instance_name ( clk, reset_n, test_expr )
```

- *severity_level* is for handling an assertion violation. The values are OVL_FATAL, OVL_ERROR, OVL_WARNING, or OVL_INFO. The default value is OVL_ERROR.

- *width* is the width of the test expression (default value is 1).

- *property_type* determines whether to use the assertion as an assert property or an assume property. The default value is OVL_ASSERT.

- *msg* is a string expression displayed whenever the assertion fails. It has a language-dependent default value.

- *coverage_level* is for enabling or disabling coverage monitoring for the checker. The default value is OVL_COVERALL.
- *clk* is a signal whose positive edge triggers the checking of the assertion.
- *reset_n* is a signal, that when deasserted, indicates that the assertion is to be monitored.
- *test_expr[width-1:0]* is the expression being monitored by this instance of **assert_zero_one_hot**.

Index

\# sign, 24, 38
$array, 156
$async$nand$array, 75, 156
$asynch, 156
$bitstoreal, 75
$display, 33, 34, 44, 67–68, 73, 201,
   203–204
$displayb, 45, 200
$fflush, 278, 280
$fgetc, 73
$fgets, 279, 280
$finish, 74, 84, 85, 106, 193, 197,
   198, 205,
$fopen, 73, 277, 280, 303, 304
$fread, 73
$frewind, 73
$fscanf, 73, 278–280
$fseek, 73, 278–280
$hold, 150
$itor, 75
$monitor, 73, 200, 201, 248, 249
$nand, 156
$nochange, 74
$period, 150
$printtimescale, 74
$random, 76, 198, 204, 205
$readmemb, 153, 156, 181
$readmemh, 73, 153–155, 187,
   263, 302–304
$realtime, 76
$realtobits, 75
$rtoi, 75
$scanf, 73
$setup, 44, 149–150

$setuphold, 150
$sformat, 302–304
$sign, 73
$sscanf, 73
$stime, 76
$stop, 6, 41, 74, 197, 202–206, 249,
   263, 277, 301
$strobe, 203, 204
$swrite, 73
$sync_async$and_or$array_plane,
   74
$time, 33, 67, 76, 200, 201, 249
$time, 57
$timeformat, 74
$timescale, 22, 23, 44, 74
$width, 150
%b, 200
%t, 200
%v, 248
@ sign, 27
`define, 172
`celldefine, 72
`celldefine, 72
`default–nettype, 71
`default–nettype, 71
`define, 71–72, 175
`else, 72
`endcelldefine, 72
`endif, 72
`ifdef, 72
`include, 71, 110
`nounconnected_drive, 72
`resetall, 71, 72
`timescale, 71, 92

`unconnected_drive`, 72
`undef`, 72

2-to-1 Multiplexer, 230, 231, 250
4-bit Adder, 95, 96,
4-bit Comparator, 107, 108, 133
4-bit Shifter, 232
4-value logic, *See* four-value logic

abstraction level, 11, 15, 102
Accellera, 208
add operation, 109
add-and-shift, 253, 255
address range, 46
Altera, 10, 184
ALU, 27–28, 121, 124, 125, 130,
    132, 192, 269, 281, 285, 286
always, 65
always block, 26, 27, 30, 31, 34, 68,
    113, 114, 116, 118, 129, 130
always statement, 21, 24–25, 34,
    113, 117, 129, 139
ambiguity, 48
ambiguous conditions, 105
ambiguous strength, 250
ambiguous value, 105, 106, 173, 176
analysis, 7
and primitive, 82, 87
and-or, 74
antecedent, 217
AOI, 229–230
application-specific integrated
    circuits, *See* ASIC
arithmetic, 48
arithmetic expressions, 108, 119
arithmetic logic unit, *See* ALU
arithmetic operators, 48, 108, 162
arithmetic shift, 51
array, 46, 153
array declaration, 46, 47
array indexing, 58
array indexing, 58, 96
array of instances, 96, 97
array_plane, 74
ASIC, 2, 3, 7, 9, 10, 223
assert_one_hot, 212, 213
assert_always, 208–210, 220
assert_always_on_edge, 208

assert_change, 208, 211
assert_cycle_sequence, 208, 213,
    214, 216
assert_decrement, 208
assert_delta, 208
assert_even_parity, 208
assert_fifo_index, 208
assert_frame, 208
assert_handshake, 208
assert_implication, 208, 217, 219
assert_increment, 208
assert_never, 208
assert_never_at_x_or_z, 208
assert_next, 208, 215–218
assert_no_overflow, 208, 218–220
assert_no_transition, 208
assert_no_underflow, 208
assert_odd_parity, 208
assert_one_cold, 208
assert_one_hot, 208, 211–213
assert_proposition, 208
assert_quiescent_state, 208
assert_range, 208
assert_time, 208
assert_transition, 208
assert_unchange, 208
assert_width, 208
assert_win_change, 208
assert_win_unchange, 208
assert_window, 208
assert_zero_one_hot, 208
assertion, 6, 138
assertion library, 207
assertion module, 207, 209
assertion monitor, 6, 207–210, 298
assertion templates, 216
assertion verification, 3, 4, 6, 191,
    207, 208,
assign statement, 20, 22, 23, 25,
    38, 39–40, 102–105, 109–114,
    128, 132
assign-deassign, 148
asynchronous, 74, 157, 180, 217
asynchronous circuit, 30
asynchronous clear, 185
asynchronous control, 146
asynchronous data signal, 30
asynchronous PLA, 75

asynchronous preset, 185
asynchronous reset, 149
asynchronous set, 145, 149
asynchronous set and reset, 144, 145,
    149, 153, 164, 187, 188, 217
auxiliary logic, 213

barrel shifter, 233, 234
base specification, 45
based numbers, 57
base-identifier, 45
basic operators, 48
BCD, 133, 209
BCD counter, 210
begin, 25
begin-end bracketing, 115, 116, 118
behavioral level, 12, 19, 112,144
behavioral description, 31, 35, 113, 186
behavioral model, 3, 142
behavioral simulation, 5
behavioral synthesis, 129
behavioral Verilog, 2
bidirectional, 12
bidirectional, 153, 160, 225, 235, 239,
    253, 259, 262
bidirectional switches, 12, 225, 250
bidirectionality, 154, 155
binary, 44
binary multiplication, 254
binding, 8–9
bit-select, 58–59
bitwise operators, 50, 51, 102, 103
blocking assignment, 24, 65–67,
    78–79, 117
blocking procedural assignments,
    67, 110, 116, 130
blocking procedural assignments, 110
boolean operations, 48, 50
bracketing, 114
buf primitive, 86, 225, 244–245
buffered data application, 205
Bufif1, 83, 84, 86
built-in gate, 242
bus contention, 110
bus resolution, 12
bus specifications, 22
bus structure, 110
bussing, 129

CAD tools, 1
Cadence Design Systems, 11
capacitive, 42, 55
capacitive model, 136
capacitive net, 242, 250
capacitive network, 243
capacitive storage, 136
capacitive wire, 55
cascadable comparator, 107, 108
cascading, 107
case, 118, 120, 122, 125
case alternative, 26, 32, 120, 122, 172
case default, 130
case equality operators, 49
case expression, 120–122, 172, 273
case inequality operators, 49
case statement, 26–28, 31, 32, 120,
    121, 125, 131, 160, 273, 279,
    288, 295
casex, 122
casez, 121, 122
charge, 241
charge storage, 12
charge strength, 242–243, 247
clocked SR-latch, 138, 139, 140
clocking, 171
CMOS, 224, 225
CMOS flip-flop, 56
CMOS gates, 226, 227
CMOS NAND, 229
code format, 41
combinational, 181
combinational block, 177
combinational circuit, 81, 129
combinational circuit testing, 192
combinational UDP, 87, 139
comments, 16, 41, 208
comparator, 104, 107
compare, 124, 125, 128
compilation, 7
compiler directives, 44, 71–72, 76
complimentary metal oxide
    semiconductor (CMOS), 12
component description, 22
concatenation, 157, 181
concatenation operator, 26, 48, 52,
    104, 122
concurrency, 39, 40

concurrent, 13, 26, 33, 39, 97
concurrent assignments, 13, 129
concurrent body, 26, 39, 40, 61
concurrent statements, 40, 43, 62,
    97, 103, 178
concurrent subcomponents, 39
conditional assignment, 106
conditional expression, 20, 155, 291
conditional jump, 307
conditional operation, 20, 105–106
conditional operator, 52–53, 107
conditional statements, 128
conflict, 42
conflict x, 232
connection list, 94
consequence, 217
console, 34
constant part-select, 58
continuous assignments, 3, 43,
    53, 56, 60, 61, 64, 65, 69, 70,
    128, 140
control/data partitioning, 16
control part/unit, 16, 17
controller, 29, 253, 256, 270, 287
converting reals, 75
counter, 24–25, 161
cross-couple SRAM memory, 237
curly brackets, 26
current state, 171
custom IC, 2, 3, 7, 9, 10

data/control partitioning, 265
data components, 22
data files, 263
data part, 16, 17
data part/path, *See* datapath
data types, 37, 42, 54, 59, 60, 65,
    70, 73, 76,
Datapath, 16, 17, 253, 256–261,
    265–276, 287
deassign, 70, 148, 149
decimal, 44
default, 28, 120, 121, 160, 172,
    176, 248
default delay, 226
default **else**, 145
default **net** type, 82
default strength, 238, 242
defparam, 91

delay, 5, 12, 23–25, 38, 64, 66
delay control, 66, 115, 116, 147
delay control statement, 66, 115, 116
delay expression, 115
delay formats, 88, 92, 100, 103
delay parameters, 56, 61, 85–87, 100
delay paths, 38
delay specification, 61–62, 87, 88, 93,
    99, 101, 116
delay3, 88
design entry, 3
design flow, 2
design validation, 4
design verification, 3, 206
designer discipline, 208
dirty page, 302
display format specifications, 200
display tasks, 73
distribuited delay, 99–102
distributed delay, gate, 100
DoD, 11
Drain, 226
drive, 241
drive strength, 242–243
D-type flip-flop, 144–150, 184,
    187, 240,
dynamic arguments, 209, 210
dynamic cell, 239
dynamic memory, 234
dynamic RAM, 239

EDA, 1, 3, 11, 208
EDA environments, 3
electronic design automation,
    See EDA
Elements of Verilog, 18
else part, 123
end keywords, 114
endfunction, 109
endmodule, 18, 42
endprimitive, 86
endspecify, 99
endtable, 88
endtask, 277
equality operator, 48–50, 107
even control statement, 144, 146, 199
event control, 65, 66, 113–115,
    119, 199
event driven simulation, 62

event expression, 144
event sequence, 216
executable comments, 208
external data files, 153, 220
external files, 263
external memory file, 162

fall delay, 88, 90, 91, 103
falling edge, 23, 211
fault simulation, 11, 13, 19
feedback model, 136
feedback path, 237
Fetch, 270, 293
field programmable logic device,
    See FPLD
FIFO queue, 167
file I/O, 276
file I/O tasks, 73, 276
file output, 73, 281
finite state machine, 259
flip-flop, 22–24, 42, 139, 150, 240
flip-flop modeling, 144
flow control, 113, 115
for loop, 118, 123
for procedural statement, 123
for statement, 98, 122, 123
force, 70
force and **release**, 70
forcing 0, 42
forcing 1, 42
forever loop, 199
fork-join, 146–147
formal verification, 1, 3–4, 6, 13,
    206–208
four value logic, 12, 41, 62, 81, 241
FPGA, 7
FPLD, 3, 7, 9, 10
full adder, 25, 39, 40, 57, 93, 94,
    97–100, 104, 117, 118, 127,
full adder tester, 41
full-path, 100
function, 34, 44, 72, 75, 109–110
functional description, 78
functional registers, 157
functional specification, 7
functional verification, 33, 127

Gate, 227
gate capacitance, 234–235

gate delays, 5, 38
gate level, 12, 127, 132, 234
gate level delays, 111, 142
gate level logic, 85
gate level modeling, 223
gate level primitives, 12, 137
gate level simulation, 12
gate level synthesis, 127
gate level timing, 7
gate output, 241
gate primitives, 85, 86, 102, 133
Gateway Design Automation, 11
generate loops, 233
generate statement, 96–98
genvar, 98
glitch, 38
Gray code, 162, 213
Gray code counter, 162, 163, 213

half-register, 235, 240
hardware description languages
    (HDL), 11, 18, 103, 223
hardware generation, 10
hardware modules, 18
hardware/software environments, 73
HDL, *See* hardware description
    languages
hexadecimal, 44
hierarchical design, 13
hierarchical fashion, 3
hierarchical naming, 91, 139,
    200, 202
hierarchical structures, 93
hierarchy, 13, 91, 93, 207
high-impedance, 42, 153, 272, 302
high-level synthesis, 1
high-to-low propagation time (tphl),
    25
hold time, 150
Huffman model, 176, 178–179,
    187, 291
identifier, 41, 43–44
IEEE std 1364–1995, 11
IEEE std 1364–2001, 41
IEEE std 754–1985, 45
if statement, 123
if-condition, 120
if-else, 21, 25, 31, 113, 118–119, 161
immediate data, 266, 282

implication, 217
implicit model, 136
inactive, 259
inactive values, 130
indexed part-select, 58
indexing memories, 59
infinite loop, 113, 147
initial, 65
initial block, 67–68
initial reseting, 217
initial statement, 34, 85, 122, 153,
    183, 185, 192, 199, 200, 204
initial value, 42, 56
initializing **reg**, 197
inout, 43, 81–82, 154
inout ports, 225, 239
input, 43, 81–82
input/output specification, 23–24
instance name, 29, 33
instance name, 94, 96
instantiation, 19–20
instantiation statement, 4
instantiation statements, 4
instruction register, 281
integer, 44, 45, 57
interactive testbench, 201
interconnections, 26, 28–29, 82
intermediate format, 7
intermediate signal values, 57
intermediate wires, 20, 23
intra-assignment delay, 66, 68, 116,
    204, 116–117
intra-assignment event, 116
invalid state, 218
iterative structure, 96

Join, *See* fork-join

large, 63
latch, 137, *See also* SR-latch
level (RTL), 3
LFSR and MISR, 163
limiting data sets, 198
linear feedback shift register (LFSR),
    163–166
load, 157
localparam, 90, 171–172, 176
logic optimization, 8

logic value system, 41
logical operation, 22, 27, 48, 50, 287
logical shift, 51
low to high propagation time (tplh),
    25

majority circuit, 86, 90
majority UDP, 88
master-slave flip-flop, 141–142,
    240–241
maximum delay, 88
Mealy machine, 174, 181, 217
medium, 63
memory, 59, 110, 153
memory buffering, 276
memory initialization, 153
memory modeling, 276
memory read, 266, 293
min:typ:max, 88, 100, 103
minimum delay, 88
MISR, 164–164, 167–169
missing delay values, 226
mixbed synthesis, 132
mnemonic, 266
module, 18, 42, 81
module instantiation, 21, 72, 85, 91,
    94, 113, 124, 132, 209
module name, 18
module path delay, 99–100
module ports, 43
module with no ports, 196
module-under-test, *See* MUT
modulus operation, 48
monitoring, 73, 191
monitors, 6
Moore machine, 31, 171, 174,
    176, 178
multidimensional arrays, 46–47
multidimensional memories, 59, 137
multiple assignments, 68–69
multiple drivers, 64
multiple input signature register, *See*
    MISR
multiplexer, 19–22, 35, 82, 83, 105,
    111, 118, 119, 132, 230–232, 254,
    257–259
multiply operation, 48
MUT, 85, 191

named connection, 29, 94
named parameter assignment, 91
nand primitive, 242
n-bit adder, 97
negative edge, 24, 25
negedge, 23, 145
nested condition operators, 106
nested generate loops, 233
nested if-then-else, 106, 120
net, 42, 43, 55–56, 236
net assignments, 63–64
net declaration, 54, 56
net declaration assignment, 63–64,
    111, 242–243
net strength, 242
netlist, 10, 12
nmos, 224–226, 232
NMOS-PMOS inverter, 235
nonblocking assignment, 24, 67–68,
    117
nonblocking procedural assignments,
    67, 68, 116
not, 83, 84
number specification, 45
numbers, 44–45

observability, 208, 220
octal, 44
octal latch, 157
odd-even parity, 104
opcode, 271
Open Verification Library, *See* OVL
operations, 48
operator precedence, 53
operators, 13, 48
ordered connection, 29, 94
ordered parameter assignment,
    90–91, 94
ordered port connection, 29
output, 43, 81–82, 243
output latch, 129
output ports, 18
overdriving, 238
overflow, 124, 132
overriding parameters, 139
OVI (Open Verilog International),
    11
OVL, 6, 11, 207–209, 220

parallel path, 101
parameter, 90, 91, 111
parameter declaration, 31
parameter override, 91
parameterized module instantiation,
    91
parameters, 18, 57, 91
parity, 124–125
part-select, 58
pass gate logic, 230–231
path delay, 100–102
path delay specification, 99, 101, 102
period check, 150
pin-to-pin delay, 12, 93, 100
PLA, 12, 74
PLA modeling, 155
PLA modeling tasks, 74–75
PLD, 2
PLI, 72
pmos, 224, 232
polynomial, 164
ports, 18, 43, 81–82
posedge, 144, 145
positive-edge, 30
postsynthesis, 10, 128
postsynthesis simulation, 3, 10, 199
precedence of operators, 53
predefined bus resolution functions,
    12
predefined gate primitives, 85
predefined parameters, 12
predefined wire resolution functions,
    12
presynthesis description, 127
presynthesis simulation, 4
prime numbers, 194
primitive instantiation, 19, 94
primitives, 19, 82, 83, 85, 127
procedural, 21, 24
procedural assignments, 65, 66, 115,
    117, 131
procedural assignments, 65–66,
    130
procedural block, 85, 113
procedural block, 12, 13, 20, 40, 43,
    112–118, 129, 137,
procedural blocking assignments, 66
procedural body, 40, 68

procedural case statement, 120
procedural continuous assignment,
    65, 69–70
procedural flow control, 65
procedural for statement, 122
procedural **if-else**, 118
procedural statement, 3, 34, 110, 130
procedural statement, 3
procedural **while** loop, 123
Program Counter, 281
programmable logic arrays, 155
programmable logic devices, *See* PLD
Programming Language Interface
    (PLI), 13
propagation delay, 85
property, 6, 206–207
property coverage, 7
pseudo-static d-latch, 236
pseudo-static memory, 234
pull0, 63
pull1, 63
pulldown, 225
pull-down, 228, 229, 231
pull-up, 228, 231
pulup, 225

Quartus II, 10
queues, 167

random, 204
random access, 12
random data, 199
random number generation, 163
random time intervals, 204–205
range specification, 46
reading data files, 263
real, 44, 45, 57
realtime, 57
reduction, 125
reduction operation, 50–51, 105
reduction XOR, 124
reg, 25, 42, 54, 56–57, 113
reg data type, 52, 54, 56, 66
reg declaration, 56–57
register, 26
register block, 160
register file, 187, 281, 285, 288
register transfer, 3
register transfer level, *See* RTL

registers, 22
regular structure, 96, 97, 114
relational operation, 48, 105, 107
release, 70
repeat, 263
repeat statement, 198, 202
repetition multiplier, 52
replication operation, 48, 52
reset sequence, 216
resistive 0, 1, 42
resistive switches, 247
resolution, 64, 243–244
resolution function, 12, 224, 231,
    244, 246
response observation, 207
retaining old values, 120, 121, 123,
    129, 177
rise delay, 88, 90, 91, 103
rising clock edge, 30
ROM based controller, 181
routing and placement, 9
RT level, *See* RTL
RT level simulation, 4
RTL, 3, 15, 16, 17, 253, 264–265
RTL synthesis, 7

SAYEH, 281, 282
SAYEH datapath, 283
SAYEH Verilog description, 287
seed, 163, 164, 167
seed parameters, 165
sensitive, 181
sensitivity list, 26–27, 114, 129
sequence detector, 30–31, 171,
    188, 204
sequential, 181
sequential blocks, 160, 180
sequential circuit, 70, 76, 135, 194
sequential circuit synthesis, 186
sequential circuit testing, 194, 218
sequential multiplier, 253
sequential UDP, 139–141
setup time, 149, 150
seven-segment display, 133
shadow instructions, 281
shift operation, 51, 52, 157
shift register, 26–27, 157
shift-and-add, 254
shifter, 26, 157, 232

short instructions, 281
signal strength, 223, 241
signed, 54, 57
signed data, 57
signed number, 45
simple architecture, yet enough
    hardware, *See* SAYEH
simple assignments, 61
simple procedural blocks, 113
simple tester, 33, 84
simulation, 3, 4
simulation control, 197
simulation control tasks, 74
simulation performance, 229
sizable register, 153
size mismatches, 109
size specification, 22
sized integer, 44
small, 63
Source, 226, 227
specify, 99
specify block, 90, 100–101, 149–150
specify parameters, 90
SRAM, 238
SR-latch, 137, 138
stacks, 167
standard memory, 58
state machine, 16, 31, 32, 171, 268
state machine coding, 171
state machine testing, 195
state transitions, 171
static memory element, 237
static parameters, 209
Status Register, 281
strength, 241, 243, 248
strength modeling, 231, 241, 250
strength reduction, 231, 247–248
strength specification, 62–63
strong, 238
strong0, 63
strong1, 63
supply0, 56, 63, 227
supply1, 56, 63, 227
switch, 243
switch level, 11, 13, 85, 127, 136,
    228–230, 242, 250
switch level 2-to-1 multiplexer, 231
switch level barrel shifter, 234
switch level half-register, 235

switch level memory element, 235
switch level modeling, 11, 136, 223,
    224, 241
switch level primitives, 224, 225, 227
switch level shifter, 232
switch level timing, 11
switch output, 241
sync-async, 74
synchronizer, 30
synchronous, 176, 180, 217
synchronous reset, 23, 24, 144
synchronous reset, 160–161
synchronous set, 144
synthesis, 7– 9, 13, 19, 125, 127, 128,
    131, 181, 261, 273
synthesis tool, 11
synthesizable, 183, 270
synthesizable D-type flip-flop, 184
synthesizable gate level, 128
system functions, 75, 44, 72, 73
system tasks, 12, 34, 44, 72, 73
system tasks and functions, 72–76
system utilities, 12, 75

target device, 3
target hardware, 7, 9
target library, 129, 183
target technology, 127
task, *See* system task
test benches, 33–34
testbench, 4–6, 33, 43, 72, 84, 106, 191,
    192–207, 205, 262, 266, 276, 298
testbench techniques, 195
three-state, *See* tri-state
timing, 4, 5, 7, 10, 13, 19, 34, 37–40,
    64, 85, 135, 191, 223
timing analysis, 2, 8, 10, 11
timing and concurrency, 13, 37, 40, 76
timing check tasks, 74
timing constructs, 11
timing control, 67, 68, 113, 115, 143
timing diagram, 5, 214, 232, 239
timing parameters, 127
timing simulation, 11
timing specification, 9, 13, 39, 71, 93,
    102, 149
timing-control constructs, 12
top-down design, 265
tran, 225

tranif0, 225
tranif1, 225
transcript, 34
transistor primitives, 85
transitions to **X**, 86
transparent D-latch, 151
tri, 110
tri, 54, 55, 63, 110, 243
tri0, 55, 63, 110, 244
tri1, 44, 55, 63, 110, 244
triand, 110
triand, 54–55, 63, 110, 244, 111
triand bus, 111
tri1, 55, 110
trior, 110
trior, 54–55, 63, 111, 110, 224
trior bus, 111
trireg, 54–56, 63, 236, 239, 243, 244
tri-state, 54, 55, 83, 85, 124, 125,
    127, 129, 273, 285
tri-state buffer, 82, 83, 257, 259
tri-state busses, 129, 273
tri-state gate, 85, 127, 225
tri-state output, 132, 153
tri-state structures, 127
tri-state wired logic, 54
typical delay, 88

UDP, 87
UDP delay, 88
unary minus, 48
unary operator, 45
unary plus, 48
unary plus/minus, 48
unconnected ports, 72
underscore, 43, 45, 83
undriven, 42
unidirectional, 11, 225, 235
uninitialized value, 42
unit under test (UUT), 33
universal shift register, 26, 158, 159
unknown state, 235
unknown value, 42
unknown **X** state, 241
unsigned, 57, 44
unsized number, 44
unwanted latches, 120, 121, 129,
    130, 261, 273, 288

up-down counter, 162, 187
user-defined primitives (UDP),
    87, 88

valid states, 218
vector declaration, 47, 95
vector operation, 104
vectors, 41, 52, 151
verification, 1, 15, 191
verifying state machines, 213
Verilog attributes, 11
Verilog data types, 54
Verilog description for
    synthesis, 7
Verilog evolution, 11
Verilog HDL, 1, 10, 13, 39
Verilog Simulation Model, 59
Verilog testbench, 4, 84, 266
Verilog–2001, 11
very large scale integration (VLSI),
    223
VHDL, 11
von Neumann, 265, 270

wait statement, 146, 147, 202
wand, 111
wand, 54–56, 63, 110, 111
waveform editors, 4
weak0, 63
weak1, 63
while, 118
while loop, 123, 124, 279
width check, 150
wire, 19, 33, 54–55, 82, 110
wire values and timing, 82
wires, 81
wor, 54–56, 63, 64, 71, 110, 111,
    232, 244

X value, 42
x value, 42, 55, 84, 230
xnor, 85
XNOR operation, 50
xor, 85

Z value, 42
z value, 42, 48, 55, 56, 87,
    225, 226